ENERGY

SUPPLIES, SUSTAINABILITY, AND COSTS

ISSN 1534-1585

ENERGY

SUPPLIES, SUSTAINABILITY, AND COSTS

Sandra M. Alters

INFORMATION PLUS® REFERENCE SERIES
Formerly Published by Information Plus, Wylie, Texas

Detroit • New York • San Francisco • New Haven, Conn. • Waterville, Maine • London

Energy: Supplies, Sustainability, and Costs

Sandra M. Alters
Paula Kepos, Series Editor

Project Editor
John McCoy

Permissions
Lisa Kincade, Lista Person, Tracie Richardson

Composition and Electronic Prepress
Evi Seoud

Manufacturing
Cynde Bishop

For more information, contact
Thomson Gale
27500 Drake Rd.
Farmington Hills, MI 48331-3535
Or you can visit our Internet site at
http://www.gale.com

ISBN-13: 978-0-7876-5103-9 (set)
ISBN-10: 0-7876-5103-6 (set)
ISBN-13: 978-1-4144-0751-7
ISBN-10: 1-4144-0751-3
ISSN: 1534-1585

This title is also available as an e-book.
ISBN-13: 978-1-4144-2868-0 (set),
ISBN-10: 1-4144-2868-5 (set)
Contact your Thomson Gale sales representative for ordering information.

Printed in the United States of America
10 9 8 7 6 5 4 3 2 1

TABLE OF CONTENTS

TABLE OF CONTENTS

PREFACE

Energy: Supplies, Sustainability, and Costs is part of the *Information Plus Reference Series*. The purpose of each volume of the series is to present the latest facts on a topic of pressing concern in modern American life. These topics include today's most controversial and most studied social issues: abortion, capital punishment, care for the elderly, crime, health care, energy, the environment, immigration, minorities, social welfare, women, youth, and many more. Although written especially for the high school and undergraduate student, this series is an excellent resource for anyone in need of factual information on current affairs.

By presenting the facts, it is Thomson Gale's intention to provide its readers with everything they need to reach an informed opinion on current issues. To that end, there is a particular emphasis in this series on the presentation of scientific studies, surveys, and statistics. These data are generally presented in the form of tables, charts, and other graphics placed within the text of each book. Every graphic is directly referred to and carefully explained in the text. The source of each graphic is presented within the graphic itself. The data used in these graphics are drawn from the most reputable and reliable sources, in particular from the various branches of the U.S. government and from major independent polling organizations. Every effort was made to secure the most recent information available. The reader should bear in mind that many major studies take years to conduct and that additional years often pass before the data from these studies are made available to the public. Therefore, in many cases the most recent information available in 2007 dated from 2004 or 2005. Older statistics are sometimes presented as well, if they are of particular interest and no more-recent information exists.

Although statistics are a major focus of the *Information Plus Reference Series*, they are by no means its only content. Each book also presents the widely held positions and important ideas that shape how the book's subject is discussed in the United States. These positions are explained in detail and, where possible, in the words of those who support them. Some of the other material to be found in these books includes: historical background; descriptions of major events related to the subject; relevant laws and court cases; and examples of how these issues play out in American life. Some books also feature primary documents or have pro and con debate sections giving the words and opinions of prominent Americans on both sides of a controversial topic. All material is presented in an evenhanded and unbiased manner; the reader will never be encouraged to accept one view of an issue over another.

HOW TO USE THIS BOOK

The United States is the world's largest consumer of energy in all its forms. Gasoline and other fossil fuels power the nation's cars, trucks, trains, and aircraft. Electricity generated by burning oil, coal, and natural gas—or from nuclear or hydroelectric plants—runs America's lights, telephones, televisions, computers, and appliances. Without a steady, affordable, and massive amount of energy, modern America could not exist. This book presents the latest information on U.S. energy consumption and production, compared with years past. Controversial issues such as the U.S. dependence on foreign oil, the possibility of exhausting fossil fuel supplies, and the harm done to the environment by mining, drilling, and pollution are explored.

Energy: Supplies, Sustainability, and Costs consists of nine chapters and three appendixes. Each of the major elements of the U.S. energy system—such as coal, nuclear power, renewable energy sources, and electricity generation—has a chapter devoted to it. For a summary of the information covered in each chapter, please see the synopses provided in the Table of Contents at the front of the book. Chapters generally begin with an overview of the basic facts and background information on the chapter's topic, then proceed to examine subtopics of particular interest. For example, Chapter 7: Energy Reserves—Oil, Gas, Coal, and Uranium begins with a description of the different ways in which natural resource reserves are measured and their reliability. The chapter then moves into an examination of proved U.S. reserves of oil and natural gas. Statistics on known and estimated reserves are presented, as are projections of how long these reserves will last if they continue to be consumed at current rates. The controversial possibility of drilling for oil in the Arctic National Wildlife Refuge is discussed. This is followed by an examination of the methods, costs, and consequences of oil and gas exploration. After this are sections on U.S. coal and uranium reserves. Then the chapter presents similar statistics on worldwide reserves of oil, coal, gas, and uranium. Readers can find their way through a chapter by looking for the section and subsection headings, which are clearly set off from the text. Or, they can refer to the book's extensive Index, if they already know what they are looking for.

Statistical Information

The tables and figures featured throughout *Energy: Supplies, Sustainability, and Costs* will be of particular use to the reader in learning about this topic. These tables and figures represent an extensive collection of the most recent and valuable statistics on energy production and consumption; for example: the amount of coal mined in the United States in a year, the rate at which energy consumption is increasing in the United States, and the percentage of U.S. energy that comes from renewable sources. Thomson Gale believes that making this information available to the reader is the most important way in which we fulfill the goal of this book: to help readers understand the topic of energy and reach their own conclusions about controversial issues related to energy use and conservation in the United States.

Each table or figure has a unique identifier appearing above it, for ease of identification and reference. Titles for the tables and figures explain their purpose. At the end of each table or figure, the original source of the data is provided.

In order to help readers understand these often complicated statistics, all tables and figures are explained in the text. References in the text direct the reader to the relevant statistics. Furthermore, the contents of all tables and figures are fully indexed. Please see the opening section of the Index at the back of this volume for a description of how to find tables and figures within it.

Appendixes

In addition to the main body text and images, *Energy: Supplies, Sustainability, and Costs* has three appendixes. The first is the Important Names and Addresses directory. Here the reader will find contact information for a number of organizations that study energy. The second appendix is the Resources section, which is provided to assist the reader in conducting his or her own research. In this section the author and editors of *Energy: Supplies, Sustainability, and Costs* describe some of the sources that were most useful during the compilation of this book. The final appendix is this book's Index.

ADVISORY BOARD CONTRIBUTIONS

The staff of Information Plus would like to extend its heartfelt appreciation to the Information Plus Advisory Board. This dedicated group of media professionals provides feedback on the series on an ongoing basis. Their comments allow the editorial staff who work on the project to make the series better and more user-friendly. Our top priorities are to produce the highest-quality and most useful books possible, and the Advisory Board's contributions to this process are invaluable.

The members of the Information Plus Advisory Board are:

- Kathleen R. Bonn, Librarian, Newbury Park High School, Newbury Park, California
- Madelyn Garner, Librarian, San Jacinto College—North Campus, Houston, Texas
- Anne Oxenrider, Media Specialist, Dundee High School, Dundee, Michigan
- Charles R. Rodgers, Director of Libraries, Pasco-Hernando Community College, Dade City, Florida
- James N. Zitzelsberger, Library Media Department Chairman, Oshkosh West High School, Oshkosh, Wisconsin

COMMENTS AND SUGGESTIONS

The editors of the *Information Plus Reference Series* welcome your feedback on *Energy: Supplies, Sustainability, and Costs*. Please direct all correspondence to:

Editors
Information Plus Reference Series
27500 Drake Rd.
Farmington Hills, MI, 48331-3535

CHAPTER 1

AN ENERGY OVERVIEW

Energy is essential to life. Living creatures draw on energy flowing through the environment and convert it to forms they can use. The most fundamental energy flow for living creatures is the energy of sunlight, and the most important conversion is the act of primary production, in which plants and phytoplankton convert sunlight into biomass by photosynthesis. Earth's web of life, including human beings, rests on this foundation.

—*Energy in the United States: 1635–2000*, U.S. Energy Information Administration (2001)

A HISTORICAL PERSPECTIVE

Before the Twentieth Century

People have always found ways to harness energy, such as using animals to do work or inventing machines to tap the power of wind or water. The industrialization of the modern world was accompanied by the widespread use of such fossil fuels as coal, oil, and natural gas.

Significant use and management of energy resulted in one of the most profound social changes in history within a few generations. In the early 1800s most Americans lived in rural areas and worked in agriculture. The country ran mainly on wood fuel. One hundred years later, most Americans were city dwellers and worked in industry. America had become the world's largest producer and consumer of fossil fuels, had roughly tripled its use of energy per capita, and had become a global superpower.

The United States has always been a resource-abundant nation, but it was not until the Industrial Revolution in the mid-1800s that the total work output of engines surpassed that of work animals. As the country industrialized, coal began to replace wood as a primary fuel. Then petroleum and natural gas began to replace coal for many applications. The United States has since relied heavily on these three fossil fuels—coal, petroleum, and natural gas.

The Twentieth and Early Twenty-First Centuries

For much of its history the United States has been nearly energy self-sufficient, although small amounts of coal were imported from Britain in colonial times. Through the 1950s domestic energy production and consumption were nearly equal. During the 1960s consumption slightly outpaced production. In the 1970s the gap widened considerably, narrowed somewhat in the early 1980s, and then widened year after year. By 2004 the gap between domestic energy production and consumption was quite significant. (See Figure 1.1.) Since the 1970s energy imports have been used to try to close the gap between energy production and consumption. However, America's dependence on other countries for energy has created significant problems.

OIL CRISIS IN THE 1970S. In 1973 the United States supported Israel in the Yom Kippur War, which was fought between Israel and neighboring Arab countries. In response, several Arab nations cut off exports of oil to the United States and decreased exports to the rest of the world. The embargo was lifted six months later, but the price of oil had tripled from the 1973 average to about $12 per barrel. (See Figure 1.2.) Not only did Americans (and others around the world) face sudden price hikes for products produced from oil, such as gasoline and home heating oil, but they faced temporary shortages as well. The energy problem quickly became an energy crisis, which led to occasional blackouts in cities and industries, temporary shutdowns of factories and schools, and frequent lines at gasoline service stations. The sudden increase in energy prices in the early 1970s is widely considered to have been a major cause of the economic recession of 1974 and 1975.

Oil prices increased even more in the late 1970s. A revolution in Iran resulted in a significant drop in Iranian oil production from 1978 to 1981. During this same period the Iran-Iraq war began, and many other Persian

FIGURE 1.1

Energy production and consumption, 1949–2004

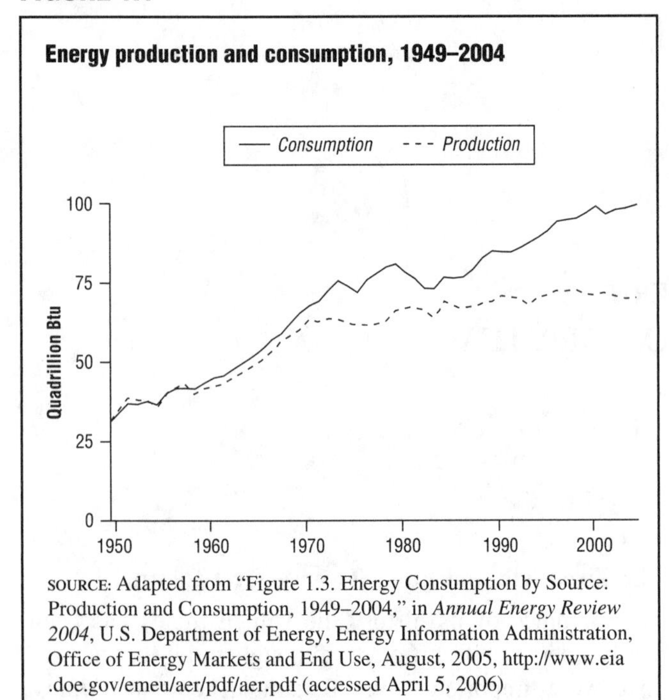

SOURCE: Adapted from "Figure 1.3. Energy Consumption by Source: Production and Consumption, 1949–2004," in *Annual Energy Review 2004*, U.S. Department of Energy, Energy Information Administration, Office of Energy Markets and End Use, August, 2005, http://www.eia.doe.gov/emeu/aer/pdf/aer.pdf (accessed April 5, 2006)

Gulf countries decreased their oil output as well. Companies and governments began to stockpile oil. As a result, prices continued to rise and reached a peak in 1981. (See Figure 1.2.)

OIL PRICES FALL IN THE 1980S. In early 1981 the U.S. government responded to the oil crisis by removing price and allocation controls on the oil industry. By no longer controlling domestic crude oil prices or restricting exports of petroleum products, it allowed the marketplace and competition to determine the price of crude oil. Therefore, domestic oil prices rose to the level of foreign oil prices.

As a result of these increasingly high prices, individuals and industry used less oil, stepped up their conservation efforts, or switched to alternative fuels. The demand for crude oil declined. However, the Organization of Petroleum Exporting Countries (OPEC), and particularly Saudi Arabia, cut its output during the first half of the 1980s to keep the price from declining dramatically. (In 2006 the member countries of OPEC were Algeria, Indonesia, Iran, Iraq, Kuwait, Libya, Nigeria, Qatar, Saudi Arabia, the United Arab Emirates, and Venezuela.)

In 1985 Saudi Arabia moved to increase its market share of crude oil exports by increasing its production. (Saudi Arabia was and still is the world's largest producer and exporter of oil and is a key member of OPEC.) Other OPEC members followed suit, which resulted in a glut of crude oil on the world market. Crude oil prices fell sharply in early 1986, and imports to the United States increased.

THE UPS AND DOWNS OF OIL PRICES FROM THE 1990S THROUGH 2002. In August 1990 Iraq invaded Kuwait. The United Nations responded by placing an embargo on all crude oil and oil products from both countries. Oil prices rose suddenly and sharply, but non-OPEC countries in Central America, western Europe, and Asia, along with the United States, stepped up their production to fill the gap in world supplies. After the United Nations (UN) approved the use of force against Iraq, starting in October 1990, prices fell quickly. (See Figure 1.2.)

The collapse of Asian economies in the mid-1990s led to a further drop in the demand for energy, and petroleum prices dipped sharply in the late 1990s. OPEC reacted by curtailing production, which boosted prices in 2000. (See Figure 1.2.) World crude oil prices then declined through 2001 as global demand dropped because of weakening economies (especially in the United States) and reduced demand for jet fuel following the September 11, 2001, terrorist attacks in the United States. Fear of an increased worldwide economic downturn also added to the decline in crude oil prices.

In late 2002 attacks and counterattacks between Palestinians and Israelis caused concerns that Iraq might halt its crude oil shipments to countries that supported the Jewish state of Israel over Islamic Palestine. Additionally, concerns existed that the Middle East region might become destabilized should the United States invade Iraq, which has the second-largest oil reserve in the world. Moreover, Venezuelan oil workers went on strike, which cut off exports from Venezuela. These three factors were the primary causes of the rise in crude oil prices by the end of 2002. (See Figure 1.2.)

VOLATILITY AND RECORD HIGHS IN OIL PRICES FROM 2003 TO 2006. In early 2003 a U.S. war with Iraq seemed imminent. In addition, because of a cold winter and the Venezuelan strike, U.S. crude oil inventories had declined. As a result the price of crude oil rose to nearly $40 per barrel in February 2003, the highest in twenty-nine months. When the war began on March 19, 2003, Iraqi oil fields were shut down. Other oil-producing countries stepped up production to offset the shortfall, however. In addition, the Venezuelan strike had ended. The price of oil declined dramatically to about $27 per barrel by the beginning of May 2003. (See Figure 1.2.)

The lower price did not prevail, however. By June 2003 the price of oil rose above $30 per barrel, largely because supplies of crude oil were low and demand was high: Summer, when Americans drive the most, was just starting. The price of crude oil continued to climb over the summer and was pushed higher in the fall when the U.S. dollar sank to a record low against the euro. By December 2003 it had risen to nearly $34 per barrel. (See Figure 1.2.)

FIGURE 1.2

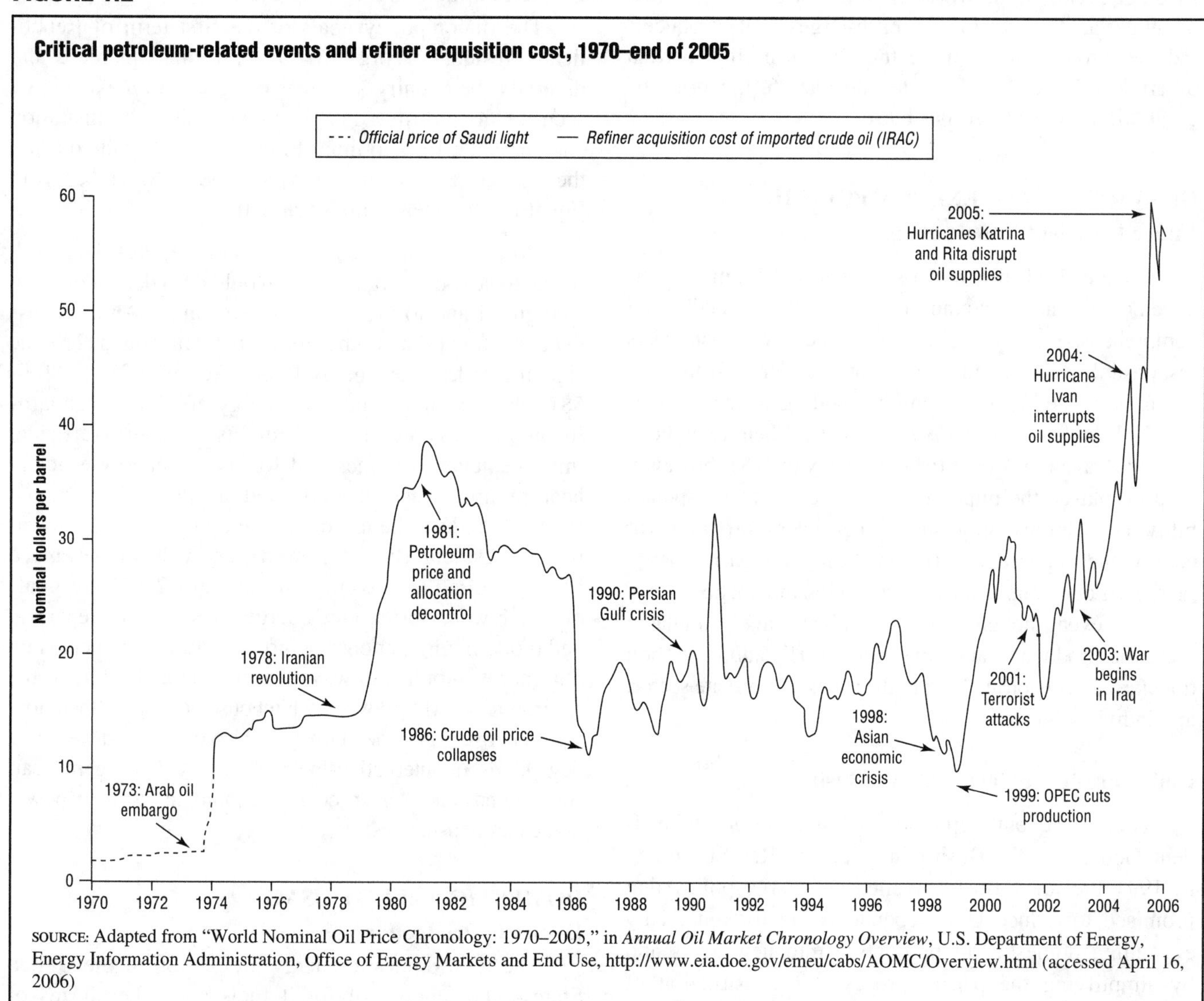

SOURCE: Adapted from "World Nominal Oil Price Chronology: 1970–2005," in *Annual Oil Market Chronology Overview*, U.S. Department of Energy, Energy Information Administration, Office of Energy Markets and End Use, http://www.eia.doe.gov/emeu/cabs/AOMC/Overview.html (accessed April 16, 2006)

In 2004 crude oil prices continued to rise because of political uncertainty, the weakened U.S. dollar, the weather, and tight supplies. Increases reflected growing demand from the world's three largest oil consumers—the United States, China, and Japan—as well as concern about terrorist attacks in Spain, Iraq, Pakistan, Saudi Arabia, and other areas. Moreover, sabotage of Iraq's northern oil pipelines prevented the country from producing the amount of oil that was expected. By March 2004 the price soared to a thirteen-year high of about $38 per barrel. By August the price of crude reached more than $45 per barrel. Then in September Hurricane Ivan hit the Gulf of Mexico, forcing the evacuation of oil workers from offshore platforms and delaying oil tankers from Venezuela. By October 2004 a barrel of crude oil cost more than $50 for the first time. (See Figure 1.2.)

In March 2005 sabotage forced Iraq to close the northern pipeline it used to export crude oil. In the following months OPEC agreed to increase its production to allow more oil into the world market to help reduce prices. In August 2005, however, Hurricane Katrina hit the Gulf Coast of the United States, severely affecting oil and natural gas production and oil refining there. (See Figure 1.2.) Hurricane Rita struck the Gulf Coast a month later. Because of the hurricanes, more than a quarter of U.S. oil refining capacity was shut down. In response, President Bush directed the Department of Energy (DOE) to release as much as 30 million barrels of crude oil from the Strategic Petroleum Reserve. (See Chapter 2.) To bring the price of oil down on the world market, the International Energy Agency released 60 million tons of crude oil to the United States. The crude oil price began to drop, falling below $60 per barrel in October and to $56 per barrel in November. However, by December 2005 the price bounced back up.

The effects of the hurricanes on U.S. refining capacity lasted into 2006. The price of oil was pushed higher by other events in the world as well. For example, unrest in Nigeria, the world's twelfth-largest oil producer, reduced production by more than half a million barrels of crude oil per day. By late April 2006, the price of crude oil surpassed $70 per barrel.

GOVERNMENTAL ENERGY POLICIES

Under President Ronald Reagan

In a televised speech in 1977, President Jimmy Carter, a Democrat, said the country could have "an effective and comprehensive energy policy only if the government takes responsibility for it and if the people understand the seriousness of the challenge and are willing to make sacrifices" (*Vital Speeches of the Day*, 1977). When Republican Ronald Reagan took over the presidency in 1981, however, he downplayed the importance of governmental responsibility. His administration sharply cut federal programs for energy and opposed governmental intervention in energy markets. For example, the administration did not tax energy imports, even though doing so might have stimulated domestic production and conservation. His administration transferred decision-making to the states, the private sector, and individuals.

Under President George H. W. Bush

The subsequent Republican administration of President George H. W. Bush continued the Reagan policy. In 1991 President Bush unveiled an energy policy that promised to reduce U.S. dependence on foreign oil by expanding domestic oil production into new areas and by simplifying the permit process for construction of nuclear power plants. Both proposals put him in conflict with conservationists, who objected to increased offshore drilling, especially in the coastal plain of the Arctic National Wildlife Refuge in Alaska. They also wanted to see automobile fuel economy improved and conservation methods stressed, rather than the use of nuclear power. President Bush's proposals did not include government-directed conservation efforts or tax incentives.

Under President Bill Clinton

The Democratic administration of President Bill Clinton sought a larger role for government in energy and environmental issues, although the major energy bills it proposed were not passed by the Republican-dominated Congress. Nevertheless, the Clinton administration did increase funds for alternative-energy research, mandate new energy-efficiency measures, and enforce emission standards. The administration also opened up several areas for oil exploration, including some Alaskan and offshore areas.

Under President George W. Bush

The major policy goals of the first term of Republican President George W. Bush were to increase and diversify the country's sources of oil and to make energy security a priority. For example, his administration encouraged efforts to import more Russian crude oil into the United States and reopened the U.S. embassy in Equatorial Guinea, an oil-rich nation.

However, in his second term in office, President Bush began to advocate policies that would curb dependence on foreign oil and to promote efforts to make America less dependent on oil and other fossil fuels. In August 2005 he signed into law the Energy Policy Act of 2005 (PL-109-58), which set new minimum energy-efficiency standards for appliances; provided tax credits for energy-efficient improvements to homes and for use of energy-efficient heat pumps, water heaters, and air conditioners; and sought to reduce the use of energy by the federal government. President Bush followed that with an Advanced Energy Initiative, unveiled in February 2006, the goals of which were "promoting energy conservation, repairing and modernizing our energy infrastructure, and increasing our energy supplies in ways that protect and improve the environment" (http://www.whitehouse.gov/stateoftheunion/2006/energy/print/index.html). He also proposed developing new kinds of alternative-fuel vehicles and using nuclear, solar, wind, and "clean coal" technologies to help power homes and businesses.

DOMESTIC ENERGY USAGE

Domestic Production

The total domestic energy production of the United States—the amount of fossil fuels and other forms of energy that were mined, pumped, or otherwise originated in the United States—has more than doubled since 1949, rising from 31.7 quadrillion British thermal units (Btu) in 1949 to 70.4 quadrillion Btu in 2004. (See Table 1.1 and Figure 1.3; Figure 1.3 shows how energy production related to energy consumption as of 2004.) One quadrillion Btu equals the energy produced by approximately 170 million barrels of crude oil. Large production and consumption figures are given in these units to make it easier to compare the various types of energy, which come in different forms.

Table 1.1 and Figure 1.4 show that after declining in the 1950s, energy produced from coal increased fairly steadily between 1960 and 2004, with its highest level recorded in 1998 (23.9 quadrillion Btu). The production of oil rose from 1949 through 1970, but by 2004 it had declined to about the level produced in 1950. Natural gas production quadrupled between 1949 and 1970, from 5.4 quadrillion Btu to 21.7 quadrillion Btu; after peaking at 22.3 quadrillion Btu in 1971, natural gas production has generally ranged between 17 and 21 quadrillion Btu since

TABLE 1.1

Energy production by source, selected years 1949–2004

[Quadrillion Btu]

	Fossil fuels						Renewable energy[a]							
Year	Coal	Natural gas (dry)	Crude oil[b]	Natural gas plant liquids	Total	Nuclear electric power	Conventional hydroelectric power	Wood, waste, alcohol[c]	Geothermal	Solar	Wind	Total	Total
1949	11.974	5.377	10.683	0.714	28.748	0.000	1.425	1.549	NA	NA	NA	2.974	31.722
1950	14.060	6.233	11.447	0.823	32.563	0.000	1.415	1.562	NA	NA	NA	2.978	35.540
1955	12.370	9.345	14.410	1.240	37.364	0.000	1.360	1.424	NA	NA	NA	2.784	40.148
1960	10.817	12.656	14.935	1.461	39.869	0.006	1.608	1.320	0.001	NA	NA	2.929	42.804
1965	13.055	15.775	16.521	1.883	47.235	0.043	2.059	1.335	0.004	NA	NA	3.398	50.676
1970	14.607	21.666	20.401	2.512	59.186	0.239	2.634	1.431	0.011	NA	NA	4.076	63.501
1971	13.186	22.280	20.033	2.544	58.042	0.413	2.824	1.432	0.012	NA	NA	4.268	62.723
1972	14.092	22.208	20.041	2.598	58.938	0.584	2.864	1.503	0.031	NA	NA	4.398	63.920
1974	14.074	21.210	18.575	2.471	56.331	1.272	3.177	1.540	0.053	NA	NA	4.769	62.372
1976	15.654	19.480	17.262	2.327	54.723	2.111	2.976	1.713	0.078	NA	NA	4.768	61.602
1978	14.910	19.485	18.434	2.245	55.074	3.024	2.937	2.038	0.064	NA	NA	5.039	63.137
1980	18.598	19.908	18.249	2.254	59.008	2.739	2.900	2.485	0.110	NA	NA	5.494	67.241
1982	18.639	18.319	18.309	2.191	57.458	3.131	3.266	2.615	0.105	NA	NA	5.985	66.574
1984	19.719	18.008	18.848	2.274	58.849	3.553	3.386	2.880	0.165	s	s	6.431	68.832
1986	19.509	16.541	18.376	2.149	56.575	4.380	3.071	2.841	0.219	s	s	6.132	67.087
1988	20.738	17.599	17.279	2.260	57.875	5.587	2.334	2.937	0.217	s	s	5.489	68.951
1990	22.456	18.326	15.571	2.175	58.529	6.104	3.046	2.662	0.336	0.060	0.029	6.133	70.765[R]
1992	21.629	18.375	15.223	2.363	57.590	6.479	2.617	2.847	0.349	0.064	0.030	5.907	69.976[R]
1994	22.111	19.348	14.103	2.391	57.952	6.694	2.683	2.939	0.338	0.069	0.036	6.065	70.711[R]
1996	22.684	19.344	13.723	2.530	58.281	7.087	3.590	3.127	0.316	0.071	0.033	7.137	72.504[R]
1998	23.935	19.613	13.235	2.420	59.204	7.068	3.297	2.835	0.328	0.070	0.031	6.561	72.833[R]
1999	23.186	19.341	12.451	2.528	57.505	7.610	3.268	2.885	0.331	0.069	0.046	6.599	71.714[R]
2000	22.623	19.662	12.358	2.611	57.254	7.862	2.811	2.907	0.317	0.066	0.057	6.158	71.274[R]
2001	23.490[R]	20.205	12.282	2.547	58.523[R]	8.033	2.242[R]	2.640	0.311	0.065	0.070[R]	5.328[R]	71.884[R]
2002	22.622[R]	19.439[R]	12.163	2.559	56.783[R]	8.143	2.689[R]	2.648[R]	0.328	0.064	0.105	5.835[R]	70.762[R]
2003	21.970[R]	19.626[R]	12.026[R]	2.346[R]	55.968[R]	7.959[R]	2.825[R]	2.740[R]	0.339[R]	0.064[R]	0.115[R]	6.082[R]	70.009[R]
2004[P]	22.686	19.339	11.528	2.468	56.021	8.232	2.725	2.845	0.340	0.063	0.143	6.116	70.369

[a]Electricity net generation from conventional hydroelectric power, geothermal, solar, and wind; consumption of wood, waste, and alcohol fuels; geothermal heat pump and direct use energy; and solar thermal direct use energy.
[b]Includes lease condensate.
[c]"Alcohol" is ethanol blended into motor gasoline.
R=Revised. P=Preliminary. NA=Not available. s=Less than 0.0005 quadrillion Btu.
Notes: This table no longer shows energy consumption by hydroelectric pumped storage plants. The change was made because most of the electricity used to pump water into elevated storage reservoirs is generated by plants other than pumped-storage plants; thus, the associated energy is already accounted for in other data columns in this table (such as "conventional hydroelectric power," "coal," "natural gas," and so on). Totals may not equal sum of components due to independent rounding.

SOURCE: Adapted from "Table 1.2. Energy Production by Source, Selected Years, 1949–2004 (Quadrillion Btu)," in *Annual Energy Review 2004*, U.S. Department of Energy, Energy Information Administration, Office of Energy Markets and End Use, August, 2005, http://www.eia.doe.gov/emeu/aer/pdf/aer.pdf (accessed April 5, 2006)

that time. Energy produced from nuclear power rose dramatically during the 1970s, from 0.2 quadrillion Btu in 1970 to 3 quadrillion Btu in 1978; it had doubled again by the early 1990s and reached a peak of 8.2 quadrillion Btu in 2004. Hydroelectric and biofuel power production (wood, waste, alcohol) reached their highest levels during the mid-1990s but have not shown dramatic changes over the past several decades. At 22.7 quadrillion Btu, more energy was produced from coal in the United States during 2004 than from any other energy source. Energy produced from natural gas was second (19.3 quadrillion Btu), followed by oil (11.5 quadrillion Btu), and nuclear electric power (8.2 quadrillion Btu).

Domestic Consumption

Although total domestic energy production more than doubled from 1949 to 2004, total domestic energy consumption—the amount of energy used in the United States—more than tripled during that time. (See Figure 1.1.) This increase did not happen in an even progression over those years. A doubling of domestic energy consumption occurred from 1949 to 1973, increasing from 30 quadrillion Btu to 74 quadrillion Btu. However, after huge oil price increases in 1973, energy consumption fell, then rose, and then fell again, eventually returning to 1973 levels by 1984. The level of domestic energy consumption remained relatively stable through 1986, but following a drop in the price of crude oil that year, imports of oil began to rise and energy consumption increased quite steadily. In 2004 domestic energy consumption reached an all-time high of 99.7 quadrillion Btu.

One of the reasons that energy consumption has grown in the United States is the number of people who use it: According to U.S. Census Bureau data, the U.S. population grew from 151.3 million in 1950 to 281.4

FIGURE 1.3

Energy flow (overview), 2004

[In quadrillion Btu]

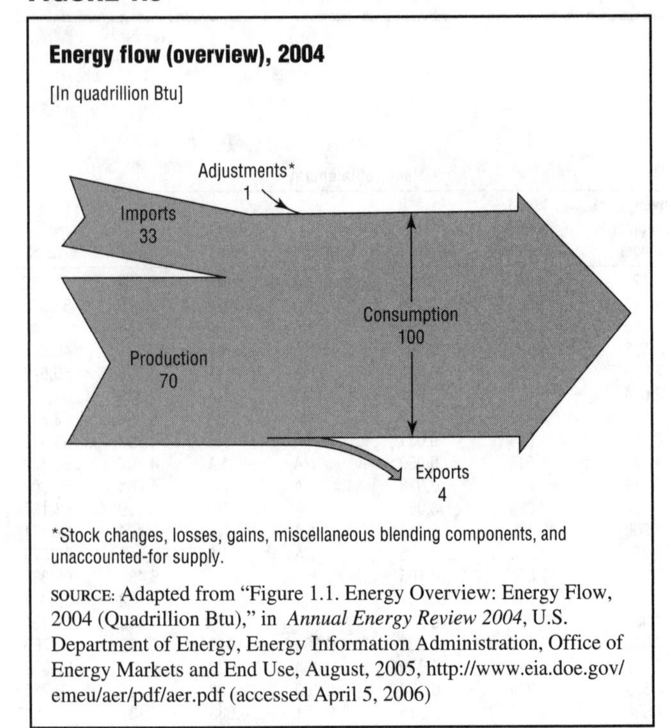

*Stock changes, losses, gains, miscellaneous blending components, and unaccounted-for supply.

SOURCE: Adapted from "Figure 1.1. Energy Overview: Energy Flow, 2004 (Quadrillion Btu)," in *Annual Energy Review 2004*, U.S. Department of Energy, Energy Information Administration, Office of Energy Markets and End Use, August, 2005, http://www.eia.doe.gov/emeu/aer/pdf/aer.pdf (accessed April 5, 2006)

million in 2000, an increase of 88%. However, energy consumption rose by 181% during the same period. While consumption grew fairly steadily, it slowed somewhat after the 1973 oil crisis. (See Figure 1.1.) For a time, Americans became more efficient and used less energy per capita.

Figure 1.5 shows energy production and consumption flows, including types of energy sources, in 2004. Coal, which in 1973 accounted for 17% of all energy consumed, accounted for 22.4 quadrillion Btu in 2004, or 22.5% of the total consumed. Nuclear electric power, which contributed barely 1% of the nation's consumption in 1973, had grown considerably by 2004 to 8.2 quadrillion Btu, or 8.3% of the total energy consumed.

ENERGY IMPORTS AND EXPORTS

As indicated in Figure 1.1, since the late 1950s energy consumption in the United States has outpaced energy production, and the difference has been made up by importing energy sources. Imports (mainly oil) grew rapidly from 1953 through 1973 as the U.S. economy expanded using inexpensive oil. In 1973 net imports of petroleum reached almost 13 quadrillion Btu.

Although the Arab oil embargo of 1973–74—coupled with increased oil prices—momentarily slowed growth in petroleum imports, the general increase continued, with imports exceeding 19 quadrillion Btu in 1977 through 1979. During those years U.S. dependence on petroleum imports rose to more than 45% of the

FIGURE 1.4

Energy production by major source, 1949–2004

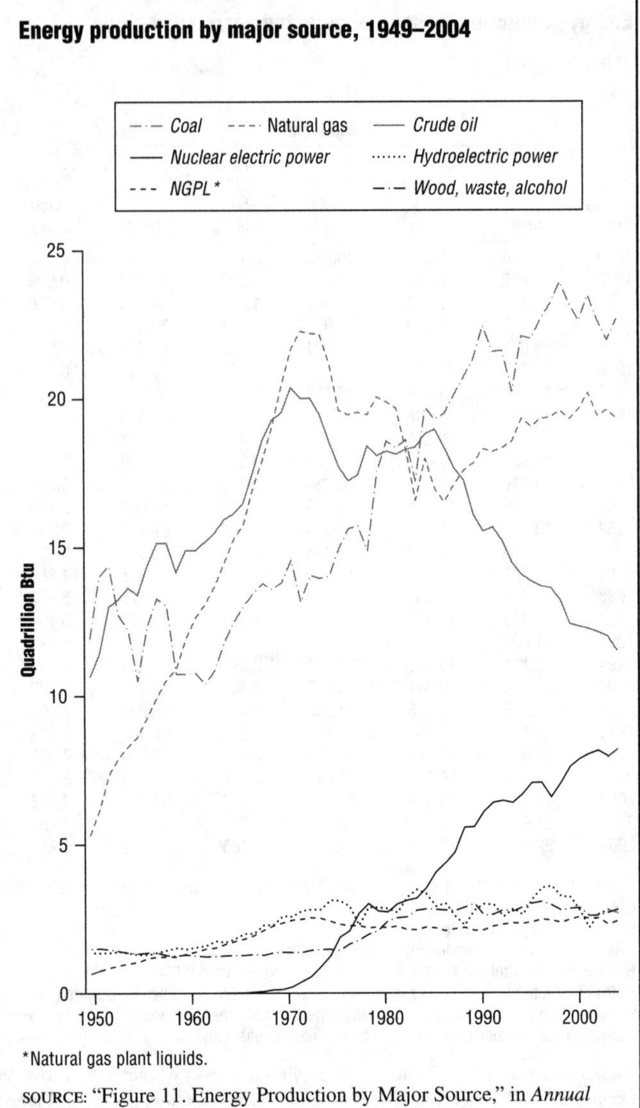

*Natural gas plant liquids.

SOURCE: "Figure 11. Energy Production by Major Source," in *Annual Energy Review 2004*, U.S. Department of Energy, Energy Information Administration, Office of Energy Markets and End Use, August, 2005, http://www.eia.doe.gov/emeu/aer/pdf/aer.pdf (accessed April 5, 2006)

nation's oil consumption. Despite the lesson of 1973, it took a second round of price increases in 1979–80, accompanied by lengthy and frustrating lines at gas stations, to persuade Americans to become less dependent on imported oil, conserve resources, or both. By 1985 U.S. dependence on foreign oil had decreased sharply, to 27.3% of oil consumption. (See Figure 1.6.)

After 1985 U.S. dependence on foreign sources of oil increased gradually, as a drop in the price of crude oil drove up demand. When Iraq invaded Kuwait in 1990, the potential threat to the flow of oil to America and other industrialized nations was one of the reasons the United States challenged Saddam Hussein in Operation Desert Storm. However, after the terrorist attacks of September 11, 2001, and the Bush administration's subsequent "war on terror," the concept of energy independence, or at

FIGURE 1.5

Energy flow (detail), 2004

[Quadrillion Btu]

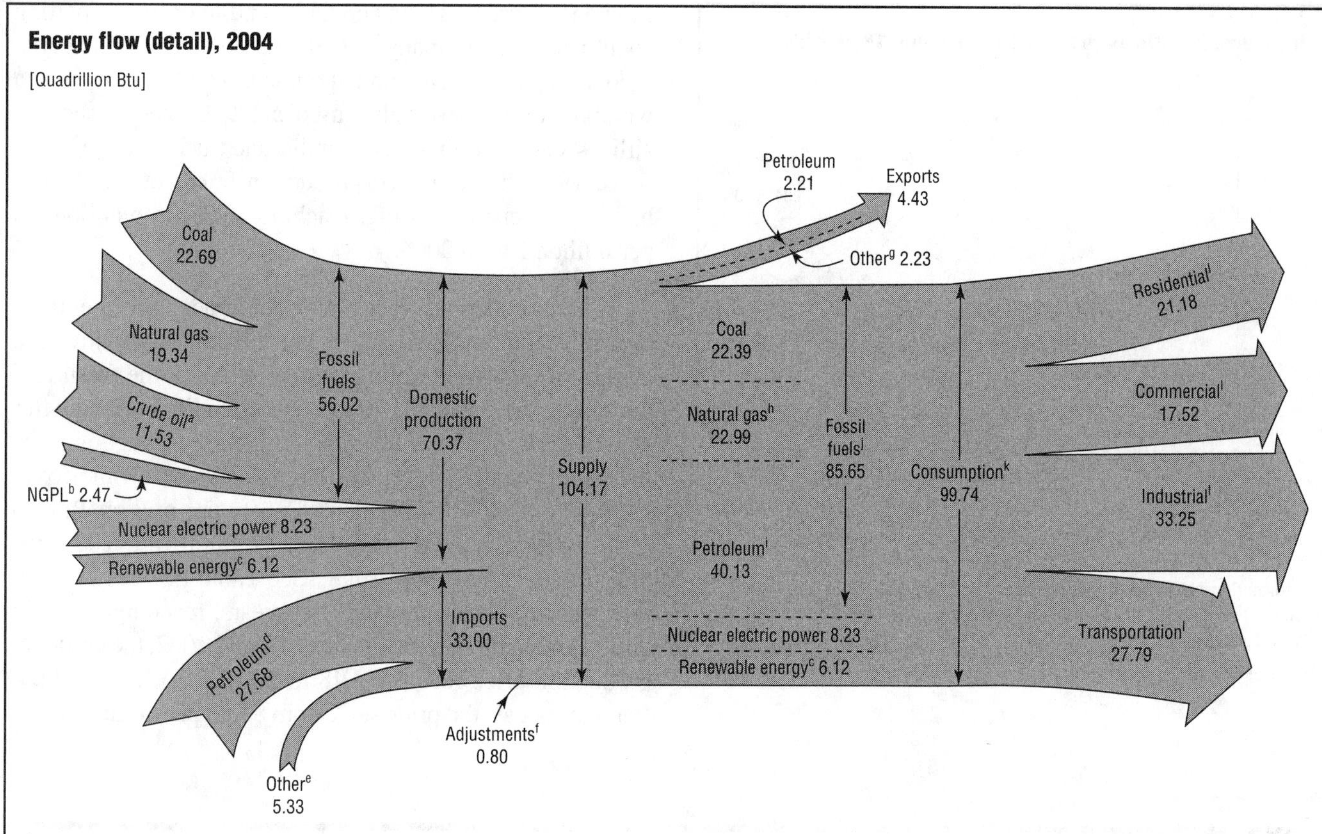

[a]Includes lease condensate.
[b]Natural gas plant liquids.
[c]Conventional hydroelectric power, wood, waste, ethanol blended into motor gasoline, geothermal, solar, and wind.
[d]Crude oil and petroleum products. Includes imports into the Strategic Petroleum Reserve.
[e]Natural gas, coal, coal coke, and electricity.
[f]Stock changes, losses, gains, miscellaneous blending components, and unaccounted-for supply.
[g]Coal, natural gas, coal coke, and electricity.
[h]Includes supplemental gaseous fuels.
[i]Petroleum products, including natural gas plant liquids.
[j]Includes 0.14 quadrillion Btu of coal coke net imports.
[k]Includes, in quadrillion Btu, 0.30 ethanol blended into motor gasoline, which is accounted for in both fossil fuels and renewable energy but counted only once in total consumption; and 0.04 electricity net imports.
[l]Primary consumption, electricity retail sales, and electrical system energy losses, which are allocated to the end-use sector's in proportion to each sector's share of total electricity retail sales.
Notes: Data are preliminary. Totals may not equal sum of components due to independent rounding.

SOURCE: "Diagram 1. Energy Flow, 2004 (Quadrillion Btu)," in *Annual Energy Review 2004*, U.S. Department of Energy, Energy Information Administration, Office of Energy Markets and End Use, August, 2005, http://www.eia.doe.gov/emeu/aer/pdf/aer.pdf (accessed April 5, 2006)

least reduced energy dependence, became increasingly important. In 2003 the United States was again at war with Iraq, and by 2004 imported oil accounted for a record 57.8% of U.S. oil consumption. (See Figure 1.6.) That year net imports of petroleum (that is, total imports minus total exports) reached 25.5 quadrillion Btu. (See Table 1.2.)

Although the United States imports energy, primarily in the form of oil, it exports small amounts of energy in the form of coal and oil. In 2004 coal exports totaled 1.3 quadrillion Btu, about 28% of U.S. energy exports. (See Figure 1.7 and Table 1.2.) The United States also exports some oil (petroleum). The reasons for exporting oil are complicated but have to do with the cost of transporting Alaskan oil; the sale of certain types of petroleum used to make steel; and exchanges with Canada and Mexico of crude products for refined products.

FOSSIL FUEL PRODUCTION PRICES

Production prices are the value of fuel produced. The combined production prices of fossil fuels (crude oil, natural gas, and coal) slowly declined from 1949 through 1972. (See fossil fuel composite, Figure 1.8.) These prices then increased dramatically from 1973 through 1981, and fell just as dramatically through 1998: The composite value of all fossil fuel prices (in real dollars, which account for inflation) dropped from $4.64 per million Btu in 1981 to $1.46 per million Btu in 1998, a decrease of more than two-thirds. (See Table 1.3.) This huge drop created economic problems in fuel-producing

FIGURE 1.6

Petroleum imports as share of consumption, 1960–2004

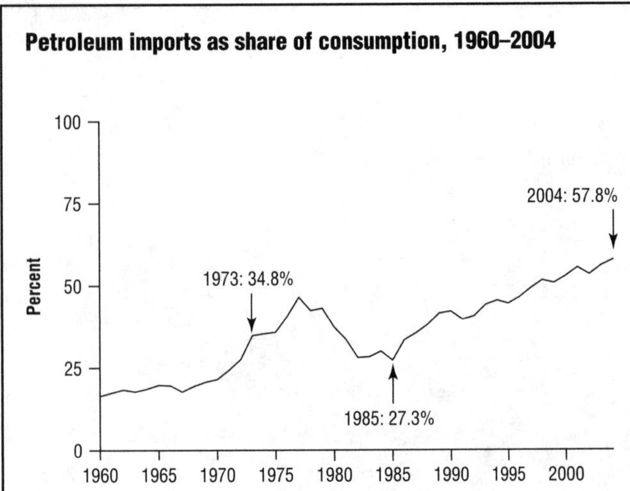

SOURCE: Adapted from "Figure 5.7. Petroleum Net Imports by Country of Origin, 1960–2004: Total Net Imports as Share of Consumption," in *Annual Energy Review 2004*, U.S. Department of Energy, Energy Information Administration, Office of Energy Markets and End Use, August, 2005, http://www.eia.doe.gov/emeu/aer/pdf/aer.pdf (accessed April 5, 2006)

American states, such as Texas, Louisiana, Oklahoma, Montana, West Virginia, and Ohio, and in energy-exporting countries, such as many Middle Eastern nations, Nigeria, Indonesia, Venezuela, and Trinidad. However, it was a windfall for industries that used a lot of energy, such as airlines, electric utilities, steel mills, and trucking companies. Since 1998 the combined production prices of fossil fuels have been generally rising, reaching $3.35 (in real dollars) per million Btu in 2004.

The production prices of both crude oil (the most expensive of the fossil fuels) and natural gas followed a pattern of rising and falling similar to that of the fossil fuel composite price from 1949 to 2000. (See Figure 1.8.) After slowly declining from 1949, the crude oil production price rose dramatically—the most dramatically of all the fossil fuels—from 1972, when it was $1.94 per million Btu, to 1981, when it topped at $9.27 per million Btu. The price then tumbled to $1.94 in 1998. (See Table 1.3.) However, it rose sharply during the next two years, reaching $4.61 in 2000. After a bit of a decline in 2001 and 2002, the crude oil production price rose to $4.48 in 2003 and $5.86 in 2004. For natural gas, the price sank from $3.55 per million Btu in

TABLE 1.2

Energy imports, exports, and net imports, selected years 1949–2004

	Imports					Exports					Net imports				
Year	Coal	Natural gas	Petroleum[a]	Other[b]	Total	Coal	Natural Gas	Petroleum[b]	Other[b]	Total	Coal	Natural gas	Petroleum[a]	Other[b]	Total
1949	0.01	0.00	1.43	0.01	1.45	0.88	0.02	0.68	0.01	1.59	−0.87	−0.02	0.75	s	−0.14
1950	0.01	0.00	1.89	0.02	1.91	0.79	0.03	0.64	0.01	1.47	−0.78	−0.03	1.24	0.01	0.45
1955	0.01	0.01	2.75	0.02	2.79	1.46	0.03	0.77	0.01	2.29	−1.46	−0.02	1.98	s	0.50
1960	0.01	0.16	4.00	0.02	4.19	1.02	0.01	0.43	0.01	1.48	−1.02	0.15	3.57	0.01	2.71
1965	s	0.47	5.40	0.01	5.89	1.38	0.03	0.39	0.03	1.83	−1.37	0.44	5.01	−0.02	4.06
1970	s	0.85	7.47	0.02	8.34	1.94	0.07	0.55	0.08	2.63	−1.93	0.77	6.92	−0.05	5.71
1972	s	1.05	10.30	0.04	11.39	1.53	0.08	0.47	0.04	2.12	−1.53	0.97	9.83	s	9.27
1974	0.05	0.99	13.13	0.14	14.30	1.62	0.08	0.46	0.04	2.20	−1.57	0.91	12.66	0.10	12.10
1976	0.03	0.99	15.67	0.07	16.76	1.60	0.07	0.47	0.04	2.17	−1.57	0.92	15.20	0.03	14.59
1978	0.07	0.99	17.82	0.21	19.11	1.08	0.05	0.77	0.02	1.92	−1.00	0.94	17.06	0.19	17.19
1980	0.03	1.01	14.66	0.10	15.80	2.42	0.05	1.16	0.07	3.69	−2.39	0.96	13.50	0.04	12.10
1982	0.02	0.95	10.78	0.12	11.86	2.79	0.05	1.73	0.04	4.61	−2.77	0.90	9.05	0.08	7.25
1984	0.03	0.85	11.43	0.16	12.47	2.15	0.06	1.54	0.03	3.79	−2.12	0.79	9.89	0.12	8.68
1986	0.06	0.75	13.20	0.15	14.15	2.25	0.06	1.67	0.04	4.02	−2.19	0.69	11.53	0.11	10.13
1988	0.05	1.30	15.75	0.20	17.30	2.50	0.07	1.74	0.05	4.37	−2.45	1.22	14.01	0.15	12.93
1990	0.07	1.55	17.12	0.08	18.82	2.77	0.09	1.82	0.07	4.75	−2.70	1.46	15.29	0.01	14.06
1992	0.10	2.16	16.97	0.15	19.37	2.68	0.22	2.01	0.03	4.94	−2.59	1.94	14.96	0.12	14.44
1994	0.22	2.68	19.24	0.24	22.39	1.88	0.16	1.99	0.03	4.06	−1.66	2.52	17.26	0.21	18.33
1996	0.20	3.00	20.29	0.21	23.70	2.37	0.16	2.06	0.05	4.63	−2.17	2.85	18.23	0.16	19.07
1998	0.22	3.22	22.91	0.23	26.58	2.09	0.16	1.97	0.07	4.30	−1.87	3.06	20.94	0.16	22.28
1999	0.23	3.66	23.13	0.23	27.25	1.53	0.16	1.95	0.07	3.71	−1.30	3.50	21.18	0.16	23.54
2000	0.31	3.87	24.53	0.26	28.97	1.53	0.25	2.15	0.08	4.01	−1.21	3.62	22.38	0.18	24.97
2001	0.49	4.07	25.40	0.19	30.16	1.27	0.38	2.04	0.09	3.77	−0.77	3.69	23.36	0.10	26.39
2002	0.42	4.10	24.68	0.20	29.41	1.03	0.52	2.04	0.07	3.66	−0.61	3.58	22.63	0.14	25.74
2003	0.63	4.10[R]	26.22[R]	0.17	31.11[R]	1.12	0.70	2.15[R]	0.10	4.07[R]	−0.49	3.40[R]	24.07	0.07	27.05[R]
2004[P]	0.68	4.36	27.68	0.29	33.00	1.25	0.86	2.21	0.11	4.43	−0.57	3.49	25.47	0.18	28.57

[a]Includes imports into the Strategic Petroleum Reserve, which began in 1977.
[b]Coal coke and small amounts of electricity transmitted across U.S. borders with Canada and Mexico.
R=Revised. P=Preliminary. s=Less than 0.005 quadrillion Btu and greater than 90.005 quadrillion Btu.
Notes: Includes trade between the United States (50 states and the District of Columbia) and its territories and possessions. Totals or net import items may not equal sum of components due to independent rounding. For data not shown for 1951–1969, see http://www.eia.doe.gov/emeu/aer/overview.html.

SOURCE: Adapted from "Table 1.4. Energy Imports, Exports, and Net Imports, Selected Years, 1949–2004 (Quadrillion Btu)," in *Annual Energy Review 2004*, U.S. Department of Energy, Energy Information Administration, Office of Energy Markets and End Use, August 2005, http://www.eia.doe.gov/emeu/aer/pdf/aer.pdf (accessed April 5, 2006)

FIGURE 1.7

Energy exports, 1949–2004

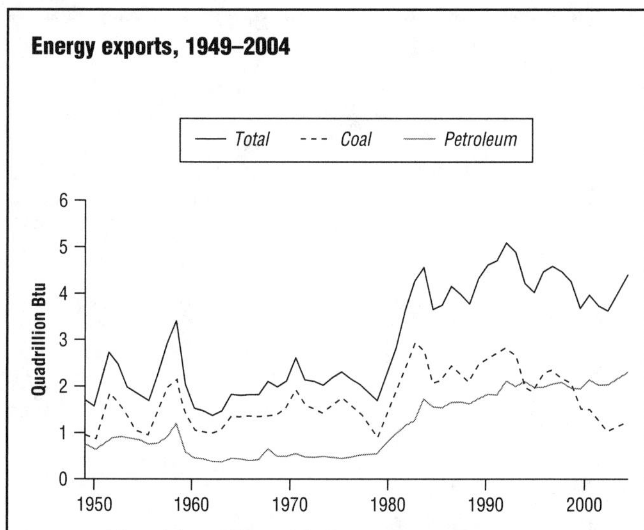

SOURCE: Adapted from "Figure 1.4. Energy Imports, Exports, and Net Imports, 1949–2004: Energy Exports," in *Annual Energy Review 2004*, U.S. Department of Energy, Energy Information Administration, Office of Energy Markets and End Use, August, 2005, http://www.eia.doe.gov/emeu/aer/pdf/aer.pdf (accessed April 5, 2006)

FIGURE 1.8

Fossil fuel production prices, 1949–2004

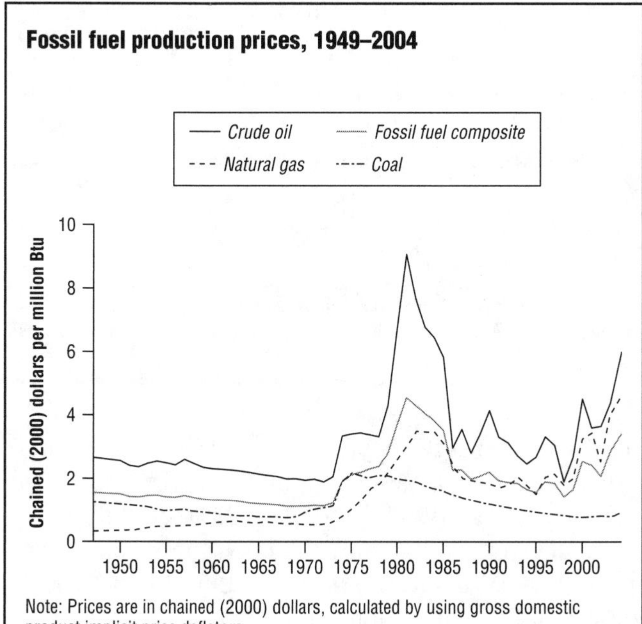

Note: Prices are in chained (2000) dollars, calculated by using gross domestic product implicit price deflators.

SOURCE: Adapted from "Figure 3.1. Fossil Fuel Production Prices: Prices, 1949–2004," in *Annual Energy Review 2004*, U.S. Department of Energy, Energy Information Administration, Office of Energy Markets and End Use, August 2005, http://www.eia.doe.gov/emeu/aer/pdf/aer.pdf (accessed April 5, 2006)

1984 to $1.83 in 1998. It rose to an all-time high of $4.59 per million Btu by 2004. (See Table 1.3.)

The story of coal prices is a bit different from that of the other fossil fuels. Coal production prices rose from 1970 ($0.97 per million Btu) to 1975 ($2.22), but then—unlike the production prices for natural gas and crude oil—declined steadily through 2000 ($0.80). Coal prices rose a bit in 2004 to $0.90 per million Btu. (See Figure 1.8 and Table 1.3.)

ENERGY USE BY SECTOR

Energy use can be classified into four main "end-use" sectors: commercial, industrial, residential, and transportation. Historically, industry has been the largest energy-consuming sector of the economy. In 2004 industry used about 33 quadrillion Btu, compared with approximately 28 quadrillion Btu in the transportation sector, 21 quadrillion Btu in the residential sector, and 18 quadrillion Btu in the commercial sector. (See Figure 1.9.)

Within sectors, energy sources have changed over time. For example, in the commercial sector, coal was the leading energy source through 1953 but declined dramatically in favor of petroleum (through 1962) and then natural gas. In 1990 energy in the form of electricity pulled ahead as the leading energy source. (See Figure 1.10.) Similarly, coal was the leading energy source in the residential sector in 1949. (See Figure 1.11.) Natural gas quickly took over, with petroleum in second place. In 1979 electricity took over second place from petroleum. Industry used more coal than natural gas or petroleum through 1957, but after that natural gas and petroleum took over as nearly equally preferred energy sources. (See Figure 1.12.) In transportation, reliance on petroleum has been increasing since 1949. (See Figure 1.13.)

Not included in the four main sectors of energy consumption is the electric power sector. This sector includes electric utilities that generate, transmit, distribute, and sell electricity for use by the other four sectors—in homes, businesses, and industry. Figure 1.14 shows that most of the electricity for the United States is generated by burning coal. The electric power sector consumed approximately 20 quadrillion Btu of coal in 2004, almost all the coal used in the United States. Electricity is also generated by nuclear power, natural gas, renewable energy sources (hydroelectric, wood, waste, geothermal, solar, and wind), and petroleum. As Figure 1.14 shows, renewable sources generate very little of the nation's electricity.

INTERNATIONAL ENERGY USAGE

World Production

World production of energy rose from 215.4 quadrillion Btu in 1970 to 417.1 quadrillion Btu in 2003. (See Table 1.4.) The Energy Information Administration (EIA) of the U.S. Department of Energy stated in its report *Annual Energy Review 2004* (2005) that the world's total output of primary energy increased by 94% from 1970 to 2003. In 2002 fossil fuels were the most heavily produced fuel, accounting for 86% of all energy produced worldwide. Renewable energy

TABLE 1.3

Fossil fuel production prices, selected years, 1949–2004

[Dollars per million Btu]

Year	Coal[a]		Natural gas[b]		Crude oil[c]		Fossil fuel composite[d]		
	Nominal	Real[e]	Nominal	Real[e]	Nominal	Real[e]	Nominal	Real[e]	Percent change[f]
1949	0.21	1.29[R]	0.05	0.33	0.44	2.68	0.26	1.60	—
1950	0.21	1.25	0.06	0.38	0.43	2.62	0.26	1.54	−3.6[R]
1955	0.19	0.99	0.09	0.48	0.48	2.55	0.27	1.45	−3.7[R]
1960	0.19	0.92	0.13	0.60	0.50	2.36	0.28	1.35	−2.3[R]
1965	0.18	0.82	0.15	0.65[R]	0.49	2.19	0.28	1.23	−1.5[R]
1970	0.27	0.97	0.15	0.56	0.55	1.99	0.32	1.15[R]	0.8[R]
1972	0.33	1.09	0.17	0.57	0.58	1.94	0.35	1.16	−1.4
1974	0.69	1.98	0.27	0.79	1.18	3.41	0.68	1.95	55.8
1975	0.85	2.22	0.40	1.06	1.32	3.48	0.82	2.16	10.9
1976	0.86	2.13	0.53	1.32	1.41	3.51	0.90	2.24	3.8[R]
1978	0.98	2.15	0.84	1.83	1.55	3.39	1.12	2.44	3.4[R]
1980	1.10	2.04	1.45	2.68	3.72	6.89	2.04	3.78	32.1
1981	1.18	2.00	1.80	3.04	5.48	9.27	2.75	4.64	22.9
1982	1.23	1.95	2.22	3.54	4.92	7.84	2.76	4.40	−5.3
1984	1.16	1.72	2.40	3.55	4.46	6.60	2.65	3.91	−5.6
1986	1.09	1.52	1.75	2.45	2.16	3.03	1.65	2.32	−35.6
1988	1.01	1.34	1.52	2.01	2.17	2.87	1.53	2.03	−12.8
1990	1.00	1.22	1.55	1.90	3.45	4.23	1.84	2.26	6.2
1992	0.97	1.12	1.57	1.82	2.76	3.19	1.66	1.92	−3.0
1994	0.91	1.01	1.67	1.86	2.27	2.52	1.53	1.69	−10.4[R]
1996	0.87	0.92[R]	1.96	2.09	3.18	3.39	1.82	1.94	21.4
1998	0.83	0.86	1.77	1.83	1.87	1.94	1.41	1.46	−22.8
1999	0.79	0.81	1.98	2.02	2.68	2.74	1.65	1.69	15.4
2000	0.80	0.80	3.32	3.32	4.61	4.61	2.60	2.60	54.3[R]
2001	0.83	0.82[R]	3.62	3.54	3.77	3.68	2.53	2.47	−5.0[R]
2002	0.87	0.84[R]	2.67	2.56	3.88	3.73	2.21	2.12	−14.1[R]
2003	0.87[R]	0.82	4.41[R]	4.16[R]	4.75	4.48[R]	3.10[R]	2.92[R]	37.5[R]
2004[P]	0.97	0.90	4.96	4.59	6.34	5.86	3.63	3.35	14.8

[a]Free-on-board (f.o.b.) rail/barge prices, which are the f.o.b. prices of coal at the point of first sale, excluding freight or shipping and insurance costs.
[b]Wellhead prices.
[c]Domestic first purchase prices.
[d]Derived by multiplying the price per Btu of each fossil fuel by the total Btu content of the production of each fossil fuel and dividing this accumulated value of total fossil fuel production by the accumulated Btu content of total fossil fuel production.
[e]In chained (2000) dollars, calculated by using gross domestic product implicit price deflators.
[f]Based on real values.
R=Revised. P=Preliminary. —=Not applicable.
Note: For data not shown for 1951–1969, see http://www.eia.doe.gov/emeu/aer/finan.html.

SOURCE: Adapted from "Table 3.1. Fossil Fuel Production Prices, Selected Years, 1949–2004 (Dollars per Million Btu)," in *Annual Energy Review 2004*, U.S. Department of Energy, Energy Information Administration, Office of Energy Markets and End Use, August, 2005, http://www.eia.doe.gov/emeu/aer/pdf/aer.pdf (accessed April 5, 2006)

accounted for 8% of all energy produced worldwide that year, and nuclear power accounted for 6%.

In 2003 the United States, Russia, and China were by far the leading producers of energy, followed by Saudi Arabia, Canada, and Iran. (See Figure 1.15.) According to the EIA, almost all the energy produced in the Middle East is in the form of oil or natural gas, while coal is a major source in China. Canada is the leading producer of hydroelectric power and alone accounted for 12.6% of world production of this form of power in 2003. France produces the highest percentage of its energy from nuclear power—more than 75% in 2004.

World Consumption

Table 1.5 shows world consumption of energy by region from 1980 to 2003. The five countries singled out in the table—the United States, China, Russia, Japan, and Germany—together consumed about 50% of the world's total energy supply in 2003. The United States, by far the world's largest consumer of energy, used about 98.8 quadrillion Btu in 2003, or 23% of the energy consumed worldwide. This amount was more than twice the 45.5 quadrillion Btu consumed in China. Russia consumed 29.1 quadrillion Btu in 2003.

FUTURE TRENDS IN ENERGY CONSUMPTION, PRODUCTION, AND PRICES

The EIA's *Annual Energy Outlook*, which forecasts energy supply, demand, and prices, is used by decision makers in the public and private sectors. The EIA's projections, through 2030, are based on current U.S. laws, regulations, and economic conditions.

Total energy consumption in the United States is projected to increase from 99.7 quadrillion Btu in 2004 to 133.9 quadrillion Btu in 2030, an average annual increase of 1.1%. (See Table 1.6.) That projection may

FIGURE 1.9

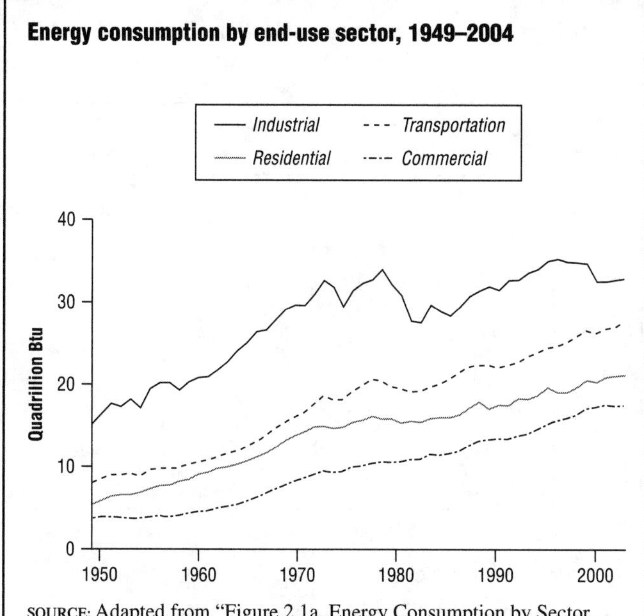

Energy consumption by end-use sector, 1949–2004

SOURCE: Adapted from "Figure 2.1a. Energy Consumption by Sector Overview: Total Consumption by End-Use Sector, 1949–2004," in *Annual Energy Review 2004*, U.S. Department of Energy, Energy Information Administration, Office of Energy Markets and End Use, August, 2005, http://www.eia.doe.gov/emeu/aer/pdf/aer.pdf (accessed April 5, 2006)

FIGURE 1.10

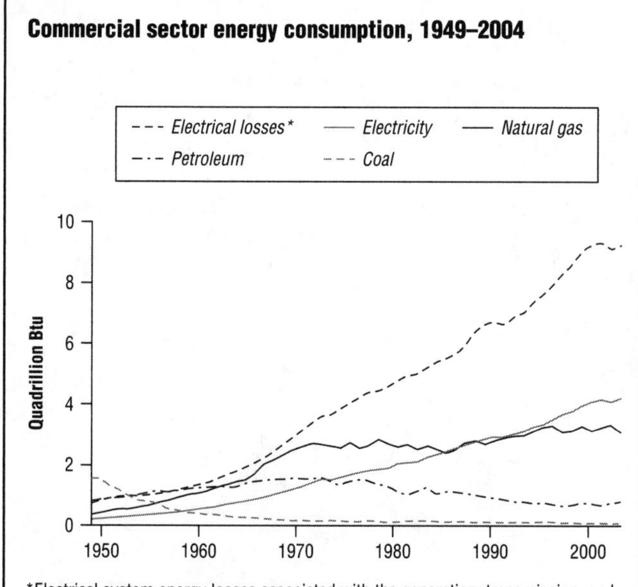

Commercial sector energy consumption, 1949–2004

*Electrical system energy losses associated with the generation, transmission, and distribution of energy in the form of electricity.

SOURCE: Adapted from "Figure 2.1b. Energy Consumption by End-Use Sector, 1949–2004: Commercial, by Major Sector," in *Annual Energy Review 2004*, U.S. Department of Energy, Energy Information Administration, Office of Energy Markets and End Use, August 2005, http://www.eia.doe.gov/emeu/aer/pdf/aer.pdf (accessed April 5, 2006)

FIGURE 1.11

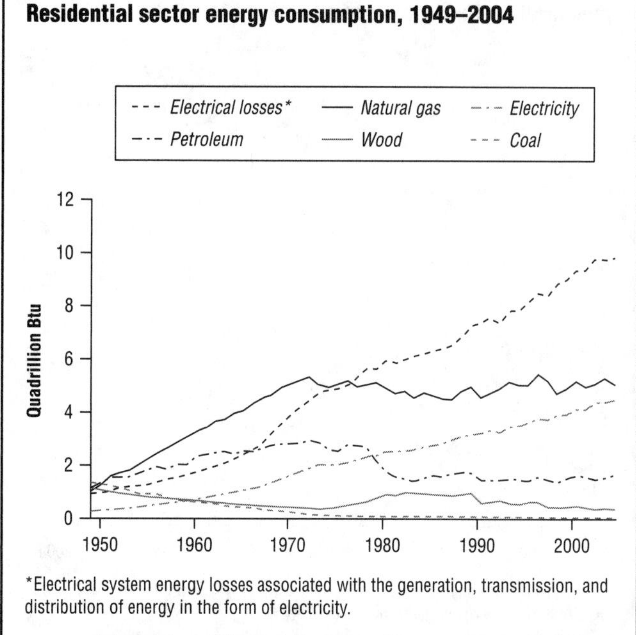

Residential sector energy consumption, 1949–2004

*Electrical system energy losses associated with the generation, transmission, and distribution of energy in the form of electricity.

SOURCE: Adapted from "Figure 2.1b. Energy Consumption by End-Use Sector, 1949–2004: Residential, by Major Sector," in *Annual Energy Review 2004*, U.S. Department of Energy, Energy Information Administration, Office of Energy Markets and End Use, August, 2005, http://www.eia.doe.gov/emeu/aer/pdf/aer.pdf (accessed April 5, 2006)

be greatly affected by such factors as economic growth and world oil prices.

The consumption of petroleum, natural gas, coal, and nonhydroelectric renewable energy sources is expected to rise significantly from 2004 to 2030. (See Figure 1.16.) The EIA projects that total petroleum demand will increase from 20.8 million barrels of oil per day in 2004 to 27.6 million barrels per day in 2030. Coal, natural gas, and renewable fuels consumption is projected to grow in part to meet the increased demand for electricity. Consumption of hydroelectric power and electricity generated from nuclear power will remain steady.

The rising consumption of petroleum by Americans is projected to lead to increasing petroleum imports by the United States through 2030. (See Figure 1.17.) Gross oil imports are projected to increase from 13.1 million barrels per day in 2004 to 18.3 million barrels per day in 2030. The EIA projects that gross petroleum imports will account for 64% of the total U.S. petroleum supply in 2030.

Electricity prices in the United States are projected to decline slightly through 2015 (after an initial jump) because of falling natural gas prices, but will then increase slowly as natural gas prices go up. (See Figure 1.18.) From 2004 to 2030 coal prices will remain relatively stable, but petroleum prices will mirror quite closely the rise, then fall, then slow rise of natural gas prices.

FIGURE 1.12

Industrial sector energy consumption, 1949–2004

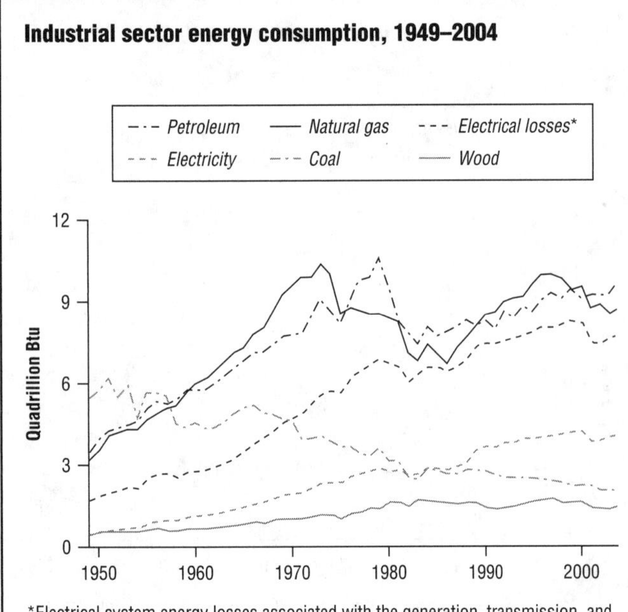

*Electrical system energy losses associated with the generation, transmission, and distribution of energy in the form of electricity.

SOURCE: Adapted from "Figure 2.1b. Energy Consumption by End-Use Sector, 1949–2004: Industrial, by Major Sector," in *Annual Energy Review 2004*, U.S. Department of Energy, Energy Information Administration, Office of Energy Markets and End Use, August, 2005, http://www.eia.doe.gov/emeu/aer/pdf/aer.pdf (accessed April 5, 2006)

FIGURE 1.13

Transportation sector energy consumption, 1949–2004

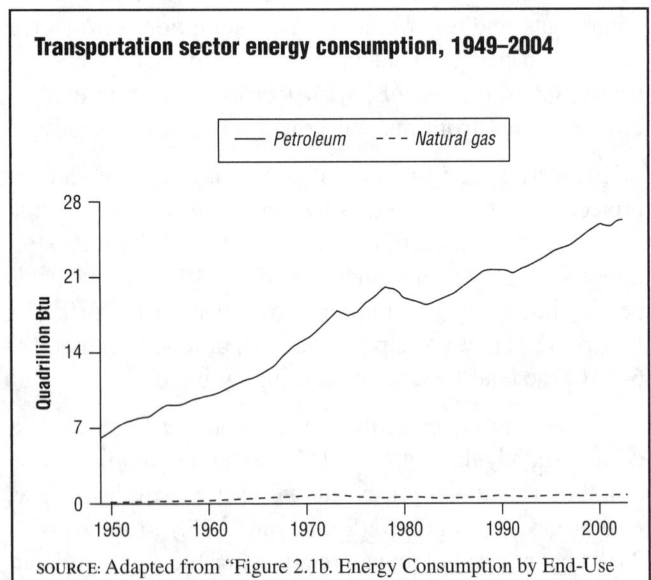

SOURCE: Adapted from "Figure 2.1b. Energy Consumption by End-Use Sector, 1949–2004: Transportation, by Major Sector," in *Annual Energy Review 2004*, U.S. Department of Energy, Energy Information Administration, Office of Energy Markets and End Use, August, 2005, http://www.eia.doe.gov/emeu/aer/pdf/aer.pdf (accessed April 5, 2006)

FIGURE 1.14

Electric power sector energy consumption, 1949–2004

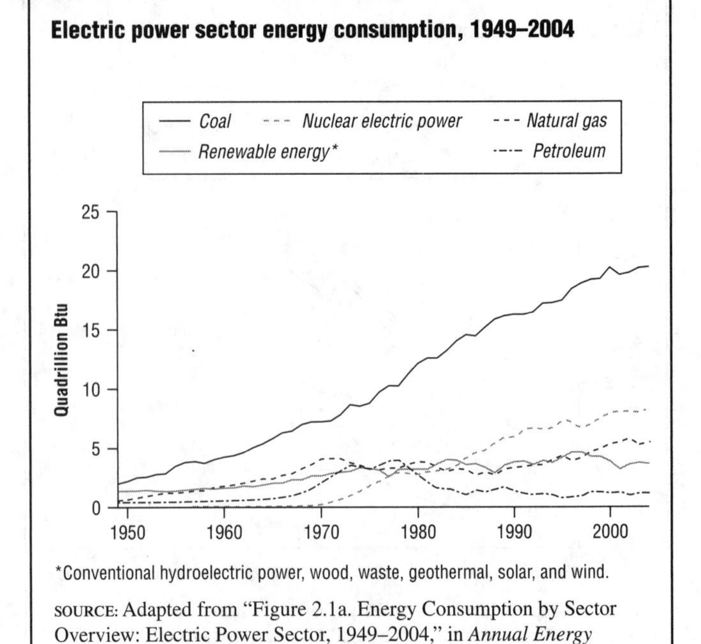

*Conventional hydroelectric power, wood, waste, geothermal, solar, and wind.

SOURCE: Adapted from "Figure 2.1a. Energy Consumption by Sector Overview: Electric Power Sector, 1949–2004," in *Annual Energy Review 2004*, U.S. Department of Energy, Energy Information Administration, Office of Energy Markets and End Use, August, 2005, http://www.eia.doe.gov/emeu/aer/pdf/aer.pdf (accessed April 5, 2006)

TABLE 1.4

World primary energy production by source, selected years 1970–2003

[Quadrillion Btu]

Year	Coal	Natural gas[a]	Crude oil[b]	Natural gas plant liquids	Nuclear electric power[c]	Hydroelectric power[c]	Geothermal[c] and other[d]	Total
1970	62.96	37.09	97.09	3.61	0.90	12.15	1.59	215.39
1972	63.65	42.08	108.52	4.09	1.66	13.31	1.68	234.99
1974	63.79	45.35	117.82	4.22	2.86	14.84	1.76	250.64
1976	67.32	47.62	122.92	4.24	4.52	15.08	1.97	263.67
1978	69.56	50.26	128.51	4.55	6.42	16.80	2.32	278.41
1980	71.24[R]	54.73	128.12	5.10	7.58	17.90[R]	2.95	287.62[R]
1982	74.25[R]	55.49	114.51	5.34	9.51	18.71[R]	3.24	281.06[R]
1984	78.38[R]	61.78	116.86	5.71	12.99	20.18[R]	3.64	299.54[R]
1986	84.28[R]	65.32	120.24	6.12	16.25	20.88[R]	3.74	316.83[R]
1988	87.94[R]	71.80	125.93	6.63	19.23	21.50[R]	3.94[R]	336.98[R]
1990	90.93[R]	75.87	129.50	6.85	20.36[R]	22.38[R]	3.96	349.85[R]
1992	85.94[R]	76.90	129.13	7.38	21.28[R]	22.76[R]	4.32	347.71[R]
1994	86.27[R]	79.18	130.46	8.25[R]	22.41[R]	24.19[R]	4.53[R]	355.29[R]
1996	88.92[R]	83.94	136.64	8.76[R]	24.11[R]	25.84[R]	4.89[R]	373.10[R]
1998	90.88[R]	85.94[R]	143.15	9.19[R]	24.32[R]	26.13[R]	4.93[R]	384.54[R]
1999	90.42[R]	87.89[R]	140.79	9.45[R]	25.10[R]	26.62[R]	5.16[R]	385.44[R]
2000	91.30[R]	91.39[R]	146.50	9.88[R]	25.68[R]	27.05[R]	5.40[R]	397.20[R]
2001	97.02[R]	93.73[R]	145.10[R]	10.51[R]	26.41[R]	26.45[R]	5.29[R]	404.51[R]
2002	96.85[R]	95.47[R]	142.73[R]	10.78[R]	26.71[R]	26.64[R]	5.56[R]	404.74[R]
2003[P]	99.69	98.70	147.82	11.35	26.52	27.18	5.87	417.12

[a]Dry production.
[b]Includes lease condensate.
[c]Net generation, i.e., gross generation less plant use.
[d]Includes net electricity generation from wood, waste, solar, and wind. Data for the United States also include other renewable energy.
R=Revised. P=Preliminary.
Notes: Totals may not equal sum of components due to independent rounding. For related information, see http://www.eia.doe.gov/international.

SOURCE: Adapted from "Table 11.1. World Primary Energy Production by Source, 1970–2003 (Quadrillion Btu)," in *Annual Energy Review 2004*, U.S. Department of Energy, Energy Information Administration, Office of Energy Markets and End Use, August, 2005, http://www.eia.doe.gov/emeu/aer/pdf/aer.pdf (accessed April 5, 2006)

FIGURE 1.15

Top energy producing countries, 2003

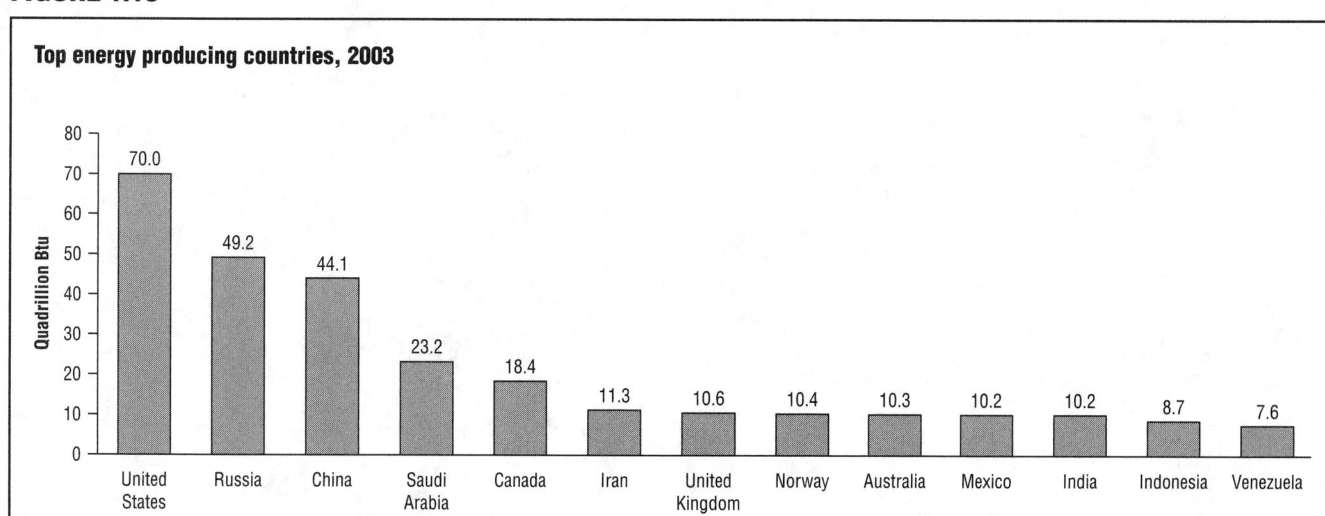

SOURCE: Adapted from "Figure 11.2. World Primary Energy Production by Region and Country: Top Producing Countries, 2003," in *Annual Energy Review 2004*, U.S. Department of Energy, Energy Information Administration, Office of Energy Markets and End Use, August, 2005, http://www.eia.doe.gov/emeu/aer/pdf/aer.pdf (accessed April 5, 2006)

TABLE 1.5

World primary energy consumption by region, selected years 1980–2003

[Quadrillion Btu]

Region/country	1980	1985	1990	1995	2000	2001	2002	2003
United States	78.321	76.450	84.642	91.249	98.962	96.510	98.095	98.843
North America	**91.765**	**91.152**	**100.836**	**108.822**	**118.302**	**115.615**	**117.487**	**119.136**
Central & South America	**11.543**	**12.367**	**14.526**	**17.599**	**20.884**	**21.229**	**21.265**	**21.876**
Germany	NA	NA	NA	14.310	14.261	14.620	14.369	14.241
Germany, East	3.602	3.859	3.357	NA	NA	NA	NA	NA
Germany, West	11.270	11.020	11.460	NA	NA	NA	NA	NA
Western Europe	**58.468**	**59.566**	**64.063**	**66.274**	**71.593**	**72.714**	**72.695**	**73.555**
Russia	NA	NA	NA	27.942	27.458	27.703	27.928	28.763
Eastern Europe & former U.S.S.R.	**59.983**	**69.179**	**73.037**	**52.740**	**50.745**	**51.248**	**52.085**	**53.834**
Middle East	**5.880**	**8.590**	**11.288**	**13.895**	**17.351**	**18.093**	**19.085**	**19.643**
Africa	**6.796**	**8.508**	**9.450**	**10.642**	**12.029**	**12.503**	**12.761**	**13.327**
China	17.288	22.193	26.985	35.146	38.829	40.913	42.082	45.482
Japan	15.224	15.716	18.404	20.766	22.321	22.320	22.173	22.421
Asia & Oceania	**48.977**	**59.275**	**74.147**	**95.639**	**108.850**	**112.588**	**114.930**	**120.141**
World total	**283.412**	**308.638**	**347.347**	**365.611**	**399.753**	**403.991**	**410.307**	**421.513**

NA=Data not available

SOURCE: Adapted from "Table E.1. World Total Primary Energy Consumption (Quadrillion Btu), 1980–2003," in *International Energy Annual 2003*, U.S. Department of Energy, Energy Information Administration, July 1, 2005, http://www.eia.doe.gov/pub/international/iealf/tablee1.xls (accessed April 10, 2006)

TABLE 1.6

Summary of projected total energy supply and disposition, selected years 2003–30

Energy and economic factors	2003	2004	2010	2015	2020	2025	2030	Average annual change, 2004–2030
Primary energy production (quadrillion Btu)								
Petroleum	14.40	13.93	14.83	14.94	14.41	13.17	12.25	−0.5%
Dry natural gas	19.63	19.02	19.13	20.97	22.09	21.80	21.45	0.5%
Coal	22.12	22.86	25.78	25.73	27.30	30.61	34.10	1.6%
Nuclear power	7.96	8.23	8.44	8.66	9.09	9.09	9.09	0.4%
Renewable energy	5.69	5.74	7.08	7.43	8.00	8.61	9.02	1.8%
Other	0.72	0.64	2.16	2.85	3.16	3.32	3.44	6.7%
Total	**70.52**	**70.42**	**77.42**	**80.58**	**84.05**	**86.59**	**89.36**	**0.9%**
Net imports (quadrillion Btu)								
Petroleum	24.19	25.88	26.22	28.02	30.39	33.11	36.49	1.3%
Natural gas	3.39	3.49	4.45	5.23	5.15	5.50	5.72	1.9%
Coal/other (−indicates export)	−0.45	−0.42	−0.58	0.20	0.90	1.54	2.02	NA
Total	**27.13**	**28.95**	**30.09**	**33.44**	**36.44**	**40.15**	**44.23**	**1.6%**
Consumption (quadrillion Btu)								
Petroleum products	38.96	40.08	43.14	45.69	48.14	50.57	53.58	1.1%
Natural gas	23.04	23.07	24.04	26.67	27.70	27.78	27.66	0.7%
Coal	22.38	22.53	25.09	25.66	27.65	30.89	34.49	1.7%
Nuclear power	7.96	8.23	8.44	8.66	9.09	9.09	9.09	0.4%
Renewable energy	5.70	5.74	7.08	7.43	8.00	8.61	9.02	1.8%
Other	0.02	0.04	0.07	0.08	0.05	0.05	0.05	0.9%
Total	**98.05**	**99.68**	**107.87**	**114.18**	**120.63**	**126.99**	**133.88**	**1.1%**
Petroleum (million barrels per day)								
Domestic crude production	5.69	5.42	5.88	5.84	5.55	4.99	4.57	−0.7%
Other domestic production	3.10	3.21	3.99	4.50	4.90	5.45	5.84	2.3%
Net imports	11.25	12.11	12.33	13.23	14.42	15.68	17.24	1.4%
Consumption	20.05	20.76	22.17	23.53	24.81	26.05	27.57	1.1%
Natural gas (trillion cubic feet)								
Production	19.11	18.52	18.65	20.44	21.52	21.24	20.90	0.5%
Net imports	3.29	3.40	4.35	5.10	5.02	5.37	5.57	1.9%
Consumption	22.34	22.41	23.35	25.91	26.92	26.99	26.86	0.7%
Coal (million short tons)								
Production	1,083	1,125	1,261	1,272	1,355	1,530	1,703	1.6%
Net imports	−18	−21	−26	5	36	63	83	NA
Consumption	1,095	1,104	1,233	1,276	1,390	1,592	1,784	1.9%
Prices (2004 dollars)								
Imported low-sulfur light crude oil (dollars per barrel)	31.72	40.49	47.29	47.79	50.70	54.08	56.97	1.3%
Imported crude oil (dollars per barrel)	28.46	35.99	43.99	43.00	44.99	47.99	49.99	1.3%
Domestic natural gas at wellhead (dollars per thousand cubic feet)	5.08	5.49	5.03	4.52	4.90	5.43	5.92	0.3%
Domestic coal at minemouth (dollars per short ton)	18.40	20.07	22.23	20.39	20.20	20.63	21.73	0.3%
Average electricity price (cents per kilowatthour)	7.6	7.6	7.3	7.1	7.2	7.4	7.5	0.0%

TABLE 1.6

Summary of projected total energy supply and disposition, selected years 2003–30 [CONTINUED]

Energy and economic factors	2003	2004	2010	2015	2020	2025	2030	Average annual change, 2004–2030
Economic indicators								
Real gross domestic product (billion 2000 dollars)	10,321	10,756	13,043	15,082	17,541	20,123	23,112	3.0%
GDP chain-type price index (index, 2000=1.000)	1.063	1.091	1.235	1.398	1.597	1.818	2.048	2.5%
Real disposable personal income (billion 2000 dollars)	7,742	8,004	9,622	11,058	13,057	15,182	17,562	3.1%
Value of manufacturing shipments (billion 2000 dollars)	5,378	5,643	6,355	7,036	7,778	8,589	9,578	2.1%
Energy intensity (thousand Btu per 2000 dollar of GDP)	**9.51**	**9.27**	**8.28**	**7.58**	**6.88**	**6.32**	**5.80**	**−1.8%**
Carbon dioxide emissions (million metric tons)	**5,785**	**5,900**	**6,365**	**6,718**	**7,119**	**7,587**	**8,114**	**1.2%**

Notes: Quantities are derived from historical volumes and assumed thermal conversion factors. Other production includes liquid hydrogen, methanol, supplemental natural gas, and some inputs to refineries. Net imports of petroleum include crude oil, petroleum products, unfinished oils, alcohols, ethers, and blending components. Other net imports include coal coke and electricity. Some refinery inputs appear as petroleum product consumption. Other consumption includes net electricity imports, liquid hydrogen, and methanol.

SOURCE: "Table 1. Total Energy Supply and Disposition in the AEO2006 Reference Case: Summary, 2003–2030," in *Annual Energy Outlook 2006*, U.S. Department of Energy, Energy Information Administration, Office of Integrated Analysis and Forecasting, February 2006, http://www.eia.doe.gov/oiaf/aeo/pdf/0383(2006).pdf (accessed April 5, 2006)

FIGURE 1.16

Energy consumption by fuel, 1980–2030

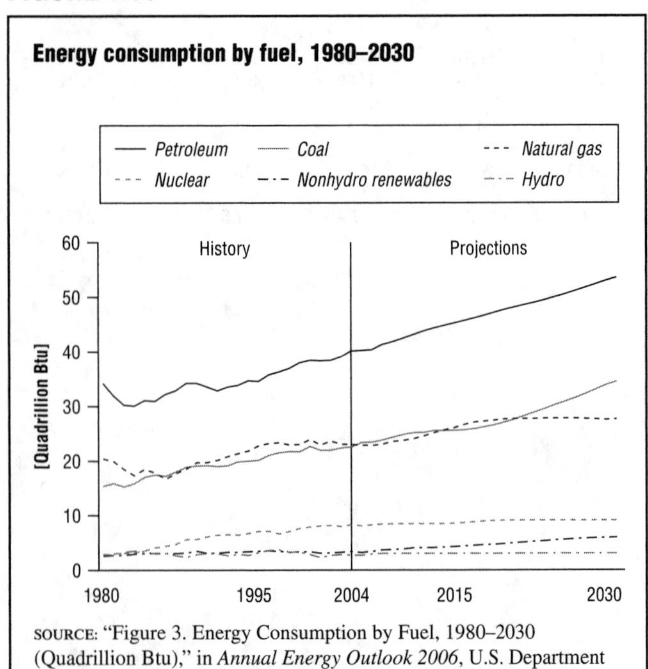

SOURCE: "Figure 3. Energy Consumption by Fuel, 1980–2030 (Quadrillion Btu)," in *Annual Energy Outlook 2006*, U.S. Department of Energy, Energy Information Administration, Office of Integrated Analysis and Forecasting, February 2006, http://www.eia.doe.gov/oiaf/aeo/pdf/0383(2006).pdf (accessed April 5, 2006)

FIGURE 1.17

Gross petroleum imports by source, 2004–30

[Million barrels per day]

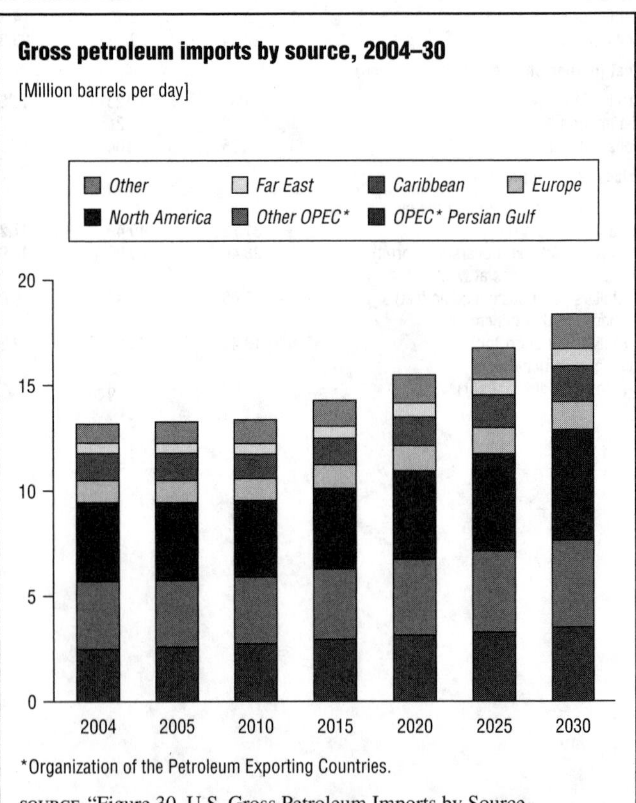

*Organization of the Petroleum Exporting Countries.

SOURCE: "Figure 30. U.S. Gross Petroleum Imports by Source, 2004–2030 (Million Barrels per Day)," in *Annual Energy Outlook 2006*, U.S. Department of Energy, Energy Information Administration, Office of Integrated Analysis and Forecasting, February 2006, http://www.eia.doe.gov/oiaf/aeo/pdf/0383(2006).pdf (accessed April 5, 2006)

FIGURE 1.18

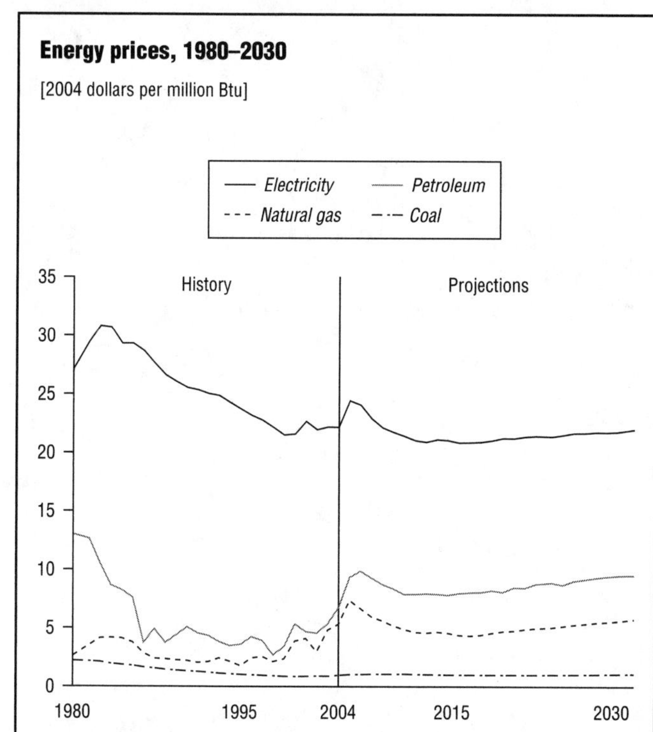

SOURCE: "Figure 1. Energy Prices, 1980–2030 (2004 Dollars per Million Btu)," in *Annual Energy Outlook 2006*, U.S. Department of Energy, Energy Information Administration, Office of Integrated Analysis and Forecasting, February 2006, http://www.eia.doe.gov/oiaf/aeo/pdf/0383(2006).pdf (accessed April 5, 2006)

CHAPTER 2
OIL

THE QUEST FOR OIL

On August 27, 1859, Edwin Drake struck oil sixty-nine feet below the surface of the earth near Titusville, Pennsylvania. This was the first successful modern oil well, which ushered in the "Age of Petroleum." Not only did petroleum help meet the growing demand for new and better fuels for heating and lighting, it also proved to be an excellent fuel for the internal combustion engine, which was developed in the late 1800s.

Sources of Oil

Almost all oil comes from underground reservoirs. The most widely accepted explanation of how oil and gas are formed within the earth is that these fuels are the products of intense heat and pressure applied over millions of years to organic (formerly alive) sediments buried in geological formations. For this reason they are called fossil fuels. They are limited (nonrenewable) resources, which means that they are formed much more slowly than they are used, so they are finite in supply.

At one time it was believed that crude oil flowed in underground streams and accumulated in lakes or caverns in the earth. Today, scientists know that a petroleum reservoir is usually a solid sandstone or limestone formation overlaid with a layer of impermeable rock or shale, which creates a shield. The petroleum accumulates within the pores and fractures of the rock and is trapped beneath the shield. Anticlines (archlike folds in a bed of rock), faults, and salt domes are common trapping formations. (See Figure 2.1.) Oil and natural gas deposits can be found at varying depths. Wells are drilled to reach the reservoirs and extract the oil.

How Oil Is Drilled and Recovered

Most oil wells are drilled with a rotary drilling system, or rotary rig, as illustrated in Figure 2.2. The rotating bit at the end of a pipe drills a hole into the ground. Drilling mud is pushed down through the pipe and the drill bit, forcing small pieces of drilled rock to the surface, as shown by the arrows in the diagram. As the well gets deeper, more pipe is added. The oil derrick above the ground supports equipment that can lift the pipe and drill bit from the well when drill bits need to be changed or replaced.

After oil reservoirs have been tapped for several years or decades, their supply of oil becomes depleted. Several techniques can be used to recover additional petroleum, including the injection of water, chemicals, or steam to force more oil from the rock. These recovery techniques can be expensive and add to the cost of producing each barrel of crude oil.

TYPES OF OIL

While crude oil is usually dark when it comes from the ground, it may also be green, red, yellow, or colorless, depending on its chemical composition and the amount of sulfur, oxygen, nitrogen, and trace minerals present. Its viscosity (thickness, or resistance to flow) can range from as thin as water to as thick as tar. Crude oil is refined, or chemically processed, into finished petroleum products; it has limited uses in its natural form.

Crude oils vary in quality. "Sweet" crudes have little sulfur, refine easily, and are worth more than "sour" crudes, which contain more impurities. "Light" crudes, which have more short molecules, yield more gasoline and are more profitable than "heavy" crudes, which have more long molecules and bring a lower price in the market.

In addition to crude oil, there are two other sources of petroleum: lease condensate and natural gas plant liquids. Lease condensate is a liquid recovered from natural gas at the well. It consists primarily of chemical compounds called pentanes and heavier hydrocarbons and is generally

FIGURE 2.1

Petroleum traps

Anticline

Fault

Salt dome

SOURCE: "Figure 3.1. Petroleum Traps," in *Petroleum: An Energy Profile 1999*, U.S. Department of Energy, Energy Information Administration, July 1999, http://www.eia.doe.gov/pub/oil_gas/petroleum/analysis_publications/petroleum_profile_1999/profile99v8.pdf (accessed April 10, 2006)

blended with crude oil for refining. Natural gas plant liquids, such as butane and propane, are recovered during the refinement of natural gas in processing plants.

USES FOR OIL

Many of the uses for petroleum are well known: gasoline, diesel fuel, jet fuel, and lubricants for transportation; heating oil, residual oil, and kerosene for heat; and heavy residuals for paving and roofing. Petroleum byproducts are also vital to the chemical industry, ending up in many different foams, plastics, synthetic fabrics, paints, dyes, inks, and even pharmaceutical drugs. Many chemical plants, because of their dependence on petroleum, are directly connected by pipelines to nearby refineries.

HOW OIL IS REFINED

Before they can be used by consumers, crude oil, lease condensate, and natural gas plant liquids must be processed into finished products. The first step in refining is distillation, in which crude oil molecules are separated according to size and weight.

FIGURE 2.2

A rotary drilling system

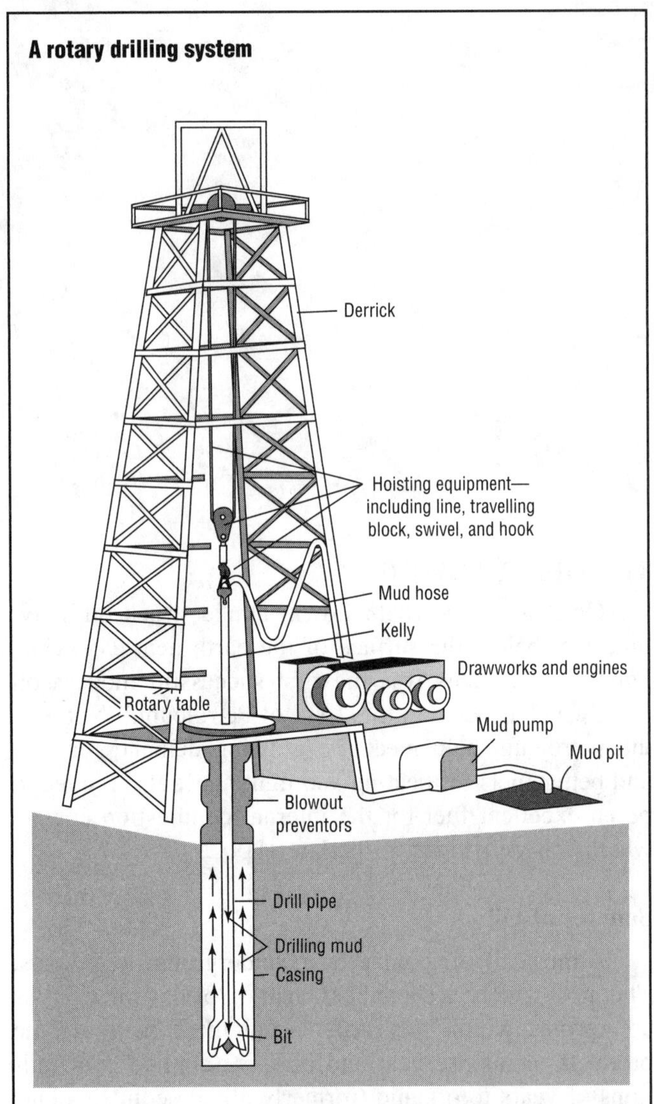

SOURCE: "Figure 4.4. A Rotary Drilling System," in *Petroleum: An Energy Profile 1999*, U.S. Department of Energy, Energy Information Administration, July 1999, http://www.eia.doe.gov/pub/oil_gas/petroleum/analysis_publications/petroleum_profile_1999/profile99v8.pdf (accessed April 10, 2006)

During distillation, crude oil is heated until it turns to vapor. (See Figure 2.3.) The vaporized crude oil enters the bottom of a distillation column, where it rises and condenses on trays. The lightest vapors, such as those of gasoline, rise to the top. The middleweight vapors, such as those of kerosene, rise about halfway up the column. The heaviest vapors, such as those of heavy gas oil, stay at the bottom. The vapors at each level condense into liquids as they are cooled. These liquids are drawn off, and processes called cracking and reforming further refine each portion. Cracking converts the heaviest fractions of separated petroleum into lighter fractions to produce jet fuel, motor gasoline, home heating oil, and less-residual fuel oils, which are heavier and used for naval ships, commercial and industrial heating, and some power generation. Reforming is used to increase the octane rating of gasoline.

FIGURE 2.3

Crude oil distillation

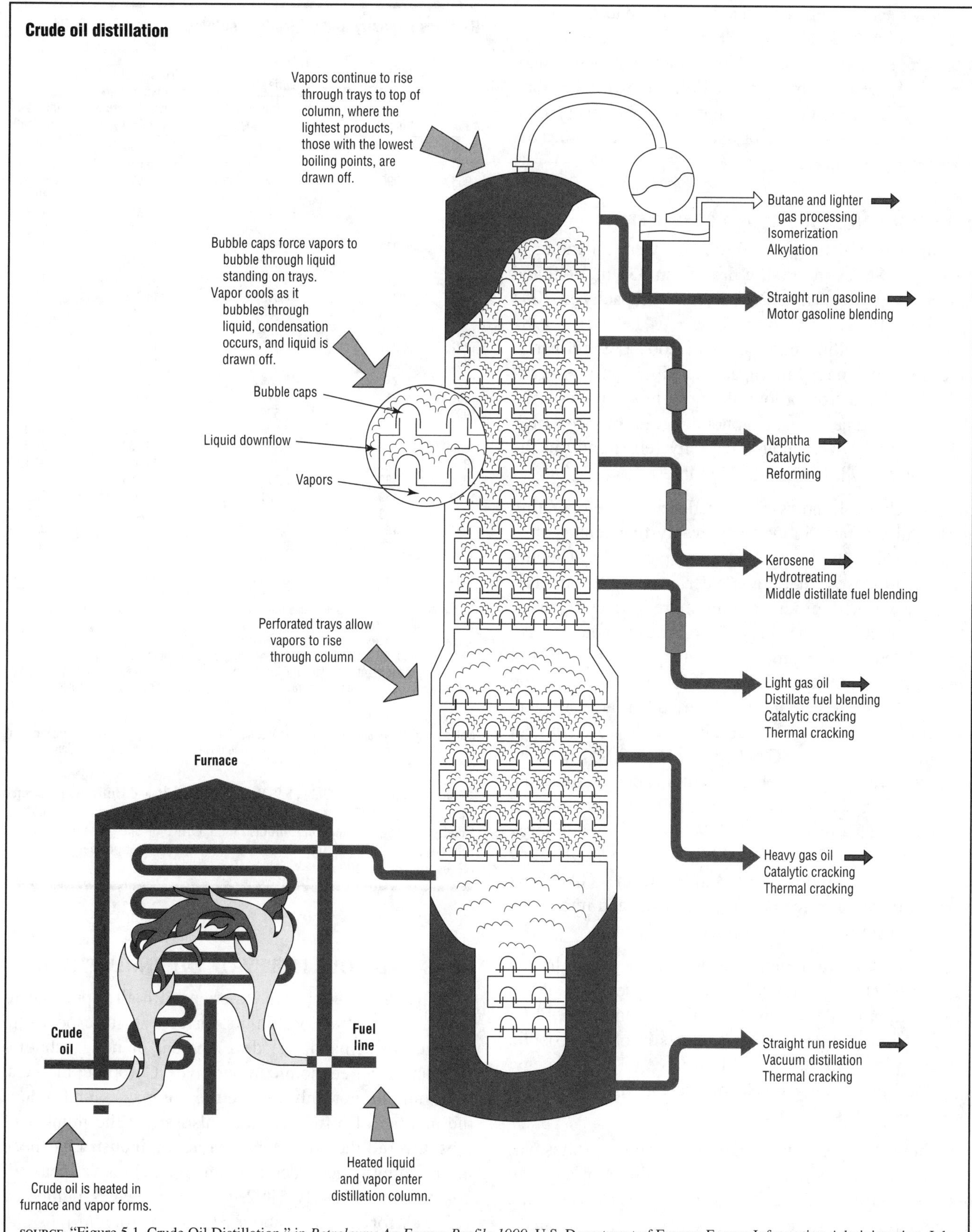

SOURCE: "Figure 5.1. Crude Oil Distillation," in *Petroleum: An Energy Profile 1999*, U.S. Department of Energy, Energy Information Administration, July 1999, http://www.eia.doe.gov/pub/oil_gas/petroleum/analysis_publications/petroleum_profile_1999/profile99v8.pdf (accessed April 10, 2006)

Refining is a continuous process, with crude oil entering the refinery at the same time that finished products leave by pipeline, truck, and train. At most refineries storage capacity is limited, so if there is a malfunction and products cannot be refined, the oil may be burned off (flared) rather than stored. While a small flare is normal at a refinery or a chemical plant, large flares or many flares likely indicate a processing problem.

REFINERY NUMBERS AND CAPACITY

One hundred forty-nine refineries were operating in the United States in 2004, a drop from 336 in 1949 and 324 in 1981. (See Table 2.1.) Refinery capacity in 2004 was about 16.9 million barrels per day, below the 1981 peak of 18.6 million barrels. As of 2004 U.S. refineries were operating near full capacity. Utilization rates generally increased from a low of 68.6% in 1981—a period of low demand because of economic recession—to a high of 95.6% in 1998. Although capacity fell slightly in the years since 1998, it was still high in 2004.

As Table 2.1 shows, fewer refineries were operating in the United States in the early twenty-first century than in the past. The number dropped partly because the petroleum industry shut down older, inefficient refineries and concentrated production in more efficient plants, which tended to be newer and larger. The industry also consolidated. For example, the merging of Gulf Oil Corporation into Chevron Corporation in 1984 led to the closing of two large refineries. Many other large mergers and closings followed. According to the Government Accountability Office (GAO), more than 2,600 mergers have occurred in the U.S. petroleum industry since the 1990s, "most frequently among firms involved in oil exploration and production" (*GAO Highlights: Energy Markets*, May 2004; http://www.gao.gov/highlights/d0496high.pdf). Industry officials told the GAO that mergers increase efficiency, reduce costs, and improve a company's ability to control prices.

An equally important reason for the drop in the number of U.S. refineries is that some members of the Organization of Petroleum Exporting Countries (OPEC) started refining their own oil products in order to obtain higher prices in the world market. This strategy, employed particularly by Saudi Arabia, was intended to maximize profits.

The last large refinery built in the United States was completed in 1976, and the last new refinery of any size began operation in Valdez, Alaska, in 1993. However, the demand for U.S.-refined petroleum continues to grow. The Energy Information Administration (EIA) stated in *Annual Energy Outlook 2006* (2006; http://www.eia.doe.gov/oiaf/aeo/) that refinery expansion is expected from 2010 to 2030. A start-up company has announced plans to open a major new refinery in Arizona in 2010.

TABLE 2.1

Refinery capacity and utilization, selected years 1949–2004

Year	Operable refineries		Gross input to distillation units	Utilization[c]
	Number[a]	Capacity[b] (thousand barrels per day)	(thousand barrels per day)	(percent)
1949	336	6,231	5,556	89.2
1950	320	6,223	5,980	92.5
1955	296	8,386	7,820	92.2
1960	309	9,843	8,439	85.1
1965	293	10,420	9,557	91.8
1970	276	12,021	11,517	92.6
1972	274	13,292	12,431	92.3
1974	273	14,362	12,689	86.6
1976	276	15,237	13,884	87.8
1978	296	17,048	15,071	87.4
1980	319	17,988	13,796	75.4
1981	324	18,621	12,752	68.6
1982	301	17,890	12,172	69.9
1984	247	16,137	12,216	76.2
1986	216	15,459	12,826	82.9
1988	213	15,915	13,447	84.7
1990	205	15,572	13,610	87.1
1992	199	15,696	13,600	87.9
1994	179	15,034	14,032	92.6
1996	170	15,333	14,337	94.1
1998	163	15,711	15,113	95.6
1999	159	16,261	15,080	92.6
2000	158	16,512	15,299	92.6
2001	155	16,595	15,369	92.6
2002	153	16,785	15,180	90.7
2003	149	16,757	15,508[R]	92.6[R]
2004[P]	149	16,894	15,701	92.8

[a]Through 1956, includes only those refineries in operation on January 1; beginning in 1957, includes all "operable" refineries on January 1.
[b]Capacity on January 1.
[c]Through 1980, utilization is derived by dividing gross input to distillation units by one-half of the current year January 1 capacity and the following year January 1 capacity. Percentages were derived from unrounded numbers. Beginning in 1981, utilization is derived by averaging reported monthly utilization.
R=Revised. P=Preliminary.
Notes: For data not shown for 1951–1969, see http://www.eia.doe.gov/emeu/aer/petro.html. For related information, see http://www.eia.doe.gov/oil_gas/petroleum/info_glance/petroleum.html.

SOURCE: Adapted from "Table 5.9. Refinery Capacity and Utilization, Selected Years, 1949–2004," in *Annual Energy Review 2004*, U.S. Department of Energy, Energy Information Administration, Office of Energy Markets and End Use, August, 2005, http://www.eia.doe.gov/emeu/aer/pdf/aer.pdf (accessed April 5, 2006)

LOSS AND VOLATILITY OF OIL INDUSTRY JOBS

The oil industry has experienced both rapid expansions and contractions in its history as it responds to changes in demand, the development of new technologies, and other economic imperatives. Corporate mergers, for example, not only restructure companies and reduce the number of refineries; they also affect the number of jobs. Overall the oil and gas extraction industry has been in an employment decline since reaching a peak of 264,500 in 1982. By 1992, when employees numbered 182,200, for example, about 31% of petroleum jobs had been lost, according to the U.S. Bureau of Labor Statistics. As of 2005, employment in the oil and gas extraction industry segment was 125,900, up 4.7% from the all-time low of 120,200 experienced in 2003, but 52.4% below the 1982 figure.

Oil prices have created some of the volatility experienced in oil employment. In 1996, after a decade of low oil prices, drilling slowed and the demand for rigs collapsed. New rig construction stopped altogether. Thousands of rigs were left idle, sold for scrap metal, or shipped overseas, and their crews were put out of work. Idle rigs became a source of spare parts for those still operating. In 1997, following a rise in oil prices, the demand for rigs soared, but by 1998 the market for rigs had once again dwindled as oil prices sank. Rising oil prices in 1999 eventually boosted the demand for drilling equipment. Although crude oil prices declined in 2001, they rose again in 2003 and reached an all-time high in April 2006, again spurring demand for oil rigs. According to Baker Hughes Inc., a company that has tallied weekly U.S. drilling activity since 1940, domestic oil drilling has rebounded sharply since late April 1999.

Jobs have also been lost in other parts of the oil industry. The number of seismic land crews and marine vessels searching for oil in the United States and its waters decreased sharply after 1981. In Texas, once the center of the U.S. oil industry, jobs plummeted from 80,000 in 1981 to 25,000 in 1996. In 1998 the Texas Comptroller of Public Accounts estimated that for every $1 drop in the price of oil per barrel, 10,000 jobs were lost in the Texas economy. That translated into 100,000 jobs lost in Texas from October 1997 to December 1998. However, domestic oil drilling rebounded sharply between mid-1999 and early 2006, creating an upswing in numbers of oil industry jobs.

DOMESTIC PRODUCTION

U.S. production of petroleum reached its highest level in 1970 at 11.3 million barrels per day. (See Table 2.2.) Of that amount, 9.6 million barrels per day were crude oil. After 1970 domestic production of petroleum first declined, then rose from 1977 through 1985, and finally declined fairly steadily. (See Figure 2.4.) By 2004 U.S. domestic production averaged about 7.2 million barrels per day. Of that amount, 5.4 million barrels per day were crude oil. Figure 2.5 shows the overall flow of petroleum in the United States for 2004.

According to the *Annual Energy Review 2004* (Energy Information Administration, 2005; http://tonto.eia.doe.gov/FTPROOT/multifuel/038404.pdf), the 510,000 producing wells in the United States in 2004 produced less than eleven barrels per day per well, significantly below peak levels of more than eighteen barrels per day per well in the early 1970s. Of the country's thirteen largest oil fields, seven are at least 80% depleted.

The United States is considered to be in a "mature" oil development phase, meaning that much of its oil has already been found. The amount of oil discovered per foot of exploratory well in the United States has fallen to less than half the rate of the early 1970s. Geological studies have estimated that 34% of the country's untapped oil resources are in Alaska.

Most domestic oil is produced in only a few states. Texas, Alaska, Louisiana, California, and the offshore areas around these states produce about 75% of the nation's oil. Most domestic oil (3.6 million barrels per day, or about 66%) comes from onshore drilling, while the remaining 1.9 million barrels come from offshore sources. (See Figure 2.6.) Supplies from Alaska, which increased with the construction of a pipeline in the late 1970s, have begun to decline. Notice the gap between "Total" and "48 States" in Figure 2.7; the difference is Alaska's share of U.S. oil production.

Unless protected wildlife refuges in Alaska are opened for drilling, U.S. oil production will likely continue to decline. The Alaskan government and the administration of President George W. Bush strongly support drilling for oil in Alaska's Arctic National Wildlife Refuge. But this proposal is highly controversial for environmental reasons. Legislation to allow the drilling has been stalled in Congress several times.

Domestic production is also affected by another factor: the expense of drilling and recovering oil. U.S. producers spend about $14 to produce a barrel of oil, not counting royalty payments and taxes, according to estimates by the U.S. Department of Energy. Middle Eastern producers, by contrast, drill and extract crude oil from enormous, easily accessible reservoirs for around $2 per barrel. When the cost of extraction severely reduces the profit margin on a barrel of oil, U.S. producers may shut down their most expensive wells.

DOMESTIC CONSUMPTION

In 2004 most petroleum was used for transportation (13.6 million barrels per day, or 66%), followed by industrial use (5.1 million barrels per day; 25%), residential use (0.9 million barrels per day; 4%), electric utilities (0.5 million barrels per day; 3%), and commercial use (0.4 million barrels per day; 2%). (See Figure 2.8.)

Most petroleum used in the transportation sector is for motor gasoline. In the residential and commercial sectors, distillate fuel oil (refined fuels used for space heaters, diesel engines, and electric power generation) accounts for most petroleum use. Liquid petroleum gas (LPG) is the primary oil used in the industrial sector. In electric utilities residual fuel oils are used the most.

A modest decline in residual fuel oil consumption resulted when electric utilities and plants were converted to coal or natural gas. An initial decline in the amount of motor gasoline used, beginning in 1978, was attributed to the federal Corporate Average Fuel Economy regulations, which required increased miles-per-gallon efficiency in

TABLE 2.2

Petroleum production, selected years 1949–2004

[Thousand barrels per day]

Year	Production[a]					Processing gain[c]	Trade			Stock change[d]	Adjust-ments[e]	Petroleum products supplied
	Crude oil			Natural gas plant liquids	Total		Imports	Exports	Net imports			
	48 states[b]	Alaska	Total									
1949	5,046	0	5,046	430	5,477	−2	645	327	318	−8	−38	5,763
1950	5,407	0	5,407	499	5,906	2	850	305	545	−56	251	6,458
1955	6,807	0	6,807	771	7,578	34	1,248	368	880	s	237	8,455
1960	7,034	2	7,035	929	7,965	146	1,815	202	1,613	−83	28	9,797
1965	7,774	30	7,804	1,210	9,014	220	2,468	187	2,281	−8	210	11,512
1970	9,408	229	9,637	1,660	11,297	359	3,419	259	3,161	103	216	14,697
1972	9,242	199	9,441	1,744	11,185	388	4,741	222	4,519	−232	43	16,367
1974	8,581	193	8,774	1,688	10,462	480	6,112	221	5,892	179	22	16,653
1976	7,958	173	8,132	1,604	9,736	477	7,313	223	7,090	−58	101	17,461
1978	7,478	1,229	8,707	1,567	10,275	496	8,363	362	8,002	−94	220	18,847
1980	6,980	1,617	8,597	1,573	10,170	597	6,909	544	6,365	140	64	17,056
1982	6,953	1,696	8,649	1,550	10,199	531	5,113	815	4,298	−147	121	15,296
1984	7,157	1,722	8,879	1,630	10,509	553	5,437	722	4,715	280	228	15,726
1986	6,814	1,867	8,680	1,551	10,231	616	6,224	785	5,439	202	197	16,281
1988	6,123	2,017	8,140	1,625	9,765	655	7,402	815	6,587	−28	249	17,283
1990	5,582	1,773	7,355	1,559	8,914	683	8,018	857	7,161	107	338	16,988
1992	5,457	1,714	7,171	1,697	8,868	772	7,888	950	6,938	−68	386	17,033
1994	5,103	1,559	6,662	1,727	8,388	768	8,996	942	8,054	15	523	17,718
1996	5,071	1,393	6,465	1,830	8,295	837	9,478	981	8,498	−151	528	18,309
1998	5,077	1,175	6,252	1,759	8,011	886	10,708	945	9,764	239	495	18,917
1999	4,832	1,050	5,881	1,850	7,731	886	10,852	940	9,912	−422	567	19,519
2000	4,851	970	5,822	1,911	7,733	948	11,459	1,040	10,419	−69	532	19,701
2001	4,839	963	5,801	1,868	7,670	903	11,871	971	10,900	325	501	19,649
2002	4,761	984	5,746	1,880	7,626	957	11,530	984	10,546	−105	527	19,761
2003	4,706[R]	974	5,681[R]	1,719[R]	7,400[R]	974[R]	12,264[R]	1,027[R]	11,238[R]	56[R]	478	20,034[R]
2004[P]	4,522	908	5,430	1,811	7,241	1,024	12,899	1,048	11,851	212	614	20,517

[a]Crude oil production on leases, and natural gas plant liquids (liquefied petroleum gases, pentanes plus, and a small amount of finished petroleum products) production at natural gas processing plants. Excludes what was previously classified as "field production" of finished motor gasoline, motor gasoline blending components, and other hydrocarbons and oxygenates; these are now included in "adjustments."

[b]United States excluding Alaska and Hawaii.

[c]Refinery output minus refinery input.

[d]A negative number indicates a decrease in stocks and a positive number indicates an increase. Distillate stocks in the "northeast heating oil reserve" are not included.

[e]An adjustment for crude oil, motor gasoline blending components, and fuel ethanol.

R=Revised. P=Preliminary. s=Less than 500 barrels per day.

Notes: Crude oil includes lease condensate. Totals may not equal sum of components due to independent rounding. For data not shown for 1951–1969, see http://www.eia.doe.gov/emeu/aer/petro.html. For related information, see http://www.eia.doe.gov/oil_gas/petroleum/info_glance/petroleum.html.

SOURCE: Adapted from "Table 5.1. Petroleum Overview, Selected Years, 1949–2004 (Thousand Barrels per Day)," in *Annual Energy Review 2004*, U.S. Department of Energy, Energy Information Administration, Office of Energy Markets and End Use, August, 2005, http://www.eia.doe.gov/emeu/aer/pdf/aer.pdf (accessed April 5, 2006)

new automobiles. However, motor gasoline use has increased steadily since then, partly because the number of users has increased and partly because vehicle efficiency has leveled off—consumers once again prefer less efficient vehicles, such as sport-utility vehicles (SUVs).

WORLD OIL PRODUCTION AND CONSUMPTION

Total world petroleum production has increased somewhat steadily, reaching 72.5 million barrels per day in 2004, after a downturn in the early 1980s. (See Table 2.3.) The largest producer in 2004 was Saudi Arabia, followed by Russia, the United States, Iran, China, Mexico, Norway, Venezuela, Nigeria, the United Arab Emirates, Canada, Kuwait, Iraq, and the United Kingdom. (See Figure 2.9.) Together, Saudi Arabia, Russia, and the United States accounted for 32% of the world's crude oil production.

Like total world petroleum production, total world petroleum consumption has increased somewhat steadily, reaching 80.1 million barrels per day in 2003. (See Table 2.4.) In 2003 the United States was by far the leading consumer, using 20 million barrels per day, followed by Japan (5.6 million barrels per day), China (5.6 million barrels per day), and Germany and Russia (2.7 million barrels per day each). Other leading petroleum consumers were India, Canada, South Korea, Brazil, France, Mexico, Italy, the United Kingdom, and Spain. (See Figure 2.10.)

OIL IMPORTS AND EXPORTS

Countries that have surplus oil (Saudi Arabia, for example) sell their excess to countries that need more than they can produce, such as the United States, China, Japan, and western European countries. They sell petroleum as both crude oil and refined products, although the

FIGURE 2.4

Petroleum production, 1949–2004

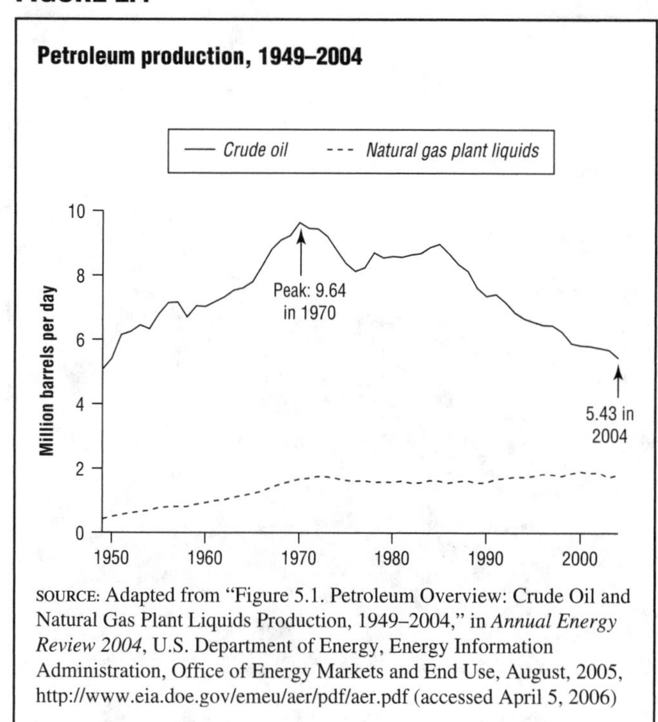

SOURCE: Adapted from "Figure 5.1. Petroleum Overview: Crude Oil and Natural Gas Plant Liquids Production, 1949–2004," in *Annual Energy Review 2004*, U.S. Department of Energy, Energy Information Administration, Office of Energy Markets and End Use, August, 2005, http://www.eia.doe.gov/emeu/aer/pdf/aer.pdf (accessed April 5, 2006)

trend has been moving toward refined products because they bring higher profits. According to the *Annual Energy Review 2004*, the leading supplier of petroleum to the United States in 2004 was Canada, followed by Mexico, Saudi Arabia, Venezuela, Nigeria, Iraq, the United Kingdom, and Norway.

Though the United States produces a significant amount of petroleum, it has been importing oil since World War II (1939–45). Initially, the imported oil was cheap and available, suiting the demands of a growing American population and economy. Furthermore, relatively low world crude oil prices often resulted in reduced domestic oil production: When the world price was lower than the cost of producing oil from some U.S. wells, domestic oil became unprofitable and was not produced. So more oil was imported.

The amount of oil imported to the United States has fluctuated over the years. Total net imports—imports minus exports—rose from 4.3 million barrels per day in 1982 to 11.9 million barrels per day by 2004. (See Table 2.2.) In 1985 imported oil supplied only 27.3% of American oil consumption. Just five years later, in 1990, the proportion had risen to 42% and by 2004 to 57.8%. (See Figure 1.6 in Chapter 1.)

CONCERN ABOUT OIL DEPENDENCY

In the 1970s U.S. leaders were concerned that so much of the country's economic structure, based heavily on imported oil, was dependent upon decisions in OPEC countries. Oil resources became an issue of national security, and OPEC countries, especially the Arab members, were often portrayed as potentially strangling the U.S. economy. Efforts were made to reduce imports by raising public awareness and by encouraging industry to create more energy-efficient products, such as automobiles that got better gas mileage. However, the Ronald Reagan and George H. W. Bush administrations in the 1980s and early 1990s took a different view. They saw oil supply as an economic, rather than political, issue and allowed energy issues to be handled by the marketplace. During the Clinton administration America's dependence on foreign oil continued largely because low prices throughout most of the 1990s set back energy-conservation efforts. In fact, efficiency gains in automobiles were offset by the public's growing preference for large vehicles, such as SUVs. By 2004 about 65% of the nation's crude oil supply (10 million barrels out of a total 15.5 million barrels per day) came from outside the country. (See Figure 2.5.) About 44% of that crude oil came from OPEC nations, according to the *Annual Energy Review 2004*.

Other factors diverted the public's attention from America's dependence on foreign oil. For example, a "comfort level" had been achieved through oil reserves, such as the Strategic Petroleum Reserve in the United States and government-required reserves in Europe, and the knowledge that, in emergencies, non-OPEC oil producers such as the United Kingdom and Norway could increase their output. Furthermore, increased use of pipelines across Saudi Arabia and Turkey lessened concern about the disruption of supplies. The pipelines allow tankers to load oil in the Red Sea or the Mediterranean Sea, rather than in the potentially dangerous Persian Gulf. Ships do not have to navigate through the narrow Strait of Hormuz, where an enemy might be able to stop the flow of oil.

Concern that America's dependence on foreign oil, especially OPEC oil, represented a threat to national security or national stability changed somewhat after the terrorist attacks of September 11, 2001. It was heightened by the "war on terror," the war with Iraq, and the consequent unrest in Middle Eastern nations. Still, demand for the product persisted. By early 2006 demand had grown, oil prices had increased dramatically, and oil supplies and reserves were tight.

PROJECTED OIL SUPPLY AND CONSUMPTION

The Energy Information Administration, in its *Annual Energy Outlook 2006*, projected that domestic crude oil production would increase from 5.4 million barrels per day in 2004 to a peak of 5.9 million barrels per day in 2014, with the peak attributed to offshore production, particularly in the deep water of the Gulf of Mexico. After 2014, production was projected to fall to

FIGURE 2.5

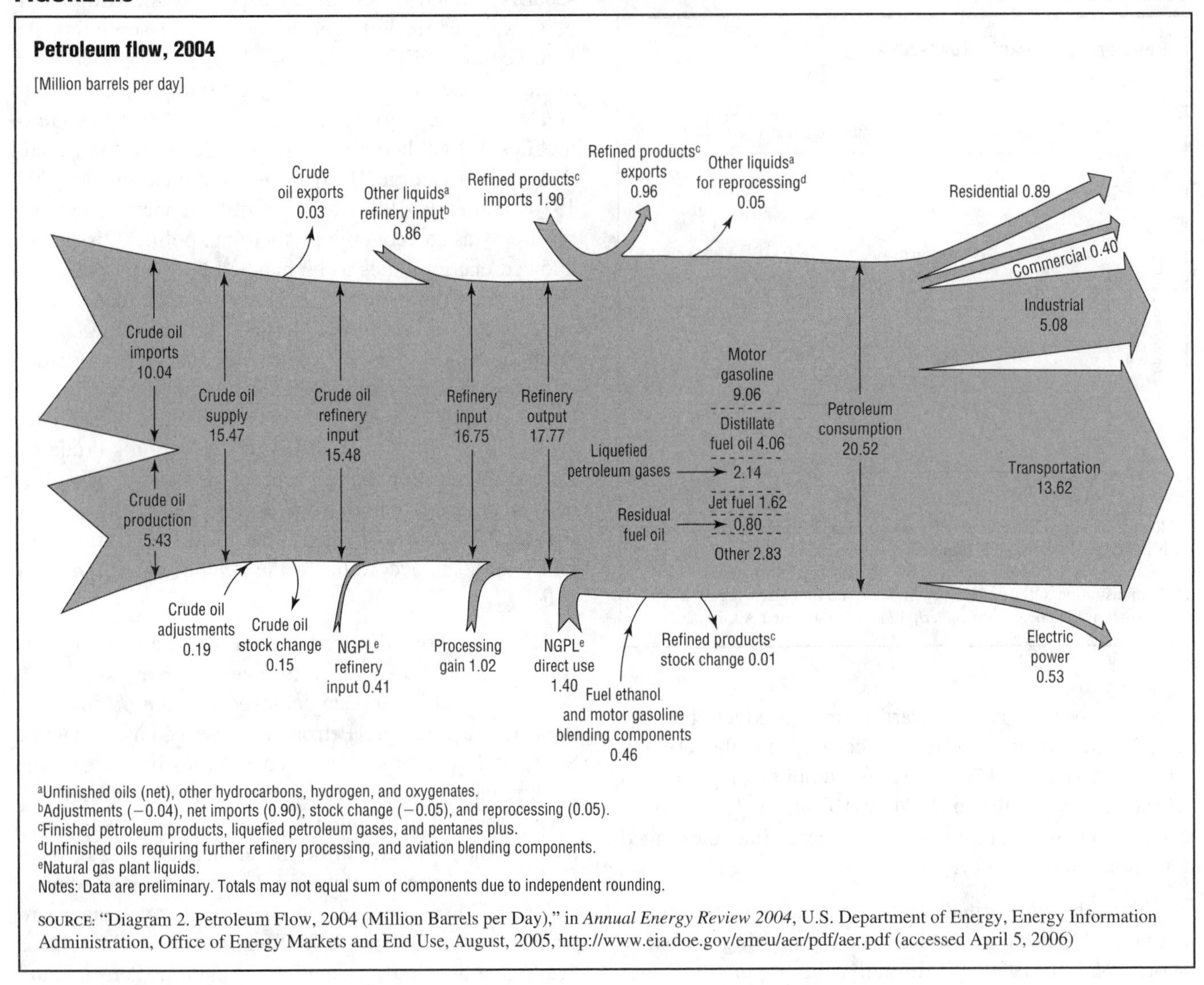

Petroleum flow, 2004

[Million barrels per day]

[a]Unfinished oils (net), other hydrocarbons, hydrogen, and oxygenates.
[b]Adjustments (−0.04), net imports (0.90), stock change (−0.05), and reprocessing (0.05).
[c]Finished petroleum products, liquefied petroleum gases, and pentanes plus.
[d]Unfinished oils requiring further refinery processing, and aviation blending components.
[e]Natural gas plant liquids.
Notes: Data are preliminary. Totals may not equal sum of components due to independent rounding.

SOURCE: "Diagram 2. Petroleum Flow, 2004 (Million Barrels per Day)," in *Annual Energy Review 2004*, U.S. Department of Energy, Energy Information Administration, Office of Energy Markets and End Use, August, 2005, http://www.eia.doe.gov/emeu/aer/pdf/aer.pdf (accessed April 5, 2006)

4.6 million barrels per day in 2030. However, while domestic supply was expected to remain relatively constant though 2030, demand was projected to increase significantly. (See Figure 2.11.) Thus, the agency foresees an increasing dependence on petroleum imports.

STRATEGIC PETROLEUM RESERVE

Early in the twentieth century a Naval Petroleum Reserve was established to ensure that the U.S. Navy would have adequate fuel in the event of war. Large tracts of government land with known deposits of oil were set aside. In 1975, in response to the growing concern over America's energy dependence, Congress expanded the concept and created the Strategic Petroleum Reserve. Oil and refined products are stored in forty-one deep salt caverns in Louisiana and Texas. (The caverns are used because oil does not dissolve salt the way water does.) If the United States suddenly finds its supplies cut off, the reserves can be connected to existing pipelines and the oil pumped out.

At the end of 2004 the Strategic Petroleum Reserve contained 676 million barrels of oil (see Figure 2.12), equal to fifty-seven days' worth of imported oil. Figure 2.13 shows a decline in the reserves in terms of days' worth of net imports, from a high of 115 days in 1985 to 57 days in 2002 through 2004. This decline reflects the country's increasing reliance on imports since 1985. As the nation has imported a greater amount of oil, the days of net import replacement represented by the amount of oil in the reserves have dropped.

OIL PRICES

The law of supply (availability) and demand (need) usually explains the change in the oil price. Higher prices lead to increased production—it becomes profitable to operate more expensive wells—and reduced demand—consumers lower usage and increase conservation efforts. The factors also work the other way: Reduced demand or increased supply generally cause the price of oil to drop.

FIGURE 2.6

Crude oil production and crude oil well productivity, by site, 1954–2004

Note: Crude oil includes lease condensate.

SOURCE: Adapted from "Figure 5.2. Crude Oil Production and Crude Oil Well Productivity, 1954–2004: By Site," in *Annual Energy Review 2004*, U.S. Department of Energy, Energy Information Administration, Office of Energy Markets and End Use, August, 2005, http://www.eia.doe.gov/emeu/aer/pdf/aer.pdf (accessed April 5, 2006)

FIGURE 2.7

Crude oil production and crude oil well productivity, by geographic location, 1954–2004

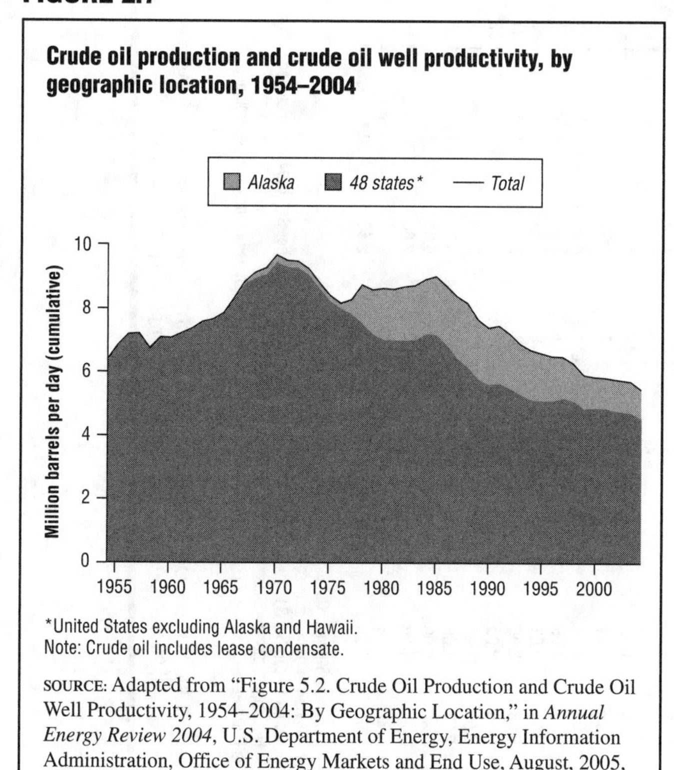

*United States excluding Alaska and Hawaii.
Note: Crude oil includes lease condensate.

SOURCE: Adapted from "Figure 5.2. Crude Oil Production and Crude Oil Well Productivity, 1954–2004: By Geographic Location," in *Annual Energy Review 2004*, U.S. Department of Energy, Energy Information Administration, Office of Energy Markets and End Use, August, 2005, http://www.eia.doe.gov/emeu/aer/pdf/aer.pdf (accessed April 5, 2006)

FIGURE 2.8

Estimated petroleum consumption by sector, 2004

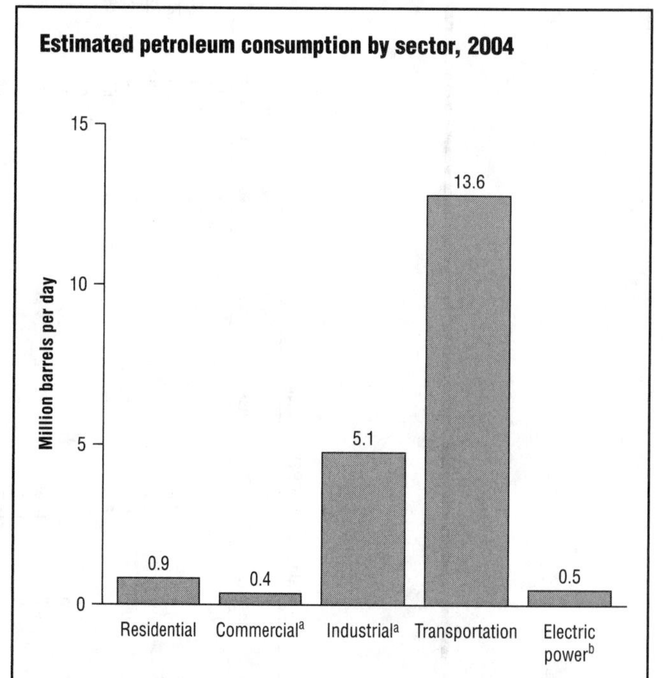

[a]Includes combined-heat-and-power plants and a small number of electricity-only plants.
[b]Electricity-only and combined-heat-and-power plants whose primary business is to sell electricity, or electricity and heat, to the public.

SOURCE: Adapted from "Figure 5.13a. Estimated Petroleum Consumption by Sector: By Sector, 2004," in *Annual Energy Review 2004*, U.S. Department of Energy, Energy Information Administration, Office of Energy Markets and End Use, August, 2005, http://www.eia.doe.gov/emeu/aer/pdf/aer.pdf (accessed April 5, 2006)

The demand for petroleum products varies. Heating oil demand rises during the winter. A cold spell, which leads to a sharp rise in demand, may result in a corresponding price increase. A warm winter may be reflected in lower prices as suppliers try to clear out their inventory. Gasoline demand rises during the summer—people drive more for recreation—so gas prices rise as a consequence. Petroleum demand also reflects the general condition of the economy. During a recession, demand for and production of petroleum products drops. Wars and other types of political unrest in oil-producing nations add volatility to petroleum prices, which fluctuate—sometimes dramatically—depending on the situation at the time.

While consumers prefer low prices that allow them to save money or get more for the same price, producers naturally prefer to keep prices high. Some oil-producing countries formed the OPEC cartel in 1960. A cartel is a group of businesses that agree to control production and marketing to avoid competing with one another. Since 1973 OPEC has tried to control the oil supply to achieve higher prices.

OPEC has faced long-term problems, however, because high prices in the late 1970s to mid-1980s encouraged conservation, reducing demand for oil and leading to a sharp decline in oil prices. As a result of the decreased demand for oil and lower prices, OPEC

TABLE 2.3

World crude oil production, selected years 1960–2004

[Million barrels per day]

		Selected OPEC producers								Selected non-OPEC producers									
Year	Persian Gulf nations[a]	Iran	Iraq	Kuwait[b]	Nigeria	Saudi Arabia[b]	United Arab Emirates	Venezuela	Total OPEC	Canada	China	Mexico	Norway	Former U.S.S.R.	Russia	United Kingdom	United States	Total non-OPEC[c]	World
1960	5.27	1.07	0.97	1.69	0.02	1.31	0.00	2.85	8.70	0.52	0.10	0.27	0.00	2.91	—	s	7.04	12.29	20.99
1962	6.19	1.33	1.01	1.96	0.07	1.64	0.01	3.20	10.51	0.67	0.12	0.31	0.00	3.67	—	s	7.33	13.84	24.35
1964	7.61	1.71	1.26	2.30	0.12	1.90	0.19	3.39	12.98	0.75	0.18	0.32	0.00	4.60	—	s	7.61	15.20	28.18
1966	9.32	2.13	1.39	2.48	0.42	2.60	0.36	3.37	15.77	0.88	0.29	0.33	0.00	5.23	—	s	8.30	17.19	32.96
1968	10.91	2.84	1.50	2.61	0.14	3.04	0.50	3.60	18.79	1.19	0.30	0.39	0.00	6.08	—	s	9.10	19.84	38.63
1970	13.39	3.83	1.55	2.99	1.08	3.80	0.78	3.71	23.30	1.26	0.60	0.49	0.00	6.99	—	s	9.64	22.59	45.89
1972	17.54	5.02	1.47	3.28	1.82	6.02	1.20	3.22	26.89	1.53	0.90	0.51	0.03	7.89	—	s	9.44	24.25	51.14
1974	21.28	6.02	1.97	2.55	2.26	8.48	1.68	2.98	30.35	1.55	1.32	0.57	0.04	8.91	—	s	8.77	25.37	55.72
1976	21.51	5.88	2.42	2.15	2.07	8.58	1.94	2.29	30.33	1.31	1.67	0.83	0.28	10.06	—	0.25	8.13	27.01	57.34
1978	20.61	5.24	2.56	2.13	1.90	8.30	1.83	2.17	29.46	1.32	2.08	1.21	0.36	11.11	—	1.08	8.71	30.70	60.16
1980	17.96	1.66	2.51	1.66	2.06	9.90	1.71	2.17	26.61	1.44	2.11	1.94	0.53	11.71	—	1.62	8.60	32.99	59.60
1982	12.16	2.21	1.01	0.82	1.30	6.48	1.25	1.90	18.78	1.27	2.05	2.75	0.52	11.91	—	2.07	8.65	34.70	53.48
1984	10.78	2.17	1.21	1.16	1.39	4.66	1.15	1.80	17.44	1.44	2.30	2.78	0.70	11.86	—	2.48	8.88	37.05	54.49
1986	11.70	2.04	1.69	1.42	1.47	4.87	1.33	1.79	18.28	1.47	2.62	2.44	0.87	11.90	—	2.54	8.68	37.95	56.23
1988	13.46	2.24	2.69	1.49	1.45	5.09	1.57	1.90	20.32	1.62	2.73	2.51	1.16	12.05	—	2.23	8.14	38.42	58.74
1990	15.28	3.09	2.04	1.18	1.81	6.41	2.12	2.14	23.20	1.55	2.77	2.55	1.70	10.98	—	1.82	7.36	37.37	60.57
1992	15.97	3.43	0.43	1.06	1.94	8.33	2.27	2.37	24.40	1.61	2.85	2.67	2.23	—	7.63	1.83	7.17	35.81	60.21
1994	16.96	3.62	0.55	2.03	1.93	8.12	2.19	2.59	25.51	1.75	2.94	2.69	2.52	—	6.14	2.37	6.66	35.48	60.99
1996	17.37	3.69	0.58	2.06	2.00	8.22	2.28	2.94	26.46	1.84	3.13	2.86	3.10	—	5.85	2.57	6.46	37.25	63.71
1998	19.34	3.63	2.15	2.09	2.15	8.39	2.35	3.17	28.77	1.98	3.20	3.07	3.02	—	5.85	2.62	6.25	38.15	66.92
1999	18.67	3.56	2.51	1.90	2.13	7.83	2.17	2.83	27.58	1.91	3.20	2.91	3.02	—	6.08	2.68	5.88	38.27	65.85
2000	19.89	3.70	2.57	2.08	2.17	8.40	2.37	3.16	29.27[R]	1.98	3.25	3.01	3.20	—		2.28	5.82	39.07[R]	68.34
2001	19.10	3.72	2.39	2.00	2.26	8.03	2.21	3.01	28.34	2.03	3.30	3.13[R]	3.12	—	6.92	2.28	5.80	39.53[R]	67.87[R]
2002	17.79	3.44	2.02	1.89	2.12	7.63	2.08	2.60	26.35[R]	2.17	3.39	3.18	2.99	—	7.41	2.29	5.75	40.43[R]	66.78[R]
2003	19.26	3.74[R]	1.31	2.18	2.24	8.85	2.35	2.34	27.98[R]	2.31	3.41	3.37	2.85	—	8.13[R]	2.09	5.68[R]	41.17[R]	69.15[R]
2004[P]	20.82	4.00	2.01	2.38	2.51	9.10	2.48	2.56	30.16	2.40	3.49	3.38	2.97	—	8.80	1.85	5.43	42.32	72.48

[a]Persian Gulf nations are Bahrain, Iran, Iraq, Kuwait, Qatar, Saudi Arabia, and United Arab Emirates.
[b]Includes about one-half of the production in the neutral zone between Kuwait and Saudi Arabia.
[c]Ecuador, which withdrew from OPEC on December 31, 1992, and Gabon, which withdrew on December 31, 1994, are included in "non-OPEC" for all years.
R=Revised. P=Preliminary. —= Not applicable. s=Less than 0.005 million barrels per day.
Notes: OPEC=Organization of the Petroleum Exporting Countries. Includes lease condensate, excludes natural gas plant liquids. Totals may not equal sum of components due to independent rounding. For related information, see http://www.eia.doe.gov/international.

SOURCE: Adapted from "Table 11.5. World Crude Oil Production, 1960–2004 (Million Barrels per Day)," in *Annual Energy Review 2004*, U.S. Department of Energy, Energy Information Administration, Office of Energy Markets and End Use, August, 2005, http://www.eia.doe.gov/emeu/aer/pdf/aer.pdf (accessed April 5, 2006)

FIGURE 2.9

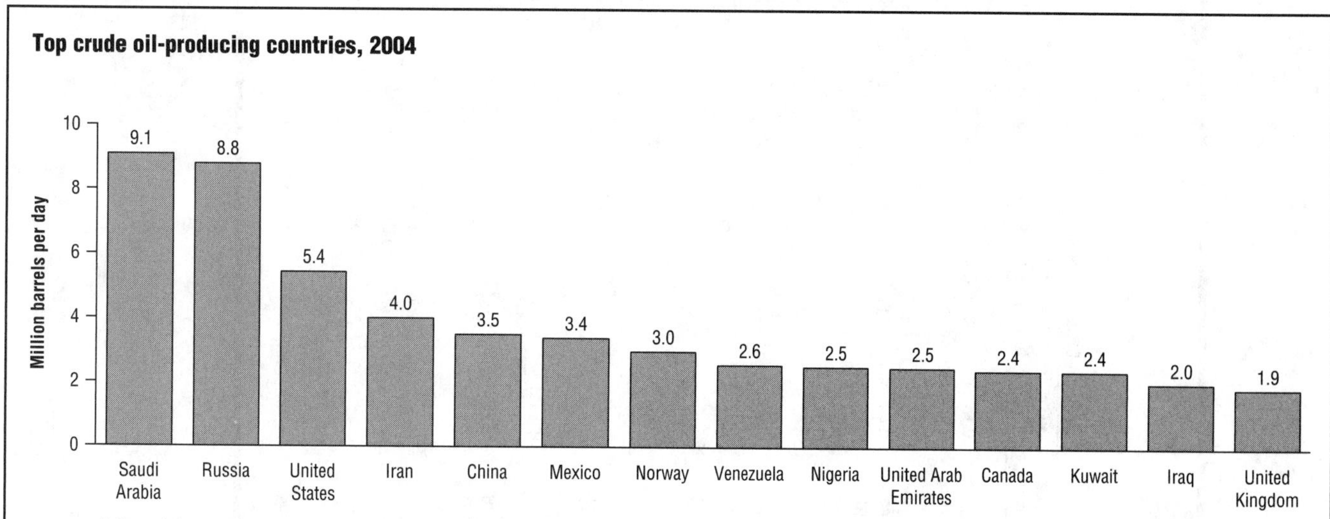

SOURCE: Adapted from "Figure 11.5. World Crude Oil Production: Top Producing Countries, 2004," in *Annual Energy Review 2004*, U.S. Department of Energy, Energy Information Administration, Office of Energy Markets and End Use, August, 2005, http://www.eia.doe.gov/emeu/aer/pdf/aer.pdf (accessed April 5, 2006)

lost some of its ability to control its members and, consequently, prices.

Nevertheless, OPEC actions can still effectively influence the petroleum market. For example, in an attempt to halt the downward slide of oil prices in 1999, Saudi Arabia, Mexico, and Venezuela agreed to cut production by 1.6% to two million barrels per day. Many other oil-producing nations also limited their production. The limitations worked: During the summer of 2000 oil prices climbed. According to *Annual Energy Review 2000* (Energy Information Administration, 2001), refiners paid $16.71 in real dollars per barrel of crude oil in 1999. In 2000 the price had risen to $26.40 per barrel.

Gasoline Prices

Many middle-aged and older Americans can remember when gas cost thirty cents per gallon. From 1972 to 1980, the price of a gallon of leaded regular gasoline (in current dollars that do not consider inflation) more than tripled, while the price in real dollars (adjusted for inflation) rose 83%. (See Table 2.5.) In 1981 the price of a gallon of regular unleaded gasoline was $2.33 in real dollars. However, after 1981, as a result of the international oil glut, real prices tumbled. In 1998 the price per gallon was only $1.10, and after price increases in 1999, the price of unleaded gas still averaged only $1.51 in 2000 and less in 2001 through 2003. By 2004, however, gasoline prices had risen to $1.74 per gallon.

By the end of April 2006, prices at the pump had soared in the United States. In many major cities, prices hovered around $3 per gallon for regular gasoline. Oil prices were rising all over the world in response to high demand, especially in the United States, China, Japan, and India. In addition, the U.S. government was changing the additives required in certain fuel blends, which caused shortages in parts of the country, pushing fuel prices up even more.

To confront the high gasoline prices, President Bush announced a four-part plan on April 25, 2006. The plan included:

1. Investigating whether the price of gas had been unfairly manipulated since Hurricanes Katrina and Rita hit the Gulf Coast in the summer 2005
2. Promoting greater fuel efficiency by providing tax credits for all hybrid and "clean diesel" vehicles sold in 2006
3. Boosting supplies of crude oil and gas by temporarily halting deposits to the Strategic Petroleum Reserve
4. Calling on Congress to support the Advanced Energy Initiative (see Chapter 1)

ENVIRONMENTAL CONCERNS ABOUT OIL TRANSPORTATION

Transporting oil carries significant environmental risks. According to the U.S. Department of the Interior, oil tanker accidents are the cause of most transportation spills.

The *Exxon Valdez* Oil Spill

While a number of events have influenced American attitudes toward oil production and use, one of the most notable occurred in March 1989, when the *Exxon Valdez*, an oil tanker, hit a reef in Alaska and spilled eleven million gallons of crude oil into the waters of Prince

TABLE 2.4

World petroleum consumption, selected years 1960–2003

[Million barrels per day]

	Selected OECD consumers											Selected non-OECD consumers						
Year	Canada	France	Germany[a]	Italy	Japan	Mexico[b]	South Korea[b]	Spain	United Kingdom	United States	Total OECD[c]	Brazil	China	India	Former U.S.S.R.	Russia	Total non-OECD	World
1960	0.84	0.56	0.63	0.44	0.66	0.30	0.01	0.10	0.94	9.80	15.78	0.27	0.17	0.16	2.38	—	5.56	21.34
1962	0.92	0.73	1.00	0.67	0.93	0.30	0.02	0.12	1.12	10.40	18.06	0.31	0.14	0.18	2.87	—	6.83	24.89
1964	1.05	0.98	1.36	0.90	1.48	0.33	0.02	0.20	1.36	11.02	21.05	0.35	0.20	0.22	3.58	—	8.03	29.08
1966	1.21	1.19	1.80	1.08	1.98	0.36	0.04	0.31	1.58	12.08	24.60	0.38	0.30	0.28	3.87	—	8.96	33.56
1968	1.34	1.46	1.99	1.40	2.66	0.41	0.10	0.46	1.82	13.39	28.56	0.46	0.31	0.31	4.48	—	10.40	38.96
1970	1.52	1.94	2.83	1.71	3.82	0.50	0.20	0.58	2.10	14.70	34.69[R]	0.53	0.62	0.40	5.31	—	12.12[R]	46.81
1972	1.66	2.32	3.13	1.95	4.36	0.59	0.23	0.68	2.28	16.37	38.95[R]	0.66	0.91	0.46	6.12	—	14.14[R]	53.09
1974	1.78	2.45	3.06	2.00	4.86	0.71	0.29	0.86	2.21	16.65	40.38[R]	0.86	1.19	0.47	7.28	—	16.30[R]	56.68
1976	1.82	2.42	3.21	1.97	4.84	0.83	0.36	0.97	1.89	17.46	41.72[R]	1.00	1.53	0.51	7.78	—	17.95[R]	59.67
1978	1.90	2.41	3.29	1.95	4.95	0.99	0.48	0.98	1.94	18.85	43.98[R]	1.11	1.79	0.62	8.48	—	20.18[R]	64.16
1980	1.87	2.26	3.08	1.93	4.96	1.27	0.54	0.99	1.73	17.06	41.76	1.15	1.77	0.64	9.00	—	21.35	63.11
1982	1.58	1.88	2.74	1.78	4.58	1.48	0.53	1.00	1.59	15.30	37.77	1.06	1.66	0.74	9.08	—	21.77	59.54
1984	1.52	1.77	2.56	1.72	4.67	1.40	0.55	0.85	1.83	15.73	37.70	1.03	1.74	0.82	8.91	—	22.13	59.83
1986	1.53	1.76	2.79	1.73	4.50	1.52	0.59	0.87	1.64	16.28	38.61	1.24	2.00	0.95	8.98	—	23.22	61.83
1988	1.68	1.80	2.72	1.83	4.85	1.60	0.75	0.98	1.69	17.28	40.68	1.30	2.28	1.08	8.89	—	24.32	65.00
1990	1.75	1.83	2.68	1.87	5.22[R]	1.75	1.05	1.01	1.78	16.99	41.52[R]	1.47	2.30	1.17	8.39	—	25.06[R]	66.58[R]
1992	1.72[R]	1.93	2.84	1.89	5.49	1.86	1.53	1.10	1.82	17.03	42.95[R]	1.52	2.66	1.27	—	4.42	24.50[R]	67.45[R]
1994	1.77	1.86[R]	2.88	1.87	5.67[R]	1.93	1.84	1.12	1.83	17.72	44.43[R]	1.67	3.16	1.41	—	3.18	24.42[R]	68.85[R]
1996	1.87	1.95	2.92	1.92	5.79[R]	1.79	2.10	1.20	1.85	18.31	46.00[R]	1.90	3.61	1.68	—	2.62	25.62[R]	71.62[R]
1998	1.94[R]	2.04	2.92	1.94	5.58[R]	1.95	1.92	1.36	1.79	18.92	46.93[R]	2.10	4.11	1.84	—	2.49	27.16[R]	74.09[R]
1999	2.03	2.03	2.84	1.89	5.70[R]	1.96	2.08	1.40	1.79	19.52	47.86[R]	2.13	4.36	2.03	—	2.54	27.97	75.83[R]
2000	2.03[R]	2.00	2.77	1.85	5.61[R]	2.04	2.14	1.43	1.76	19.70	47.97[R]	2.17	4.80	2.13	—	2.58	28.98	76.95[R]
2001	2.04	2.05	2.81	1.84	5.53[R]	1.99	2.13	1.49	1.72	19.65	48.01[R]	2.21	4.92	2.18	—	2.59[R]	29.69[R]	77.70[R]
2002	2.08[R]	1.98	2.72	1.87[R]	5.46[R]	1.94[R]	2.15[R]	1.51	1.77[R]	19.76	48.05[R]	2.13[R]	5.16	2.26[R]	—	2.64[R]	30.41[R]	78.46[R]
2003[P]	2.19	2.06	2.68	1.87	5.58	2.02	2.17	1.54	1.72	20.03	48.86	2.10	5.55	2.32	—	2.68	31.24	80.10

[a]Through 1969, the data for Germany are for the former West Germany only. For 1970 through 1990, this is East and West Germany. Beginning in 1991, this is unified Germany.
[b]Mexico, which joined the OECD on May 18, 1994, and South Korea, which joined the OECD on December 12, 1996, are included in the OECD for all years shown in this table.
[c]Hungary and Poland, which joined the OECD on May 7, 1996, and November 22, 1996, respectively, are included in Total OECD beginning in 1970, the first year that data for these countries were available. Total OECD includes Czechoslovakia from 1970–1992, and Czech Republic and Slovakia from 1993 forward.
R=Revised. P=Preliminary. —=Not applicable.
Notes: OECD=Organization for Economic Cooperation and Development. Totals may not equal sum of components due to independent rounding. For related information, see http://www.eia.doe.gov/international.

SOURCE: Adapted from "Table 11.10. World Petroleum Consumption, 1960–2003 (Million Barrels per Day)," in *Annual Energy Review 2004*, U.S. Department of Energy, Energy Information Administration, Office of Energy Markets and End Use, August, 2005, http://www.eia.doe.gov/emeu/aer/pdf/aer.pdf (accessed April 5, 2006)

William Sound. The cleanup cost Exxon $1.28 billion, a sum that does not include legal costs or any valuation for the wildlife lost. The spill was an environmental disaster for a formerly pristine area. Even measures used to clean up the spill, such as washing the beaches with hot water, caused additional damage.

The *Exxon Valdez* spill led to debate in the United States about tanker safety and design. Tankers are bigger than ever before. In 1945 the largest tanker held 16,500 tons of oil; today supertankers carry more than 550,000 tons. These ships are difficult to maneuver because of their size and are likely to spill more oil if damaged.

The Oil Pollution Act of 1990

The *Exxon Valdez* oil spill led Congress to pass the Oil Pollution Act of 1990 (PL 101-380), which increased, but still limited, oil spillers' federal liability (financial responsibility) as long as spills were not the result of "gross negligence." The bill also mandated compensation to those who were economically injured by oil spills. Damages that can be charged to oil companies were limited to $60 million for tanker accidents and $75 million for accidents at offshore facilities. The law specified that the rest of the cleanup costs were to be paid from a $500 million oil-spill fund generated by a $0.013-per-barrel tax on oil. Individual states still have the right to impose unlimited liability on spillers. Oil companies were also required to phase in double-hulled vessels by 2015. Essentially, a double-hulled vessel carries its oil in a container inside another container, providing extra protection in case of an accident.

Oil Spills Still Occur Worldwide

In November 2002 the twenty-six-year-old, single-hulled tanker *Prestige* was damaged off the coast of Spain and spilled approximately 5,000 tons of heavy fuel oil, according to Spanish government estimates. The ship continued to leak, so Spanish authorities ordered the leaking vessel towed to the open ocean. Several thousand more tons of oil were released as it sank, although much of the oil may have solidified inside the ship in the cold water at the bottom of the ocean. Scientists have estimated that the oil spilled from the *Prestige* caused the deaths of nearly 250,000 seabirds. It also killed unknown numbers of fish and dolphins and was responsible for economic damage to the Spanish fish and shellfish industries.

Smaller oil spills occur as well. For example, oil frequently washes ashore on Newfoundland's south coast. The oil often comes from bilge water—waste water contaminated with oil that accumulates in the bottoms of ships. Bilge water should be properly dumped, but it is often pumped into the ocean to save dumping costs.

Not All Oil Spills Are Due to Oil Transportation

Hurricane Katrina, which hit the coast of Louisiana and Mississippi in 2005, triggered 575 spills of petroleum and hazardous chemicals (generally refinery products), according to the Natural Resources Defense Council. The Gulf Coast is home to many oil refineries with scores of oil storage tanks and miles of oil pipeline. Some tanks and pipelines did not withstand the floodwater, releasing an estimated eight million gallons of oil and refinery products into the soil and buildings of New Orleans. Before people can safely move back into these areas, the oil and other refinery products must be cleaned up because these chemicals can be hazardous: Short-term exposure to certain chemicals in the oil causes dizziness and nausea; long-term exposure has been linked to leukemia and other serious ailments.

FIGURE 2.10

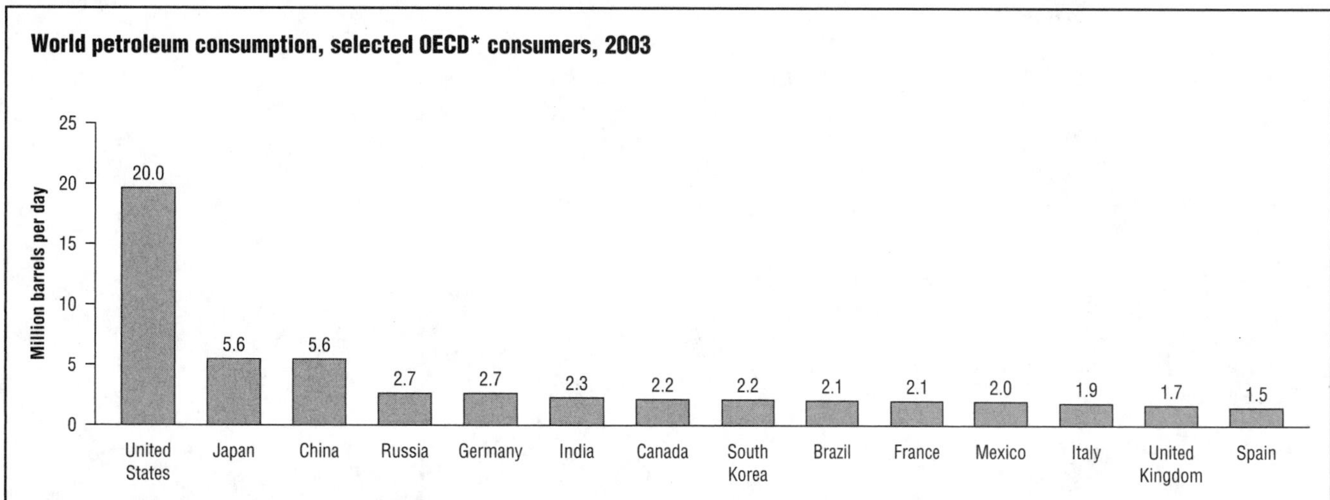

*Organization for Economic Cooperation and Development.

SOURCE: Adapted from "Figure 11.10. World Petroleum Consumption: Selected OECD Consumers, 2003," in *Annual Energy Review 2004*, U.S. Department of Energy, Energy Information Administration, Office of Energy Markets and End Use, August, 2005, http://www.eia.doe.gov/emeu/aer/pdf/aer.pdf (accessed April 5, 2006)

FIGURE 2.11

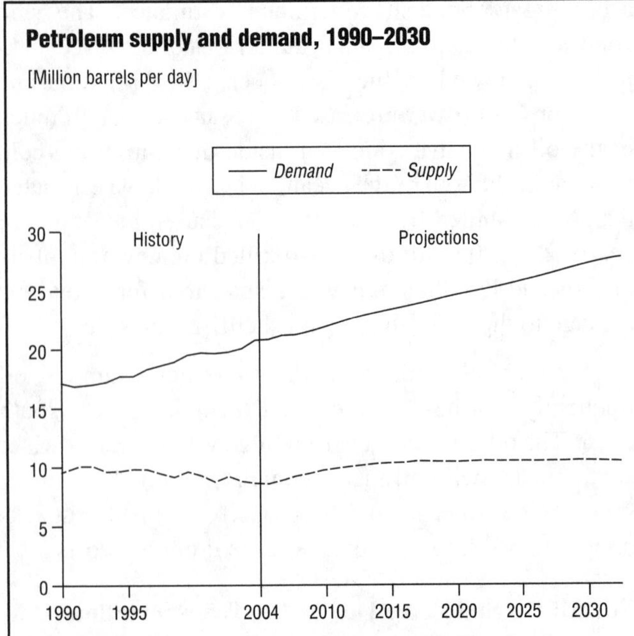

SOURCE: "Figure 93. U.S. Petroleum Product Demand and Domestic Petroleum Supply, 1990–2030 (Million Barrels per Day)," in *Annual Energy Outlook 2006*, U.S. Department of Energy, Energy Information Administration, Office of Integrated Analysis and Forecasting, February 2006, http://www.eia.doe.gov/oiaf/aeo/pdf/0383(2006).pdf (accessed April 5, 2006)

FIGURE 2.12

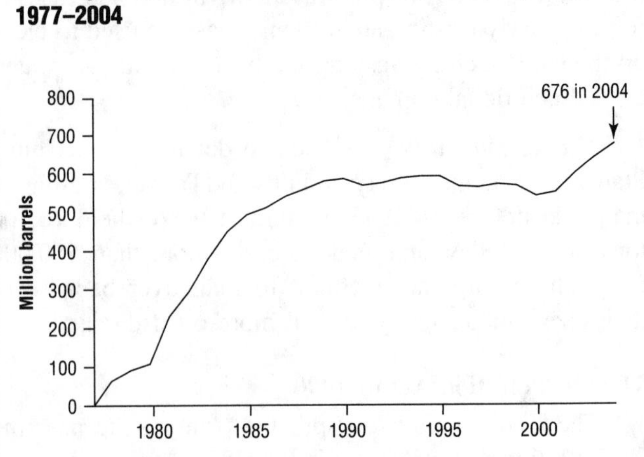

SOURCE: Adapted from "Figure 5.17. Strategic Petroleum Reserve, 1977–2004: End-of-Year Stocks in SPR," in *Annual Energy Review 2004*, U.S. Department of Energy, Energy Information Administration, Office of Energy Markets and End Use, August, 2005, http://www.eia.doe.gov/emeu/aer/pdf/aer.pdf (accessed April 5, 2006)

FIGURE 2.13

Strategic Petroleum Reserve (SPR) stocks as days of net imports, 1977–2004

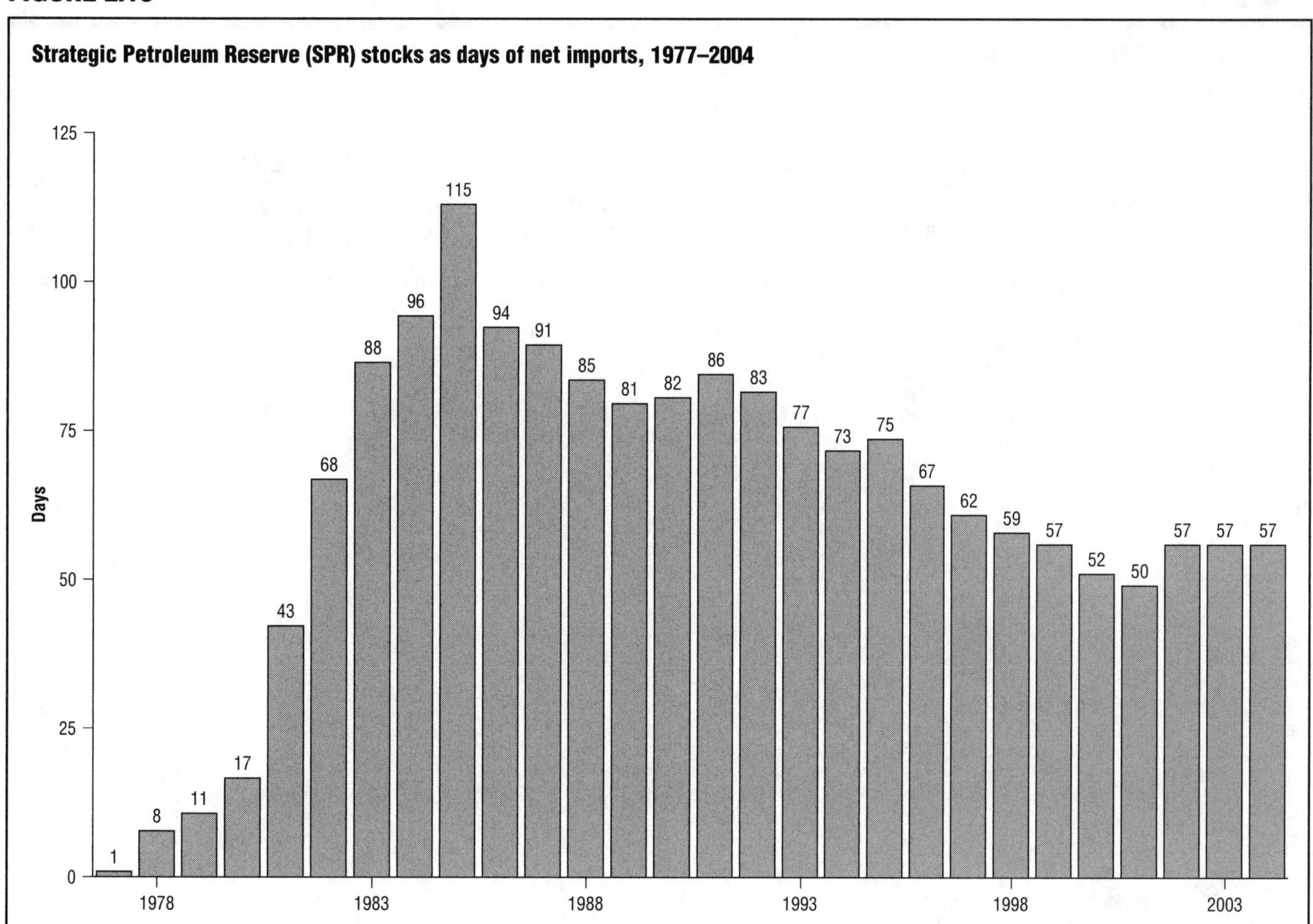

Note: Derived by dividing end-of-year SPR stocks by annual average daily net imports of all petroleum.

SOURCE: Adapted from "Figure 5.17. Strategic Petroleum Reserve, 1977–2004: SPR Stocks as Days' Worth of Net Imports," in *Annual Energy Review 2004*, U.S. Department of Energy, Energy Information Administration, Office of Energy Markets and End Use, August, 2005, http://www.eia.doe.gov/emeu/aer/pdf/aer.pdf (accessed April 5, 2006)

TABLE 2.5

Retail motor gasoline and on-highway diesel fuel prices, selected years 1949–2004

[Dollars per gallon]

	Motor gasoline by grade								Regular motor gasoline by area type[a]			
	Leaded regular		Unleaded regular		Unleaded premium		All grades					
Year	Nominal	Real[b]	Nominal	Real[b]	Nominal	Real[b]	Nominal	Real[b]	Conventional gasoline areas[c, d]	Reformulated Gasoline areas[e, f]	All areas	On-highway diesel fuel[a]
1949	0.27	1.64	NA	NA	NA	NA	NA	NA	NA	NA	NA	NA
1950	0.27	1.62	NA	NA	NA	NA	NA	NA	NA	NA	NA	NA
1955	0.29	1.55	NA	NA	NA	NA	NA	NA	NA	NA	NA	NA
1960	0.31	1.48	NA	NA	NA	NA	NA	NA	NA	NA	NA	NA
1965	0.31	1.39	NA	NA	NA	NA	NA	NA	NA	NA	NA	NA
1970	0.36	1.30	NA	NA	NA	NA	NA	NA	NA	NA	NA	NA
1972	0.36	1.20	NA	NA	NA	NA	NA	NA	NA	NA	NA	NA
1974	0.53	1.53	NA	NA	NA	NA	NA	NA	NA	NA	NA	NA
1976	0.59	1.47	0.61	1.53	NA	NA	NA	NA	NA	NA	NA	NA
1978	0.63	1.37	0.67	1.46	NA	NA	0.65	1.43	NA	NA	NA	NA
1980	1.19	2.20	1.25	2.30	NA	NA	1.22	2.26	NA	NA	NA	NA
1981	1.31	2.22	1.38	2.33	1.47	2.49	1.35	2.29	NA	NA	NA	NA
1982	1.22	1.95	1.30	2.07	1.42	2.26	1.28	2.04	NA	NA	NA	NA
1984	1.13	1.67	1.21	1.79	1.37	2.02	1.20	1.77	NA	NA	NA	NA
1986	0.86	1.20	0.93	1.30	1.09	1.52	0.93	1.31	NA	NA	NA	NA
1988	0.90	1.19	0.95	1.25	1.11	1.46	0.96	1.27	NA	NA	NA	NA
1990	1.15	1.41	1.16	1.43	1.35	1.65	1.22	1.49	NA	NA	NA	NA
1992	NA	NA	1.13	1.31	1.32	1.52	1.19	1.38	1.09	NA	1.09	NA
1994	NA	NA	1.11	1.23	1.31	1.45	1.17	1.30	1.07	NA	1.08	NA
1996	NA	NA	1.23	1.31	1.41	1.51	1.29	1.37	1.19	1.28	1.22	1.24
1998	NA	NA	1.06	1.10	1.25	1.30	1.12	1.16	1.02	1.08	1.03	1.04
1999	NA	NA	1.17	1.19	1.36	1.39	1.22	1.25	1.12	1.20	1.14	1.12
2000	NA	NA	1.51	1.51	1.69	1.69	1.56	1.56	1.46	1.54	1.48	1.49
2001	NA	NA	1.46	1.43	1.66	1.62	1.53	1.50	1.38	1.50	1.42	1.40
2002	NA	NA	1.36	1.31	1.56	1.50	1.44	1.38[R]	1.31	1.41	1.35	1.32
2003	NA	NA	1.59	1.50[R]	1.78	1.68	1.64	1.55	1.52	1.66	1.56	1.51
2004	NA	NA	1.88	1.74	2.07	1.91	1.92	1.78	1.81	1.94	1.85	1.81

[a]Nominal dollars.
[b]In chained (2000) dollars, calculated by using gross domestic product implicit price deflators.
[c]Any area that does not require the sale of reformulated gasoline.
[d]For 1993–2000, data collected for oxygenated areas are included in "conventional gasoline areas."
[e]"Reformulated gasoline areas" are ozone nonattainment areas designated by the Environmental Protection Agency that require the use of reformulated gasoline.
[f]For 1995–2000, data collected for combined oxygenated and reformulated areas are included in "reformulated gasoline areas."
R=Revised. NA=Not available.

SOURCE: Adapted from "Table 5.24. Retail Motor Gasoline and On-Highway Diesel Fuel Prices, Selected Years, 1949–2004 (Dollars per Gallon)," in *Annual Energy Review 2004*, U.S. Department of Energy, Energy Information Administration, Office of Energy Markets and End Use, August, 2005, http://www.eia.doe.gov/emeu/aer/pdf/aer.pdf (accessed April 5, 2006)

CHAPTER 3
NATURAL GAS

Natural gas is an important source of energy in the United States. Like petroleum, natural gas is composed of hydrocarbons, which are chemical compounds containing both hydrogen and carbon. The molecular structure of hydrocarbon compounds varies from the simplest, methane (CH_4), to very heavy and very complex molecules, such as those found in petroleum.

Methane, ethane, and propane are the primary constituents of natural gas, with methane making up 73% to 95% of the total. Consumer-grade natural gas is "dry gas," which means that it has been processed to remove water vapor, nonhydrocarbon gases (such as helium and nitrogen), and certain compounds that liquefy during the processing (such as lease condensate and natural gas plant liquids). Lease condensate is a liquid mix of heavy hydrocarbons recovered during natural gas processing at a lease, or field separation, facility. Natural gas plant liquids are compounds such as propane and butane recovered as liquids at other facilities later in the processing.

The natural gas industry developed out of the petroleum industry. Wells drilled for oil often produced considerable amounts of natural gas, but early oilmen had no idea what to do with it. Originally considered a waste byproduct, natural gas had no market. Even if a use for natural gas had been known at the time, there were no transmission lines to deliver it. As a result, the gas was flared, or burned off. Pictures of southeast Texas in the early twentieth century show thousands of wooden drilling rigs, each topped with a plume of flaming gas. Even today, flaring sites are sometimes the brightest spots in nighttime satellite images, outshining even the largest urban areas.

Eventually, researchers found ways to use natural gas for lighting, cooking, and heat. In 1925 the first natural gas pipeline, more than 200 miles long, was built from Louisiana to Texas. U.S. demand grew rapidly, especially after World War II. By the 1950s natural gas was providing one-quarter of the nation's energy needs. At the beginning of this century natural gas was second only to coal in the share of U.S. energy produced. Crude oil was third. (See Table 1.1 in Chapter 1.) A vast pipeline transmission system now connects production facilities in the United States, Canada, and Mexico with natural gas distributors. Figure 3.1 shows the production and consumption figures for natural gas for 2004. Figure 3.2 shows the pattern of natural gas supply and distribution in the United States in 2004.

THE PRODUCTION OF NATURAL GAS

Natural gas is produced from gas and oil wells. There is little delay between production and consumption, except for gas that is placed in storage. Changes in demand are almost immediately reflected by changes in wellhead flows, or supply.

Total U.S. natural gas production in 2004 was 18.8 trillion cubic feet (see Figure 3.3), well below the peak levels of more than 21 trillion cubic feet produced from 1970 through 1973. According to the *Annual Energy Review 2004* (Energy Information Administration, 2005), Texas, Louisiana, and Oklahoma accounted for 36% of the natural gas produced in the United States in 2004. Although production is increasing because of demand and rising prices, it continues to be outpaced by consumption. Imported gas makes up the difference between supply and demand.

Natural Gas Wells

In 2004 about 385,000 gas wells were in operation in the United States. (See Figure 3.4.) Although the number of producing wells increased steadily after 1960 and more sharply after the mid-1970s, the number of gas wells in operation fluctuates from year to year because new wells are opened and old wells are closed. Weather and economic conditions also affect well operations.

FIGURE 3.1

Natural gas flow, 2004

[Trillion cubic feet]

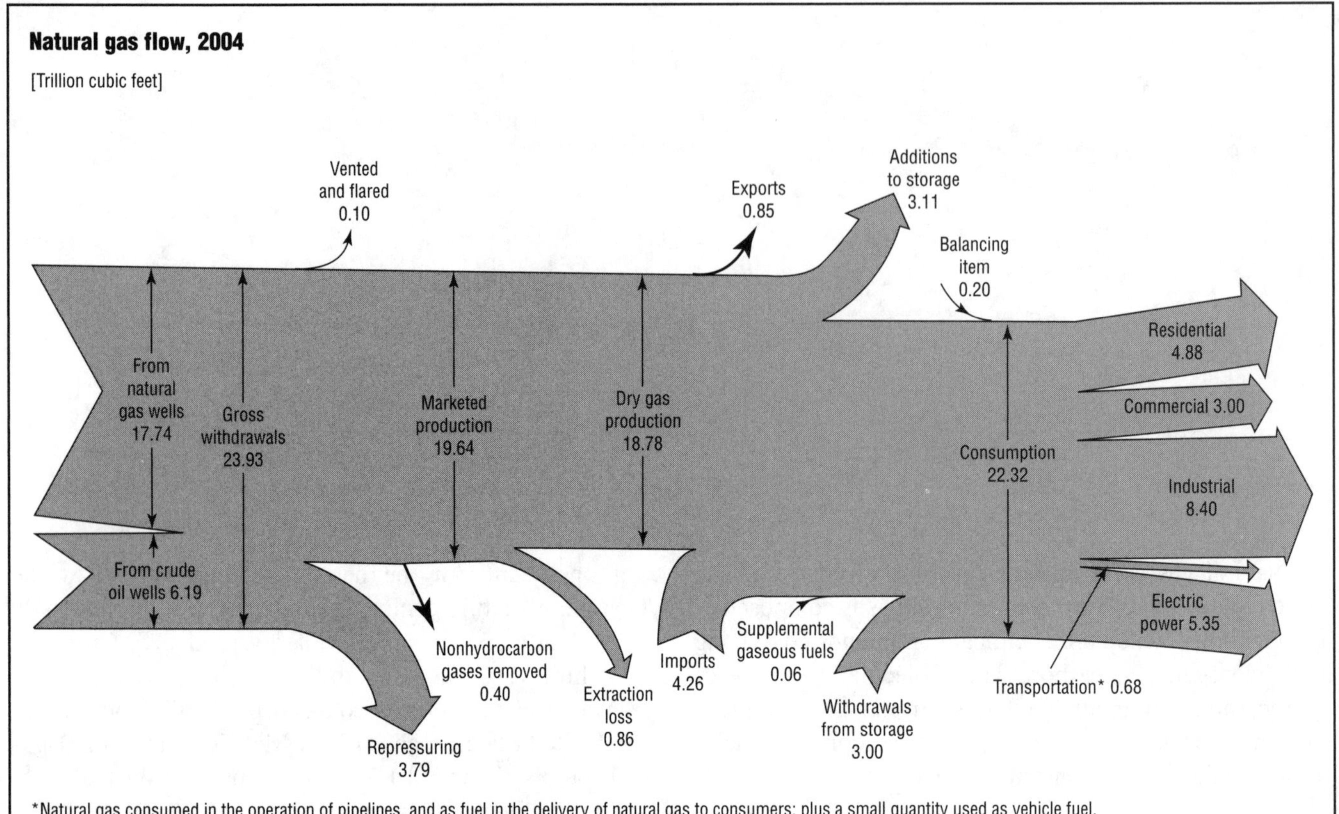

*Natural gas consumed in the operation of pipelines, and as fuel in the delivery of natural gas to consumers; plus a small quantity used as vehicle fuel.
Notes: Data are preliminary. Totals may not equal sum of components due to independent rounding.

SOURCE: "Diagram 3. Natural Gas Flow, 2004 (Trillion Cubic Feet)," in *Annual Energy Review 2004*, U.S. Department of Energy, Energy Information Administration, Office of Energy Markets and End Use, August, 2005, http://www.eia.doe.gov/emeu/aer/pdf/aer.pdf (accessed April 5, 2006)

The average productivity of natural gas wells peaked in 1971, then dropped throughout most of the 1970s and the mid-1980s; it has remained at a relatively steady low level since then. (See Figure 3.5.)

Offshore Production

Offshore wells—most are in the Gulf of Mexico and off the coast of California—accounted for nearly 4.7 trillion of the estimated 23.9 trillion cubic feet of gross withdrawals of natural gas in 2004, or nearly one-fifth of the total U.S. production. (See Figure 3.6.) Offshore production is expected to increase to meet the nation's growing need for energy. The U.S. Department of the Interior has leased more than 1.5 billion acres of offshore areas to oil companies for offshore drilling.

Offshore drilling generally occurs on the outer continental shelf, in waters up to 200 meters deep (about 650 feet). Figure 3.7 is a diagram of a continental margin. The continental shelf varies from one coastal area to another: the shelf is relatively narrow along the Pacific coast, wide along much of the Atlantic coast and the Gulf of Alaska, and widest in the Gulf of Mexico.

The development of offshore oil and gas resources began with the drilling of the Summerland oil field along the coast of California in 1896, where about 400 wells were drilled. Since then the industry has continually improved drilling technology. Today, deepwater petroleum and natural gas exploration occurs from platforms and drill ships, while shallow-water explorations occur from gravel islands and mobile units.

Even though most natural gas is transported by pipelines, rather than tanker ships, accidents such as the 1989 *Exxon Valdez* oil spill in Alaska and the *Prestige* oil spill off the coast of Spain in 2002 have focused attention on all types of offshore drilling and tanker transport. Even before the *Exxon Valdez* oil spill, environmentalists were calling for the curtailment of offshore drilling for both oil and gas. The spills from refining and storage facilities triggered by Hurricane Katrina in 2005 raised additional concerns. (See Chapter 2.)

Natural Gas Reserves

Reserves are estimated volumes of gas in known deposits that are believed to be recoverable in the future. Proved reserves are those gas volumes that geological

FIGURE 3.2

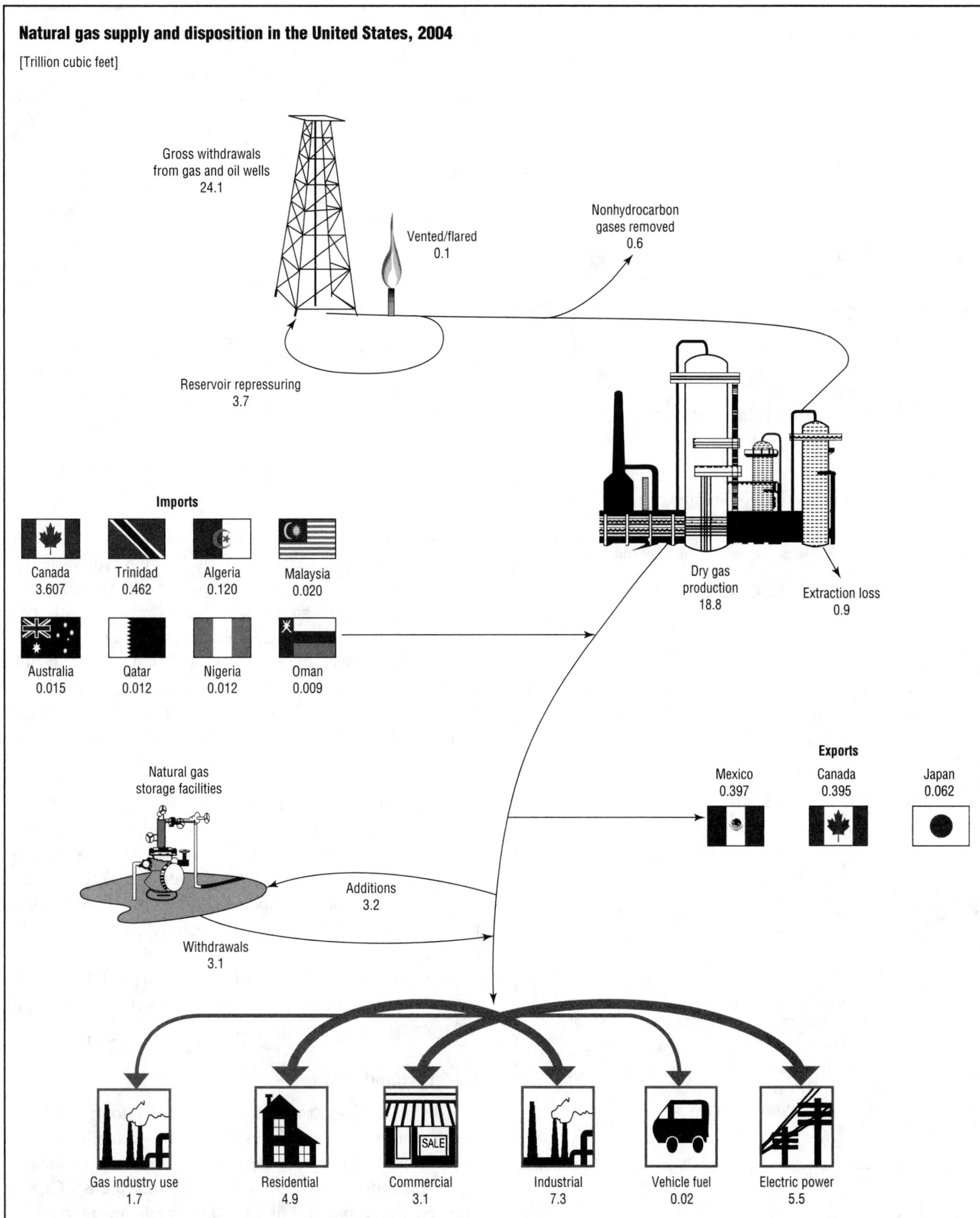

SOURCE: "Figure 2. Natural Gas Supply and Disposition in the United States, 2004," in *Natural Gas Annual 2004*, U.S. Department of Energy, Energy Information Administration, Office of Oil and Gas, December 2005, http://www.eia.doe.gov/pub/oil_gas/natural_gas/data_publications/natural_gas_annual/current/pdf/nga04.pdf (accessed April 11, 2006)

FIGURE 3.3

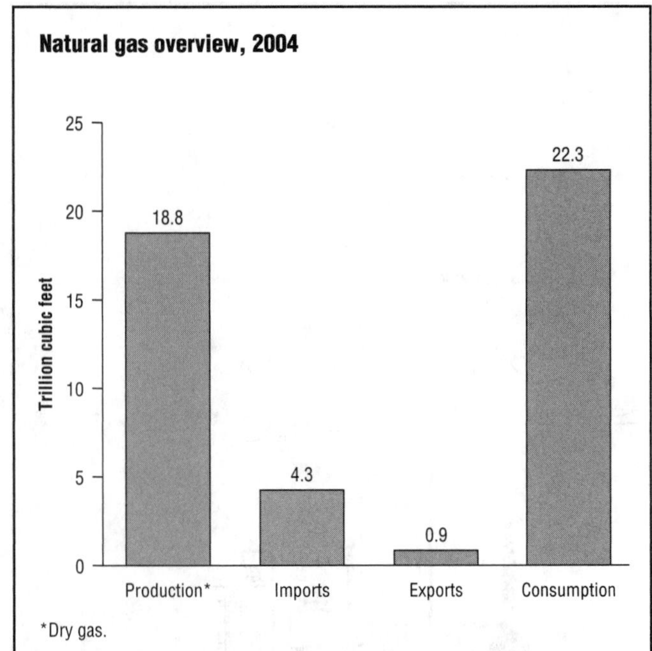

*Dry gas.

SOURCE: Adapted from "Figure 6.1. Natural Gas Overview: Overview, 2004," in *Annual Energy Review 2004,* U.S. Department of Energy, Energy Information Administration, Office of Energy Markets and End Use, August, 2005, http://www.eia.doe.gov/emeu/aer/pdf/aer.pdf (accessed April 5, 2006)

FIGURE 3.4

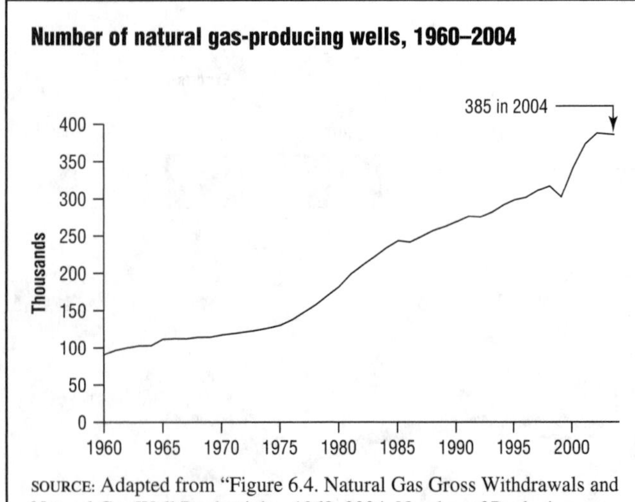

SOURCE: Adapted from "Figure 6.4. Natural Gas Gross Withdrawals and Natural Gas Well Productivity, 1960–2004: Number of Producing Wells," in *Annual Energy Review 2004,* U.S. Department of Energy, Energy Information Administration, Office of Energy Markets and End Use, August, 2005, http://www.eia.doe.gov/emeu/aer/pdf/aer.pdf (accessed April 5, 2006)

FIGURE 3.5

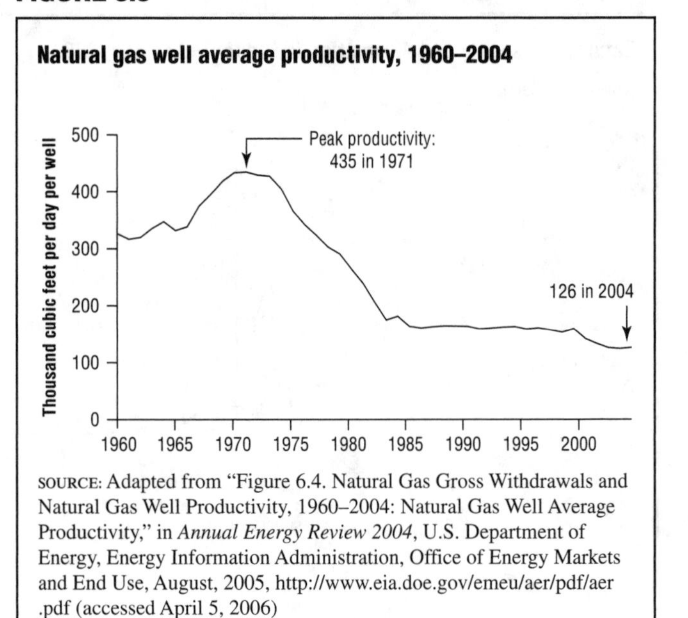

SOURCE: Adapted from "Figure 6.4. Natural Gas Gross Withdrawals and Natural Gas Well Productivity, 1960–2004: Natural Gas Well Average Productivity," in *Annual Energy Review 2004,* U.S. Department of Energy, Energy Information Administration, Office of Energy Markets and End Use, August, 2005, http://www.eia.doe.gov/emeu/aer/pdf/aer.pdf (accessed April 5, 2006)

and engineering data show with reasonable certainty to be recoverable. Proved reserves of natural gas amounted to 197.1 trillion cubic feet in 2003. (See Table 3.1.)

Natural gas reserves in North America are generally more abundant than crude oil reserves, although historically they have been difficult to assess with any accuracy. At one time the U.S. Department of Energy estimated that proven supplies of recoverable gas in the United States would last fewer than eight years. However, with new discoveries and technological improvements, it now estimates that recoverable supplies will last approximately twelve years; that is, it would take twelve years to deplete the proven reserves if current production rates are maintained and no new reserves are found (http://www.eia.doe.gov/emeu/perfpro/fig15.htm).

The North Slope fields of Alaska are estimated to contain reserves amounting to 35 trillion cubic feet. However, in early 2006 there was still no easy way to transport those reserves to the lower forty-eight states. A pipeline has been proposed but as of 2006 was not being pursued. If built, it could deliver an estimated 4.5 billion cubic feet of natural gas per day to the lower forty-eight states, or 5% of the nation's future daily natural gas consumption (http://www.ferc.gov/legal/staff-reports/angta-second.pdf).

Underground Storage

Because of seasonal, daily, and even hourly changes in demand, substantial natural gas storage facilities have been created. Many are depleted gas reservoirs located near transmission lines and marketing areas. Gas is injected into storage when market needs are lower than the available gas flow, and gas is withdrawn from storage when supplies from producing fields and the capacity of transmission lines are not adequate to meet peak demands. At the end of 2004 gas in underground storage totaled approximately 6.9 trillion cubic feet, according to the U.S. Department of Energy. (See Figure 3.8.)

FIGURE 3.6

Natural gas gross withdrawals by location, 1960–2004

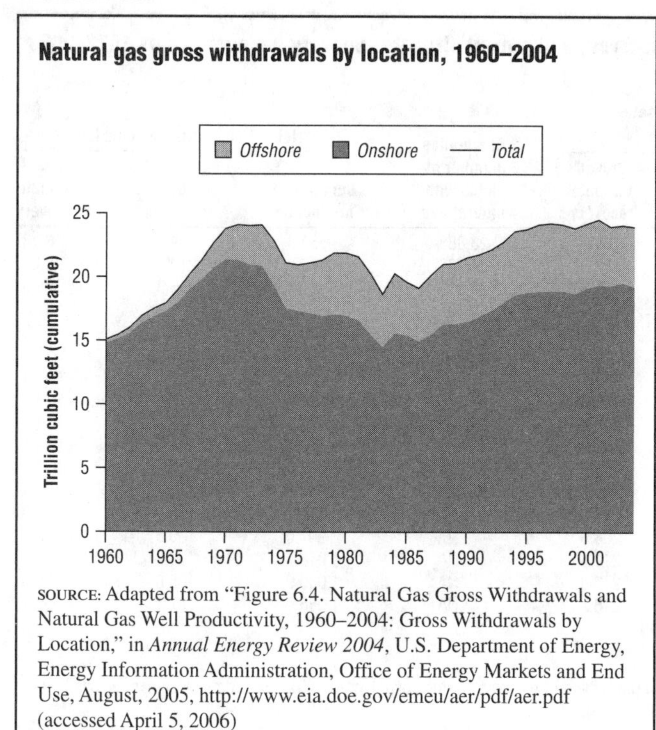

SOURCE: Adapted from "Figure 6.4. Natural Gas Gross Withdrawals and Natural Gas Well Productivity, 1960–2004: Gross Withdrawals by Location," in *Annual Energy Review 2004*, U.S. Department of Energy, Energy Information Administration, Office of Energy Markets and End Use, August, 2005, http://www.eia.doe.gov/emeu/aer/pdf/aer.pdf (accessed April 5, 2006)

TRANSMISSION OF NATURAL GAS

A vast network of natural gas pipelines crisscrosses the United States. The natural gas in this quarter-million-mile system generally flows northeastward, primarily from Texas and Louisiana, the two major gas-producing states, and from Oklahoma and New Mexico. (See Figure 3.9.) It also flows west to California.

Imports of natural gas enter the United States via pipeline from Canada into Idaho, Maine, Michigan, Montana, New Hampshire, New York, North Dakota, Washington, and Vermont. Natural gas also enters via pipeline into Texas from Mexico. According to the *Natural Gas Annual 2004* (Energy Information Administration, 2005), 85% of imported natural gas arrived in the United States by pipeline in 2004. The remainder was shipped as liquefied natural gas, arriving by tanker from Algeria, Australia, Malaysia, Nigeria, Oman, Qatar, and Trinidad. Liquefied natural gas is produced by cooling natural gas to –260° Fahrenheit; at this temperature natural gas changes from a gas to a liquid.

FIGURE 3.7

Generalized profile of the continental margin

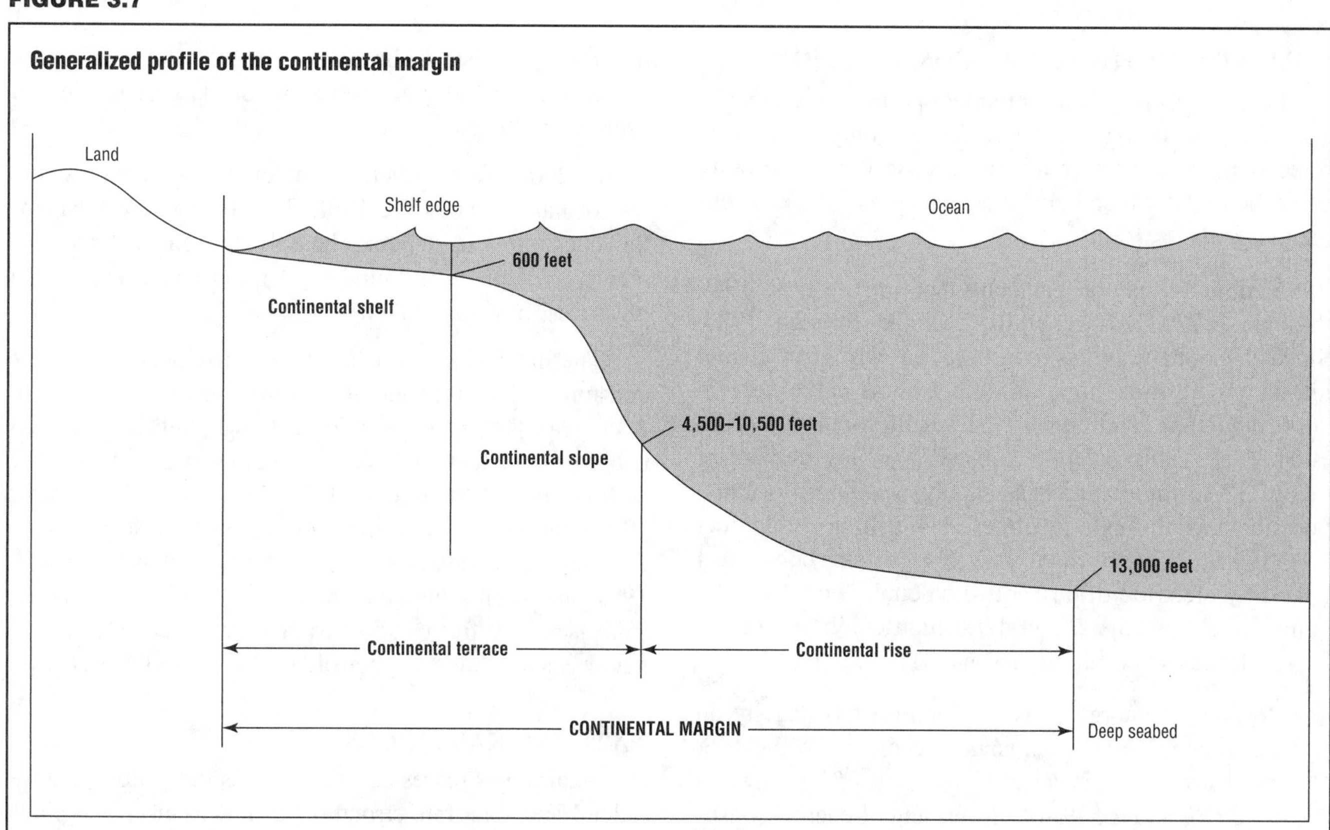

Note: Depths and gradients are approximate.

SOURCE: George Dellagiarino and Keith Meekins, "Figure 1. Profile of the Continental Margin," in *The Resource Evaluation Program: Structure and Mission on the Outer Continental Shelf*, U.S. Department of the Interior, Minerals Management Service, Resource Evaluation Division, 1998, http://www.mms.gov/itd/pubs/1998/98-0028.pdf (accessed April 11, 2006)

TABLE 3.1

Crude oil and natural gas field counts, cumulative production, proved reserves, and proved ultimate recovery, selected years 1977–2003

Year	Cumulative number of fields with crude oil and/or natural gas	Cumulative number of fields with crude oil	Crude oil and lease condensate (billion barrels)			Cumulative number of fields with natural gas	Natural gas* (trillion cubic feet)		
			Cumulative production	Proved reserves	Proved ultimate recovery		Cumulative production	Proved reserves	Proved ultimate recovery
1977	31,360	27,835	121.4	33.6	155.0	23,883	558.3	209.5	767.8
1978	32,430	28,683	124.6	33.1	157.6	24,786	578.4	210.1	788.5
1980	34,999	30,766	130.8	31.3	162.2	26,919	619.4	206.3	825.6
1982	38,123	33,375	137.1	29.5	166.6	29,375	658.1	209.3	867.4
1984	41,038	35,784	143.5	30.0	173.5	31,595	693.5	206.0	899.5
1986	43,076	37,464	150.0	28.3	178.3	33,151	727.8	201.1	928.9
1988	44,414	38,506	156.0	28.2	184.2	34,196	763.4	177.0	940.4
1990	45,385	39,244	161.5	27.6	189.0	34,975	800.4	177.6	978.0
1992	46,149	39,843	166.8	25.0	191.8	35,539	838.0	173.3	1,011.3
1994	46,922	40,417	171.7	23.6	195.3	36,142	877.1	171.9	1,049.1
1996	47,557	40,875	176.5	23.3	199.8	36,612	917.0	175.1	1,092.1
1998	47,664	35,143	181.2	22.4	203.5	32,458	957.0	172.4	1,129.4
1999	NA	NA	183.3	23.2	206.5	NA	976.8	176.2	1,153.0
2000	NA	NA	185.4	23.5	208.9	NA	997.0	186.5	1,183.5
2001	NA	NA	187.5	23.9	211.4	NA	1,016.7	191.7R	1,208.4R
2002	NA	NA	189.6	24.0	213.6	NA	1,036.9	195.6	1,232.5
2003	NA	NA	193.1	23.1	216.2	NA	1,056.0	197.1	1,253.1

*Wet, after separation of lease condensate.
NA=Not available. R=Revised.
Notes: Data are at end of year. See http://www.eia.doe.gov/oil_gas/petroleum/info_glance/petroleum.html and http://www.eia.doe.gov/oil_gas/natural_gas/info_glance/natural_gas.html for related information.

SOURCE: Adapted from "Table 4.2. Crude Oil and Natural Gas Field Counts, Cumulative Production, Proved Reserves, and Proved Ultimate Recovery, 1977–2003," in *Annual Energy Review 2004*, U.S. Department of Energy, Energy Information Administration, Office of Energy Markets and End Use, August, 2005, http://www.eia.doe.gov/emeu/aer/pdf/aer.pdf (accessed April 5, 2006)

DOMESTIC NATURAL GAS CONSUMPTION

Natural gas fulfills an important part of the country's energy needs. It is an attractive fuel not only because its price is relatively low but also because it burns cleanly and efficiently, which helps the country meet its environmental goals.

Nationally, natural gas consumption rose from 1949 through 1972, then generally declined through 1986. Since 1986 natural gas consumption has been rising, hitting an all-time high of 23.3 trillion cubic feet in 2000, and then declining to 22.3 trillion cubic feet by 2004. (See Table 3.2.) In 2004, 0.7 trillion cubic feet (about 3% of the natural gas supply) was used to transport the gas through pipelines; 8.4 trillion cubic feet (38%) was used by industry; 5.4 trillion cubic feet (24%) by electric utilities; 3 trillion cubic feet (13%) by commercial customers; and 4.9 trillion cubic feet (22%) by residences. (See Figure 3.1 and Table 3.2.)

Residential energy consumption depends heavily on weather-related heating demands. According to the U.S. Census Bureau in *Statistical Abstract of the United States: 2006*, about 57% of all residential energy consumers in the United States used gas to heat their homes in 2003 (http://www.census.gov/compendia/statab/tables/06s0961.xls). Residential consumption is also affected by conservation practices and the efficiency of gas appliances such as water heaters and stoves. The *Natural Gas Annual 2004* lists California, Illinois, Michigan, Ohio, Pennsylvania, and New Jersey as the largest residential users by volume.

In 2004 the commercial sector used 3 trillion cubic feet of natural gas. (See Table 3.2.) Its use, like residential consumption, depends heavily on seasonal requirements, as well as the number of users and conservation measures they have taken.

The industrial sector has historically been the largest consumer of natural gas. Consumption in this sector in 2004 was 8.4 trillion cubic feet, up slightly from 8.3 trillion cubic feet in 2003, but below the high of 10.2 trillion cubic feet used in 1973. (See Table 3.2.) After 1973, natural gas consumption in the industrial sector declined quite steadily through 1986, increased through 1997, and then generally declined through 2004. Industrial use grew from 1986 through 1997 because natural gas was substituted for petroleum for some purposes.

NATURAL GAS PRICES

Natural gas prices can vary across the nation because federal and state rate structures differ. Region also plays a role. For example, prices are lower in major natural gas-producing areas where transmission costs are lower. From the mid-twentieth century through the early 1970s natural gas prices were relatively stable as reported in *Annual Energy Review 2004*. (See Figure 3.10.) Then

FIGURE 3.8

Natural gas in underground storage, 1954–2004

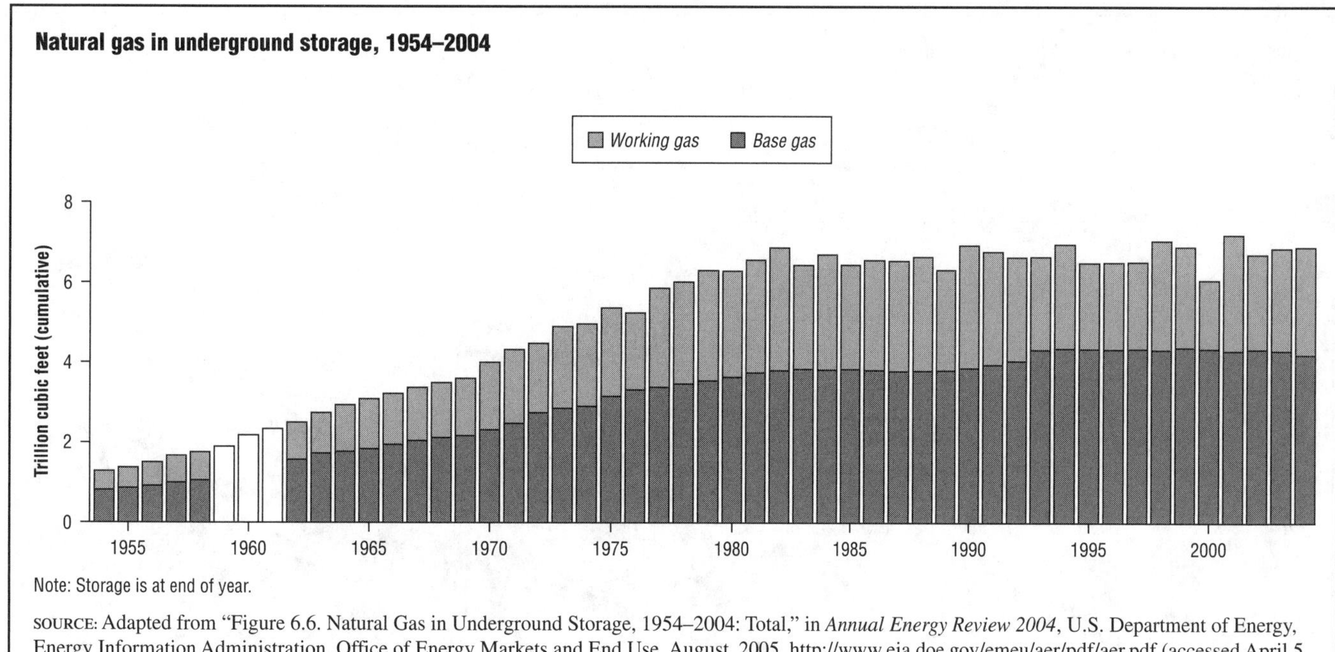

Note: Storage is at end of year.

SOURCE: Adapted from "Figure 6.6. Natural Gas in Underground Storage, 1954–2004: Total," in *Annual Energy Review 2004*, U.S. Department of Energy, Energy Information Administration, Office of Energy Markets and End Use, August, 2005, http://www.eia.doe.gov/emeu/aer/pdf/aer.pdf (accessed April 5, 2006)

governmental deregulation of the industry, as well as restructuring of companies, brought about a period of sharply rising prices, with wellhead prices (the value of natural gas at the mouth of the well) reaching a high in 1983, generally declining through 1995, and then generally increasing through 2004. The average price of natural gas at the wellhead was $5.07 in real dollars (that is, adjusted for inflation) per 1,000 cubic feet in 2004, sharply up from $2.83 in 2002.

At the retail price level (in real dollars), residential customers paid $9.92 per thousand cubic feet of natural gas in 2004, compared with $7.58 in 2002. (See Table 3.3.) Commercial consumers paid $8.56 per thousand cubic feet in 2004, while industrial consumers paid $5.91 per thousand cubic feet that year.

Much of the variation in natural gas prices through the years can be attributed to changes in the natural gas industry. The passage of the Natural Gas Policy Act of 1978 (PL 95-621) allowed prices at the wellhead to rise sharply. (See Figure 3.10.) On January 1, 1985, prices for new gas were deregulated, and additional volumes of onshore production were deregulated on July 1, 1987. In 1988 President Ronald Reagan signed legislation removing all remaining wellhead price controls by 1993.

The 1978 law not only allowed prices to go up; it also opened the market to the forces of supply and demand. Now that prices are deregulated and the industry is no longer constrained by federal controls, the natural gas industry has become more sensitive to market signals and responds more quickly to changes in economic conditions.

NATURAL GAS IMPORTS AND EXPORTS

U.S. natural gas trading was limited to the neighboring countries of Mexico and Canada until shipment of natural gas in liquefied form became a feasible alternative to pipelines. In 1969 the first shipments of liquefied natural gas were sent from Alaska to Japan, and U.S. imports of liquefied natural gas from Algeria began the following year.

In 2004 U.S. net imports of natural gas (total imports minus total exports) by all routes totaled 3.4 trillion cubic feet, or 15.3% of domestic consumption. Natural gas imports have been increasing significantly since 1986. Historically, Canada has been by far the major supplier of U.S. natural gas imports, accounting for 85% of the natural gas imported in 2004. (See Figure 3.11.)

According to the *Annual Energy Review 2004*, the United States exported 854 billion cubic feet of natural gas in 2004. Mexico bought the largest amount (397 billion cubic feet), followed by Canada (395 billion cubic feet) and Japan (62 billion cubic feet).

INTERNATIONAL NATURAL GAS USAGE

World Production

World production of dry natural gas totaled 95.2 trillion cubic feet in 2003, according to the U.S. Department of Energy. (See Table 3.4.) Russia accounted for 21.8 trillion cubic feet, while the United States produced 19 trillion cubic feet.

FIGURE 3.9

Principal interstate natural gas flow summary, 2004

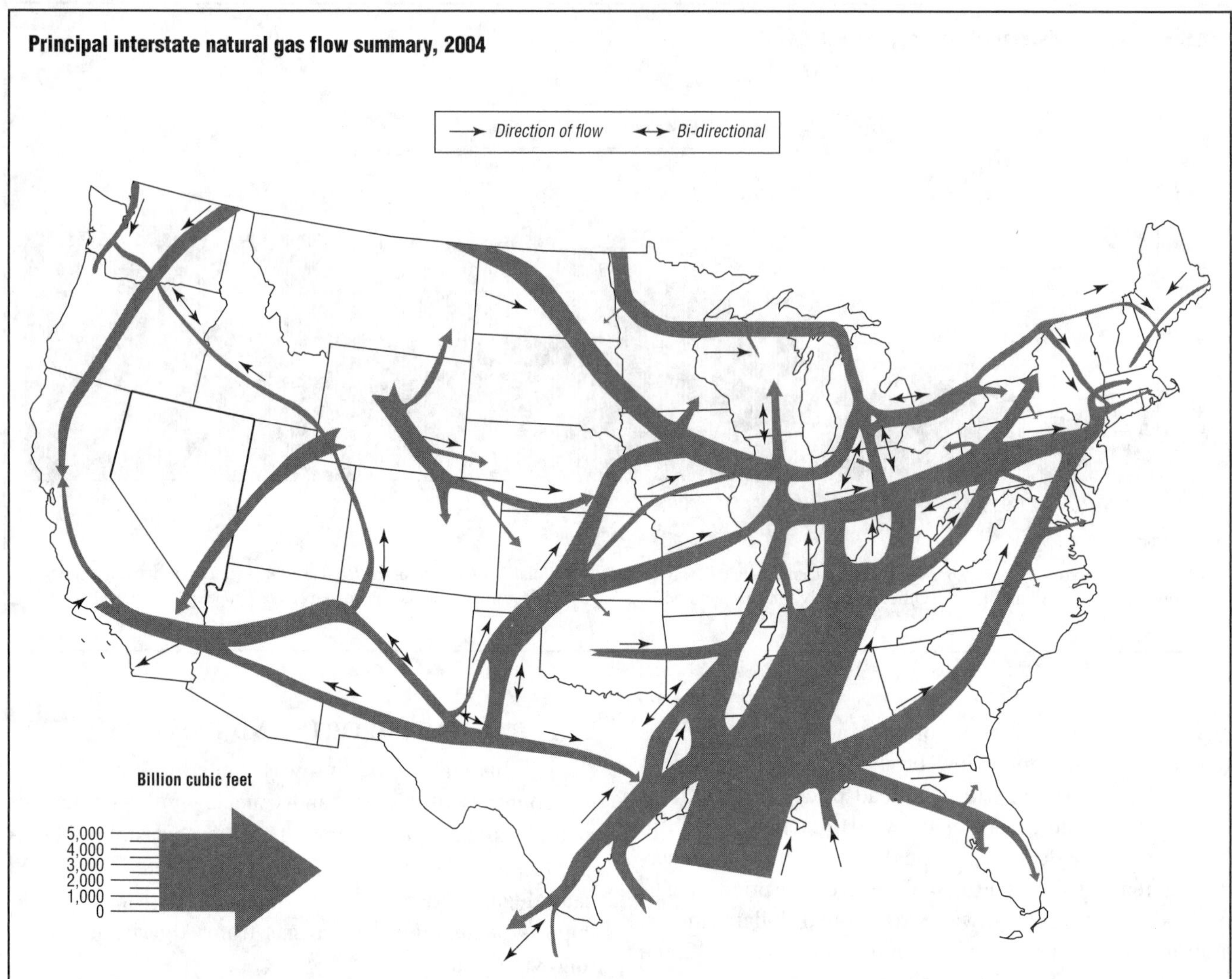

SOURCE: "Figure 7. Principal Interstate Natural Gas Flow Summary, 2004," in *Natural Gas Annual 2004*, U.S. Department of Energy, Energy Information Administration, Office of Oil and Gas, December 2005, http://www.eia.doe.gov/pub/oil_gas/natural_gas/data_publications/natural_gas_annual/current/pdf/nga04.pdf (accessed April 11, 2006)

World Consumption

The *Annual Energy Review 2004* shows that the world consumption of natural gas has increased steadily over the past several decades, from 52.9 trillion cubic feet in 1980 to 95.5 trillion cubic feet in 2003. The United States consumed the largest amount of natural gas in 2003, followed by Russia. (See Figure 3.12.) Combined, they accounted for 39% of world consumption.

FUTURE TRENDS IN THE GAS INDUSTRY

The *Annual Energy Outlook 2006* (Energy Information Administration, 2006) predicts that, by 2030, total U.S. natural gas production will decrease. However, consumption will increase, so imports will increase as well. Natural gas prices for residential customers are projected to decline from 2006 through 2016, then increase.

Domestic Production

Total domestic natural gas production, according to *Annual Energy Outlook 2006*, is projected to decrease from 2004 through 2030. Only two sources would provide increases during this period: (1) unconventional resources in the lower forty-eight states and (2) Alaska, if a natural gas pipeline becomes operational in 2015. Other sources are expected to decline. Figure 3.13 shows projected figures for these different types of natural gas production. Unconventional sources are those from which it is more difficult and less economically sound to extract natural gas because the technology to reach it has not been developed fully or is too expensive. Note that the abbreviation NA in Figure 3.13 stands for nonassociated natural gas, which is gas that is not found with crude oil. The abbreviation AD, on the other hand, stands for associated-dissolved natural gas, which is found in a dissolved state with oil, much like oxygen dissolved in aquarium water.

TABLE 3.2

Natural gas consumption by sector, selected years 1949–2004

[Billion cubic feet]

	End-use sectors													Electric power sector[a]			
		Commercial			Industrial					Transportation							
					Lease and plant fuel	Other industrial				Pipelines[f] and distribution[g]	Vehicle fuel	Total					
Year	Residential	CHP[b]	Other[c]	Total		CHP[d]	Non-CHP[e]	Total	Total				Total	Electricity only	CHP	Total	Total
1949	993	[h]	348	348	835	[i]	2,245	2,245	3,081	NA	NA	NA	4,421	550	NA	550	4,971
1950	1,198	[h]	388	388	928	[i]	2,498	2,498	3,426	126	NA	126	5,138	629	NA	629	5,767
1955	2,124	[h]	629	629	1,131	[i]	3,411	3,411	4,542	245	NA	245	7,540	1,153	NA	1,153	8,694
1960	3,103	[h]	1,020	1,020	1,237	[i]	4,535	4,535	5,771	347	NA	347	10,242	1,725	NA	1,725	11,967
1965	3,903	[h]	1,444	1,444	1,156	[i]	5,955	5,955	7,112	501	NA	501	12,959	2,321	NA	2,321	15,280
1970	4,837	[h]	2,399	2,399	1,399	[i]	7,851	7,851	9,249	722	NA	722	17,208	3,932	NA	3,932	21,139
1972	5,126	[h]	2,608	2,608	1,456	[i]	8,169	8,169	9,624	766	NA	766	18,125	3,977	NA	3,977	22,101
1973	4,879	[h]	2,597	2,597	1,496	[i]	8,689	8,689	10,185	728	NA	728	18,389	3,660	NA	3,660	22,049
1974	4,786	[h]	2,556	2,556	1,477	[i]	8,292	8,292	9,769	669	NA	669	17,780	3,443	NA	3,443	21,223
1976	5,051	[h]	2,668	2,668	1,634	[i]	6,964	6,964	8,598	548	NA	548	16,866	3,081	NA	3,081	19,946
1978	4,903	[h]	2,601	2,601	1,648	[i]	6,757	6,757	8,405	530	NA	530	16,439	3,188	NA	3,188	19,627
1980	4,752	[h]	2,611	2,611	1,026	[i]	7,172	7,172	8,198	635	NA	635	16,196	3,682	NA	3,682	19,877
1982	4,633	[h]	2,606	2,606	1,109	[i]	5,831	5,831	6,941	596	NA	596	14,776	3,226	NA	3,226	18,001
1984	4,555	[h]	2,524	2,524	1,077	[i]	6,154	6,154	7,231	529	NA	529	14,839	3,111	NA	3,111	17,951
1986	4,314	[h]	2,318	2,318	923	[i]	5,579	5,579	6,502	485	NA	485	13,619	2,602	NA	2,602	16,221
1988	4,630	[h]	2,670	2,670	1,096	[i]	6,383	6,383	7,479	614	NA	614	15,394	2,636	NA	2,636	18,030
1990	4,391	46	2,576	2,623	1,236	1,055	5,963[j]	7,018[j]	8,255	660	s	660	15,929	2,794[j]	451[j]	3,245[j]	19,174[j]
1992	4,690	62	2,740	2,803	1,171	1,107	6,420[j]	7,527[j]	8,698	588	2	590	16,780	2,829[j]	619[j]	3,448[j]	20,228[j]
1994	4,848	72	2,823	2,895	1,124	1,176	6,613	7,790	8,913	685	3	689	17,345	3,065	838	3,903	21,247
1996	5,241	82	3,076	3,158	1,250	1,289	7,146	8,435	9,685	711	6	718	18,802	2,824	983	3,807	22,609
1997	4,984	87	3,128	3,215	1,203	1,282	7,229	8,511	9,714	751	8	760	18,673	3,039	1,026	4,065	22,737
1998	4,520	87	2,912	2,999	1,173	1,355	6,965	8,320	9,493	635	9	645	17,658	3,544	1,044	4,588	22,246
1999	4,726	84	2,961	3,045	1,079	1,401	6,678	8,079	9,158	645	12	657	17,586	3,729	1,090	4,820	22,405
2000	4,996	85	3,098	3,182	1,151	1,386	6,757	8,142	9,293	642	13	655	18,127	4,093	1,114	5,206	23,333
2001	4,771	79	2,944	3,023	1,119	1,310	6,035	7,344	8,463	625	15	640	16,896	4,164	1,179[R]	5,342	22,239
2002	4,889[R]	74	3,070[R]	3,144[R]	1,113[R]	1,240	6,267[R]	7,507[R]	8,620[R]	667	15	682	17,335[R]	4,258	1,413	5,672	23,007[R]
2003	5,078[R]	58[R]	3,158[R]	3,217[R]	1,123	1,144[R]	5,995[R]	7,139[R]	8,262[R]	665[R]	18[R]	683[R]	17,240[R]	3,780[R]	1,355[R]	5,135[R]	22,375[R]
2004[P]	4,881	75	2,926	3,000	1,107	1,162	6,136	7,298	8,405	663	20	684	16,970	4,096	1,256	5,352	22,321

[a]Electricity-only and combined-heat-and-power (CHP) plants within the NAICS (North American Industry Classification System) 22 category whose primary business is to sell electricity, or electricity and heat, to the public. Through 1988, data are for electric utilities only; beginning in 1989, data are for electric utilities and independent power producers. Electric utility CHP plants are included in "electricity only."
[b]Commercial combined-heat-and-power and a small number of commercial electricity-only plants.
[c]All commercial sector fuel use other than that in "commercial CHP."
[d]Industrial combined-heat-and-power (CHP) and a small number of industrial electricity-only plants.
[e]All industrial sector fuel use other than that in "lease and plant fuel" and "industrial CHP."
[f]Natural gas consumed in the operation of pipelines, primarily in compressors.
[g]Natural gas used as fuel in the delivery of natural gas to consumers.
[h]Included in "commercial other."
[i]Included in "industrial non-CHP."
[j]For 1989–1992, a small amount of consumption at independent power producers may be counted in both "other industrial" and "electric power sector."
Notes: Totals may not equal sum of components due to independent rounding. For data not shown for 1951–1969, see http://www.eia.doe.gov/emeu/aer/natgas.html. For related information, see http://www.eia.doe.gov/oil_gas/natural_gas/info_glance/natural_gas.html.
R=Revised. P=Preliminary. NA=Not available. s=Less than 0.5 billion cubic feet.
Data are for natural gas, plus a small amount of supplemental gaseous fuels that cannot be identified separately.

SOURCE: Adapted from "Table 6.5. Natural Gas Consumption by Sector, Selected Years, 1949–2004 (Billion Cubic Feet)," in *Annual Energy Review 2004*, U.S. Department of Energy, Energy Information Administration, Office of Energy Markets and End Use, August, 2005, http://www.eia.doe.gov/emeu/aer/pdf/aer.pdf (accessed April 5, 2006)

FIGURE 3.10

Natural gas wellhead prices, 1949–2004

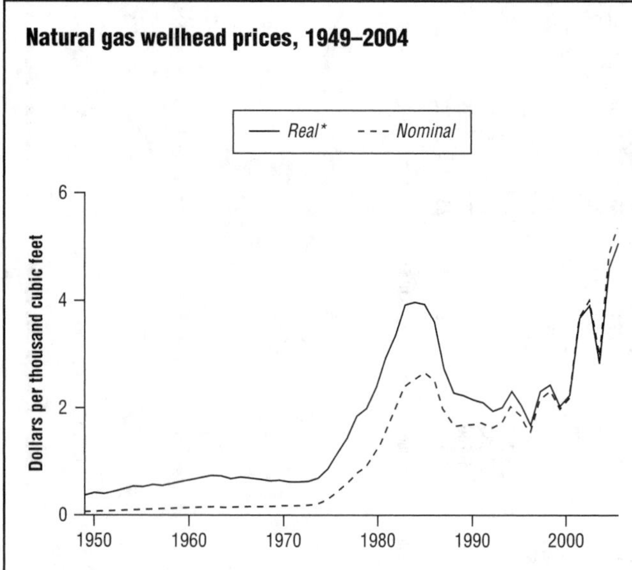

*In chained (2000) dollars, calculated by using gross domestic product implicit price deflators.

SOURCE: Adapted from "Figure 6.7. Natural Gas Wellhead, City Gate, and Imports Prices: Wellhead, 1949–2004," in *Annual Energy Review 2004*, U.S. Department of Energy, Energy Information Administration, Office of Energy Markets and End Use, August, 2005, http://www.eia.doe.gov/emeu/aer/pdf/aer.pdf (accessed April 5, 2006)

Domestic Consumption

The *Annual Energy Outlook 2006* projects that consumption of natural gas will increase from 22.4 trillion cubic feet in 2004 to 26.9 trillion cubic feet in 2030. Demand for natural gas by industrial consumers is expected to grow slowly because of high prices from 8.5 trillion cubic feet in 2004 to 10.0 trillion cubic feet in 2030. High prices will also limit growth in the residential and commercial sectors (see Figure 3.14), where natural gas use will grow from a combined 7.9 trillion cubic feet in 2004 to 9.6 trillion cubic feet in 2030. Because of its efficiency and low emissions, natural gas will be used in more electricity-generating plants, which will replace some of the nation's aging nuclear electricity-generating plants. The Energy Information Administration estimates that, to meet growing demand, natural gas pipeline capacity will have to be expanded, particularly along the corridors that move supplies to the Pacific Coast and the East. A new pipeline from the North Slope in Alaska would help meet the predicted demand.

Imports and Exports

Net imports of natural gas are projected to increase to meet demand from 2004 to 2030. Most of these imports will come from overseas in the form of liquefied natural gas. (See Figure 3.15.) A decline in Canada's non-Arctic conventional natural gas production will be only partially offset by its Arctic and unconventional production. Natural gas exports to Mexico are expected to peak in 2006 (shown as a negative import value in Figure 3.15) and then decline for a period as Mexico develops its own natural gas infrastructure. However, exports to Mexico are expected to increase again fairly steadily beginning in about 2015.

TABLE 3.3

Natural gas prices by sector, selected years 1967–2004

[Dollars per thousand cubic feet]

Year	Residential			Commercial[a]			Industrial[b]			Vehicle fuel[c]		Electric power[d]		
	Prices		Percentage of sector	Prices		Percentage of sector	Prices		Percentage of sector	Prices		Prices		Percentage of sector
	Nominal	Real[e]		Nominal	Real[e]		Nominal	Real[e]		Nominal	Real[e]	Nominal	Real[e]	
1967	1.04	4.35	NA	0.74	3.10	NA	0.34	1.42	NA	NA	NA	0.28	1.17	NA
1968	1.04	4.17	NA	0.73	2.93	NA	0.34	1.36	NA	NA	NA	0.22	0.88	NA
1970	1.09	3.96	NA	0.77	2.80	NA	0.37	1.34	NA	NA	NA	0.29	1.05	NA
1972	1.21	4.01	NA	0.88	2.92	NA	0.45	1.49	NA	NA	NA	0.34	1.13	NA
1974	1.43	4.12	NA	1.07	3.08	NA	0.67	1.93	NA	NA	NA	0.51	1.47	92.7
1976	1.98	4.93	NA	1.64	4.08	NA	1.24	3.08	NA	NA	NA	1.06	2.64	96.2
1978	2.56	5.59	NA	2.23	4.87	NA	1.70	3.72	NA	NA	NA	1.48	3.23	98.0
1980	3.68	6.81	NA	3.39	6.27	NA	2.56	4.74	NA	NA	NA	2.27	4.20	96.9
1982	5.17	8.24	NA	4.82	7.68	NA	3.87	6.17	85.1	NA	NA	3.48	5.55	92.6
1984	6.12	9.05	NA	5.55	8.20	NA	4.22	6.24	74.7	NA	NA	3.70	5.47	94.4
1986	5.83	8.18	NA	5.08	7.13	NA	3.23	4.53	59.8	NA	NA	2.43	3.41	91.7
1988	5.47	7.23	NA	4.63	6.12	90.7	2.95	3.90	42.6	NA	NA	2.33	3.08	89.6
1990	5.80	7.11	99.3	4.83	5.92	86.6	2.93	3.59	35.2	3.39	4.15	2.38	2.92	76.8[R]
1992	5.89	6.82	99.1	4.88	5.65	83.2	2.84	3.29	30.3	4.05	4.69	2.36	2.73	76.5[R]
1994	6.41	7.10	99.1	5.44	6.03	79.3	3.05	3.38	25.5	4.11	4.55	2.28	2.53	73.4[R]
1996	6.34	6.76	99.1	5.40	5.75	77.6	3.42	3.64	19.4	4.34	4.62	2.69	2.87	68.4[R]
1998	6.82	7.07	97.7	5.48	5.68	67.0	3.14	3.25	16.1	4.59	4.76	2.40	2.49	63.7[R]
1999	6.69	6.84	95.2	5.33	5.45	66.1	3.12	3.19	18.8	4.34	4.43	2.62	2.68	58.3[R]
2000	7.76	7.76	92.6	6.59	6.59	63.9	4.45	4.45	19.8	5.54	5.54	4.38	4.38	50.5[R]
2001	9.63	9.40[R]	92.4	8.43	8.23	66.0	5.24	5.12	20.8	6.60	6.45	4.61	4.50	40.2[R]
2002	7.89[R]	7.58[R]	97.9[R]	6.63[R]	6.37[R]	77.4[R]	4.02	3.86[R]	22.7[R]	5.10[R]	4.90[R]	3.68[4]	3.54[4]	83.9[4, R]
2003	9.52[R]	8.98[R]	97.6[R]	8.29[R]	7.82	77.3[R]	5.81[R]	5.48[R]	22.9[R]	6.19[R]	5.84[R]	5.54[R]	5.23[R]	90.7[R]
2004	10.74[P]	9.92[P]	96.0[E]	9.26[P]	8.56[P]	76.9[P]	6.40[P]	5.91[P]	23.3[P]	NA	NA	6.09	5.63[P]	95.0[P]

[a]Commercial sector, including commercial combined-heat-and-power (CHP) and commercial electricity-only plants.
[b]Industrial sector, including industrial combined-heat-and-power (CHP) and industrial electricity-only plants.
[c]Much of the natural gas delivered for vehicle fuel represents deliveries to fueling stations that are used primarily or exclusively by respondents' fleet vehicles. Thus, the prices are often those associated with the operation of fleet vehicles.
[d]Electricity-only and combined-heat-and-power (CHP) plants within the NAICS (North American Industry Classification System) 22 category whose primary business is to sell electricity, or electricity and heat, to the public. Through 2001, data are for electric utilities only; beginning in 2002, data are for electric utilities and independent power producers.
[e]In chained (2000) dollars, calculated by using gross domestic product implicit price deflators.
R=Revised. P=Preliminary. E=Estimate. NA=Not available.
Notes: Prices are for natural gas, plus a small amount of supplemental gaseous fuels that cannot be identified separately. The average for each end-use sector is calculated by dividing the total value of the natural gas consumed by each sector by the total quantity consumed. Prices are intended to include all taxes. See http://www.eia .doe.gov/oil_gas/natural_gas/info_glance/natural_gas.html for related information.

SOURCE: Adapted from "Table 6.8. Natural Gas Prices by Sector, 1967–2004 (Dollars per Thousand Cubic Feet)," in *Annual Energy Review 2004*, U.S. Department of Energy, Energy Information Administration, Office of Energy Markets and End Use, August, 2005, http://www.eia.doe.gov/emeu/aer/pdf/aer.pdf (accessed April 5, 2006)

FIGURE 3.11

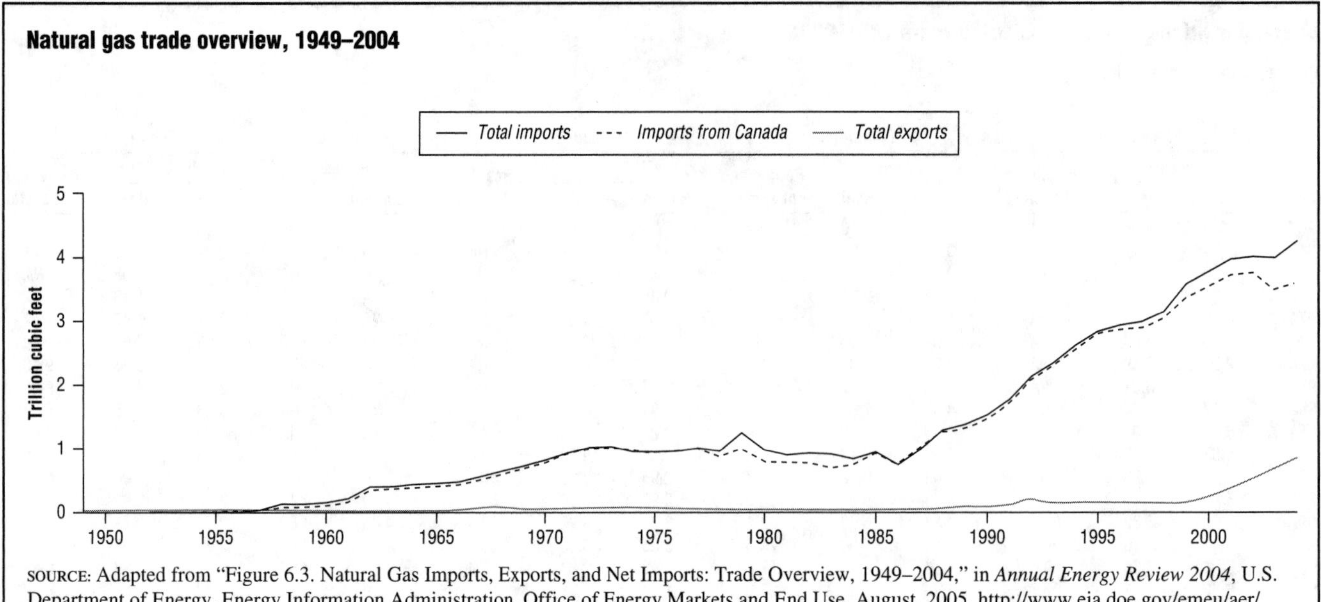

SOURCE: Adapted from "Figure 6.3. Natural Gas Imports, Exports, and Net Imports: Trade Overview, 1949–2004," in *Annual Energy Review 2004*, U.S. Department of Energy, Energy Information Administration, Office of Energy Markets and End Use, August, 2005, http://www.eia.doe.gov/emeu/aer/pdf/aer.pdf (accessed April 5, 2006)

TABLE 3.4

World dry natural gas production, 1994–2003

[Trillion cubic feet]

Region and country	1994	1995	1996	1997	1998	1999	2000	2001	2002	2003P
North, Central, and South America	**27.50**	**27.74**	**28.39**	**28.75**	**29.39**	**29.53**	**30.40R**	**31.17**	**30.62R**	**31.18**
Argentina	0.79	0.88	0.94	0.97	1.04	1.22	1.32	1.31	1.28	1.45
Canada	5.27	5.60	5.71	5.76	5.98	6.27R	6.47	6.60	6.63	6.45
Mexico	0.97	0.96	1.06	1.17	1.27	1.29	1.31	1.30	1.33	1.49
United States	18.82	18.60	18.85	18.90	19.02	18.83	19.18	19.62	18.93R	19.04
Venezuela	0.88	0.89	0.96	0.99	1.11	0.95	0.96	1.12	1.05	1.05
Other	0.78	0.81	0.86	0.96	0.96	0.98	1.15	1.22	1.39	1.71
Western Europe	**8.44**	**8.80**	**10.09**	**9.71**	**9.64**	**9.90R**	**10.19**	**10.27**	**10.55**	**10.62**
Germany	0.70	0.74	0.80	0.79	0.77	0.82	0.78	0.79	0.79	0.78
Italy	0.73	0.72	0.71	0.68	0.67	0.62	0.59	0.54	0.51	0.48
Netherlands	2.95	2.98	3.37	2.99	2.84	2.65R	2.56	2.75	2.66	2.58
Norway	1.04	1.08	1.45	1.62	1.63	1.76	1.87	1.95	2.41	2.59
United Kingdom	2.47	2.67	3.18	3.03	3.14	3.49	3.83	3.69	3.61	3.63
Other	0.55	0.61	0.59	0.60	0.58	0.57	0.57	0.57	0.57	0.55
Eastern Europe and former U.S.S.R.	**26.47**	**25.93**	**26.28**	**24.85**	**25.17**	**25.41**	**26.22**	**26.48**	**27.05**	**28.00**
Romania	0.69	0.68	0.63	0.61	0.52	0.50	0.48	0.51	0.47	0.43
Russia	21.45	21.01	21.23	20.17	20.87	20.83	20.63	20.51	21.03	21.77
Turkmenistan	1.26	1.14	1.31	0.90	0.47	0.79	1.64	1.70	1.89	2.08
Ukraine	0.64	0.62	0.64	0.64	0.64	0.63	0.64	0.64	0.65	0.69
Uzbekistan	1.67	1.70	1.70	1.74	1.94	1.96	1.99	2.23	2.04	2.03
Other	0.76	0.79	0.76	0.79	0.74	0.70	0.84	0.89	0.97	1.00
Middle East and Africa	**7.41**	**7.99**	**8.76**	**9.74**	**10.30**	**10.95**	**12.01**	**12.61**	**13.41**	**14.19**
Algeria	1.81	2.05	2.19	2.43	2.60	2.88	2.94	2.79	2.80	2.91
Egypt	0.42	0.44	0.47	0.48	0.49	0.52	0.65	0.87	0.94	0.95
Iran	1.12	1.25	1.42	1.66	1.77	2.04	2.13	2.33	2.65	2.79
Qatar	0.48	0.48	0.48	0.61	0.69	0.78	1.03	0.95	1.04	1.09
Saudi Arabia	1.33	1.34	1.46	1.60	1.65	1.63	1.76	1.90	2.00	2.12
United Arab Emirates	0.91	1.11	1.19	1.28	1.31	1.34	1.36	1.39	1.53	1.58
Other	1.34	1.33	1.53	1.67	1.79	1.76	2.15	2.39	2.45	2.74
Asia and Oceania	**7.11**	**7.50**	**8.13**	**8.47**	**8.55**	**9.14R**	**9.48**	**9.92**	**10.53R**	**11.19**
Australia	0.93	1.03	1.06	1.06	1.10	1.12R	1.16	1.19	1.23R	1.26
China	0.59	0.60	0.67	0.75	0.78	0.85	0.96	1.07	1.15	1.21
India	0.59	0.63	0.70	0.72	0.76	0.75	0.79	0.85	0.88	0.96
Indonesia	2.21	2.24	2.35	2.37	2.27	2.51	2.36	2.34	2.48	2.62
Malaysia	0.92	1.02	1.23	1.36	1.37	1.42	1.50	1.66	1.71	1.89
Pakistan	0.63	0.65	0.70	0.70	0.71	0.78	0.86	0.77	0.81	0.84
Other	1.23	1.33	1.42	1.52	1.56	1.70R	1.86	2.04	2.25	2.42
World	**76.93**	**77.96**	**81.65**	**81.52**	**83.03**	**84.93R**	**88.29R**	**90.45**	**92.15R**	**95.18**

R=Revised. P=Preliminary.

Notes: Totals may not equal sum of components due to independent rounding. For related information, see http://www.eia.doe.gov/international.

SOURCE: "Table 11.11. World Dry Natural Gas Production, 1994–2003 (Trillion Cubic Feet)," in *Annual Energy Review 2004*, U.S. Department of Energy, Energy Information Administration, Office of Energy Markets and End Use, August 2005, http://www.eia.doe.gov/emeu/aer/pdf/aer.pdf (accessed April 5, 2006)

FIGURE 3.12

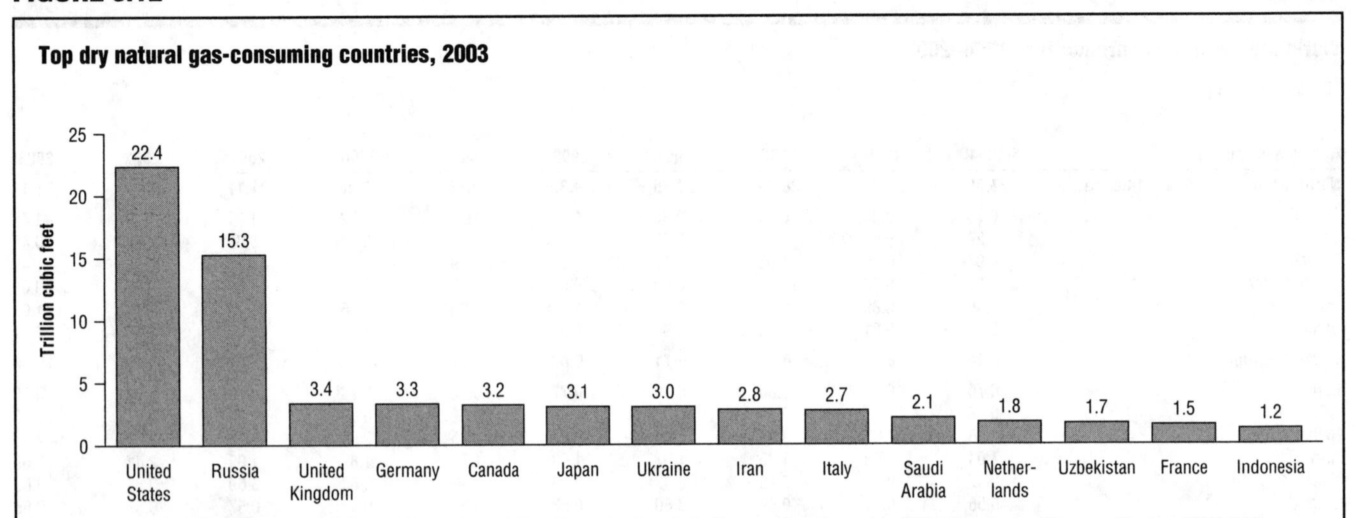

SOURCE: Adapted from "Figure 11.12. World Dry Natural Gas Consumption: Top Consuming Countries, 2003," in *Annual Energy Review 2004*, U.S. Department of Energy, Energy Information Administration, Office of Energy Markets and End Use, August, 2005, http://www.eia.doe.gov/emeu/aer/pdf/aer.pdf (accessed April 5, 2006)

FIGURE 3.13

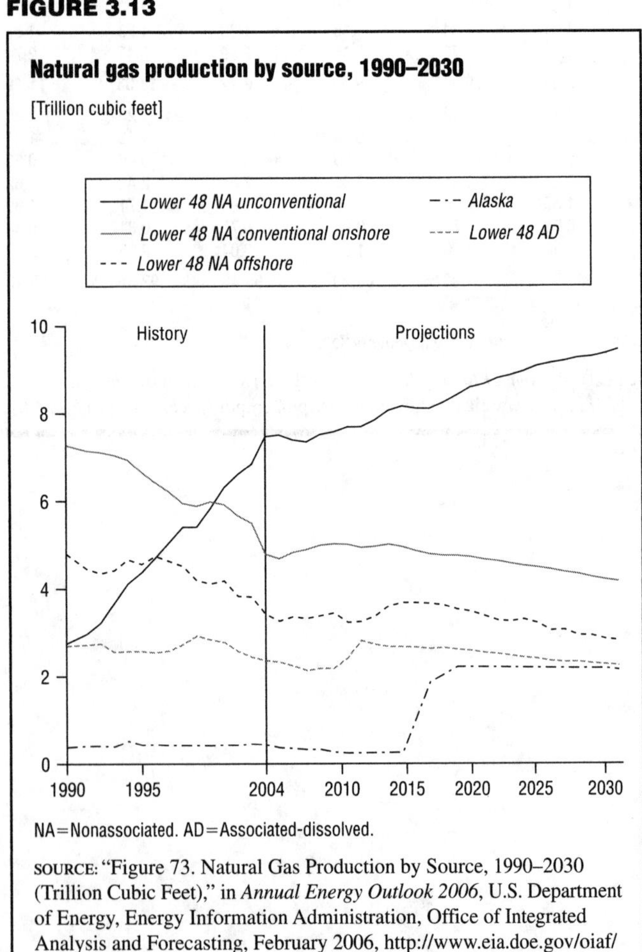

NA=Nonassociated. AD=Associated-dissolved.

SOURCE: "Figure 73. Natural Gas Production by Source, 1990–2030 (Trillion Cubic Feet)," in *Annual Energy Outlook 2006*, U.S. Department of Energy, Energy Information Administration, Office of Integrated Analysis and Forecasting, February 2006, http://www.eia.doe.gov/oiaf/aeo/pdf/0383(2006).pdf (accessed April 5, 2006)

FIGURE 3.14

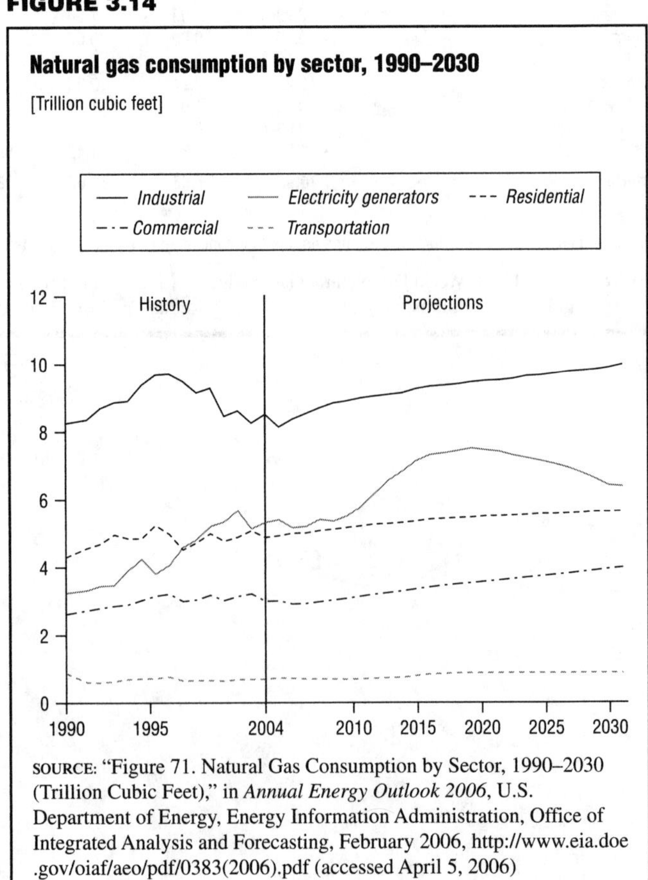

SOURCE: "Figure 71. Natural Gas Consumption by Sector, 1990–2030 (Trillion Cubic Feet)," in *Annual Energy Outlook 2006*, U.S. Department of Energy, Energy Information Administration, Office of Integrated Analysis and Forecasting, February 2006, http://www.eia.doe.gov/oiaf/aeo/pdf/0383(2006).pdf (accessed April 5, 2006)

FIGURE 3.15

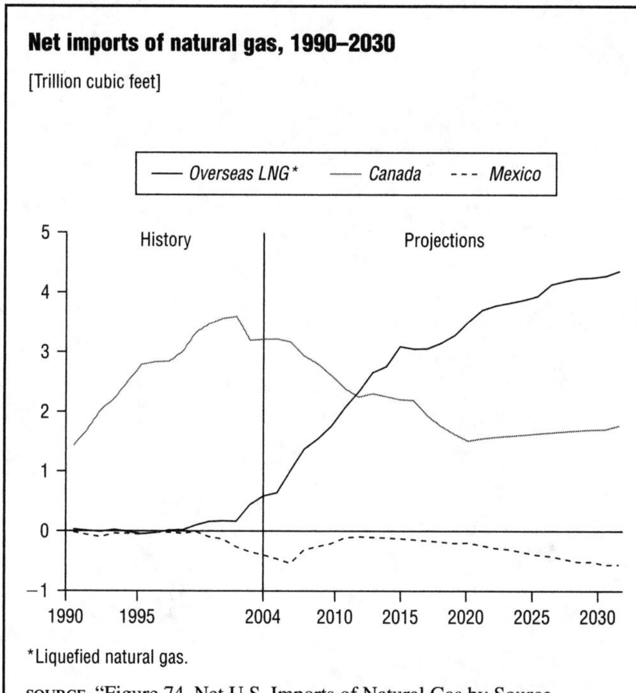

Net imports of natural gas, 1990–2030

[Trillion cubic feet]

*Liquefied natural gas.

SOURCE: "Figure 74. Net U.S. Imports of Natural Gas by Source, 1990–2030 (Trillion Cubic Feet)," in *Annual Energy Outlook 2006*, U.S. Department of Energy, Energy Information Administration, Office of Integrated Analysis and Forecasting, February 2006, http://www.eia.doe.gov/oiaf/aeo/pdf/0383(2006).pdf (accessed April 5, 2006)

CHAPTER 4
COAL

A HISTORICAL PERSPECTIVE

Although it had been a source of energy for centuries, coal was first used on a large scale during the Industrial Revolution in England. From the mid-eighteenth to the mid-nineteenth centuries, the sky was filled with billowing columns of black smoke. Soot covered the towns and cities, and workers breathed the thick coal dust swirling around them. Environmental issues, if they were considered at all, were far less important than the jobs the factories provided. Factory owners had little reason or incentive to control the smoke pouring out of their plants—the environmental and public health effects of pollution were not as well understood as they are today, so government imposed little, if any, regulation on manufacturing.

In the United States early colonists used wood to heat their homes because it was so plentiful. Coal was not as readily available and far more expensive. Before the Civil War (1861–65), some industries used coal as a source of energy, but its use expanded greatly with the building of railroads across the country. In fact, coal became such a fundamental part of American industrialization that some have called this era the Coal Age. As in England, Americans considered the development of industry a source of national pride. Photographs and postcards of the time proudly featured trains and steel mills belching dark smoke into gray skies.

By the early twentieth century coal had become the major fuel in the United States, accounting for nearly 90% of the nation's energy requirements. However, as oil—which was much cleaner—became a favored fuel for heating homes and offices, and gasoline powered the growing number of cars, coal's dominance declined. According to the Energy Information Administration, by 1945 coal accounted for approximately 40% of the energy consumed (*Annual Energy Review 2004*, 2005). It fell further out of favor in the 1950s and 1960s until, in the early 1970s, coal provided as little as 17% of the nation's energy. By then it had been overtaken by concerns about pollution, along with the emergence of nuclear power as a promising energy source.

In 1973, however, the oil embargo by the Organization of Petroleum Exporting Countries made many Americans reconsider. The embargo clearly demonstrated the nation's heavy reliance on foreign sources of energy and the potentially crippling effect that dependence could have on the U.S. economy. Consequently, the nation revived its interest in domestic coal as a plentiful and economical energy source.

In 1977 President Jimmy Carter called for an increase in domestic coal production by two-thirds—to about 1 billion tons annually by 1985. He proposed a ten-year, $10 billion program to spark that production. He also asked utility companies and other large industries to convert their operations to coal. In 2004 more coal was produced in the United States than any other form of energy: 22.7 quadrillion Btu, or 32% of all energy produced. (See Table 1.1 and Figure 1.4 in Chapter 1.) Coal was the third-largest source of energy consumed in the United States in 2004, after petroleum and natural gas. (See Figure 1.5 in Chapter 1.)

WHAT IS COAL?

Coal is a black, combustible, mineral solid. It developed over millions of years as plant matter decomposed in an airless space under increased temperature and pressure. Coal beds, sometimes called seams, are found in the earth between beds of sandstone, shale, and limestone and range in thickness from less than an inch to more than one hundred feet. Approximately five to ten feet of ancient plant material were compressed to create each foot of coal.

Coal is used as a fuel and in the production of coal gas, water gas, many coal-tar compounds, and coke (the solid substance left after coal gas and coal tar have been extracted from coal). When coal is burned, its fossil energy—sunlight converted and stored by plants over millions of years—is released. One ton of coal produces 22 million Btu on average, about the same heating value as 22,000 cubic feet of natural gas, 160 gallons of home heating oil, or a stack of seasoned firewood measuring 4′×4′×8′.

CLASSIFICATIONS OF COAL

There are four basic types of coal. Classifications, or "coal ranks," are based on how much carbon, volatile matter, and heating value are contained in the coal:

- *Anthracite*, or hard coal, is the highest ranked coal. It is hard and jet-black, with moisture content of less than 15%. It contains approximately 22 million to 28 million Btu per ton, with an ignition temperature of approximately 925° to 970° Fahrenheit. Anthracite, which is used for generating electricity and for space heating, is mined mainly in northeastern Pennsylvania. (See Figure 4.1.)
- *Bituminous*, or soft coal, is the most common. It is dense and black, with moisture content of less than 20% and an ignition range of 700° to 900° Fahrenheit. With a heating value of 19 million to 30 million Btu per ton, bituminous coal is used to generate electricity, for space heating, and to produce coke. It is mined chiefly in the Appalachian and Midwest regions of the United States. (See Figure 4.1.)
- *Subbituminous coal*, or black lignite, is dull black in color and generally contains 20% to 30% moisture. Used for generating electricity and for space heating, it contains 16 million to 24 million Btu per ton. Black lignite is mined primarily in the western United States. (See Figure 4.1.)
- *Lignite*, the lowest ranked coal, is brownish-black in color and has a high moisture content. It tends to disintegrate when exposed to weather. Lignite contains about 9 million to 17 million Btu per ton and is used mainly to generate electricity. Most lignite is mined in North Dakota, Montana, Texas, California, and Louisiana. (See Figure 4.1.)

In 2004 domestic mines produced more than 1.1 billion short tons of all types of coal. (A short ton of coal is 2,000 pounds.) About 92% of it was bituminous (546.6 million short tons) and subbituminous coal (479.6 million short tons). Lignite accounted for much of the remainder. Very little of the total was anthracite. (See Table 4.1.)

LOCATIONS OF COAL DEPOSITS

Coal is found in about 450,000 square miles, or 13%, of the total land area of the United States. Figure 4.1 shows the coal-bearing areas of the United States. Geologists have divided U.S. coalfields into the Appalachian, Interior, and Western regions. The Appalachian region is subdivided into three areas: Northern (Maryland, Ohio, Pennsylvania, and northern West Virginia); Central (eastern Kentucky, Tennessee, Virginia, and southern West Virginia); and Southern Appalachia (Alabama). The Interior region includes mines in Arkansas, Illinois, Indiana, Iowa, Kansas, western Kentucky, Louisiana, Missouri, Oklahoma, and Texas. The Western region is divided into the Northern Great Plains (northern Colorado, Montana, North and South Dakota, and Wyoming); the Rocky Mountains; the Southwest (Arizona, southern Colorado, New Mexico, and Utah); and the Northwest (Alaska and Washington).

Before 1999 most of the nation's coal was mined east of the Mississippi River. Miners had been digging deeper and deeper into the Appalachian Mountains for years before the bulldozers cut open the rich seams of eastern Montana. In 1965, according to the U.S. Department of Energy in *Annual Energy Review 2004*, western mines produced 27.4 million short tons, only 5% of the national total. By 1999, however, western production had increased more than twentyfold, to 570.8 million short tons, or 52% of the total. (See Table 4.1.) The amount of coal mined east of the Mississippi that year was 529.6 million short tons. In 2004 mines west of the Mississippi produced 627.3 million short tons—56% of the total—while eastern mines produced 484.1 million short tons.

The growth in coal production in the western states resulted, in part, because of an increased demand for low-sulfur coal, which is concentrated there. Low-sulfur coal burns cleaner and is considered less dangerous to the environment. In addition, the coal is closer to the surface, so it can be extracted by surface mining, which is cheaper and more efficient. Improved rail service has also made it easier to deliver this low-sulfur coal to electric power plants located east of the Mississippi River.

COAL-MINING METHODS

The method used to mine coal depends on the terrain and the depth of the coal. Before the early 1970s most coal was taken from underground mines. Since then coal production has shifted to surface mines. (See Table 4.1 and Figure 4.2.)

Underground mining is required when the coal lies more than two hundred feet below ground. The depth of most underground mines is less than 1,000 feet, but a few are 2,000 feet deep. In underground mines some coal must be left untouched to form pillars that prevent the mines from caving in.

Figure 4.3 shows three types of underground mines: a shaft mine, a slope mine, and a drift mine. In a shaft mine, elevators take miners and equipment up and down

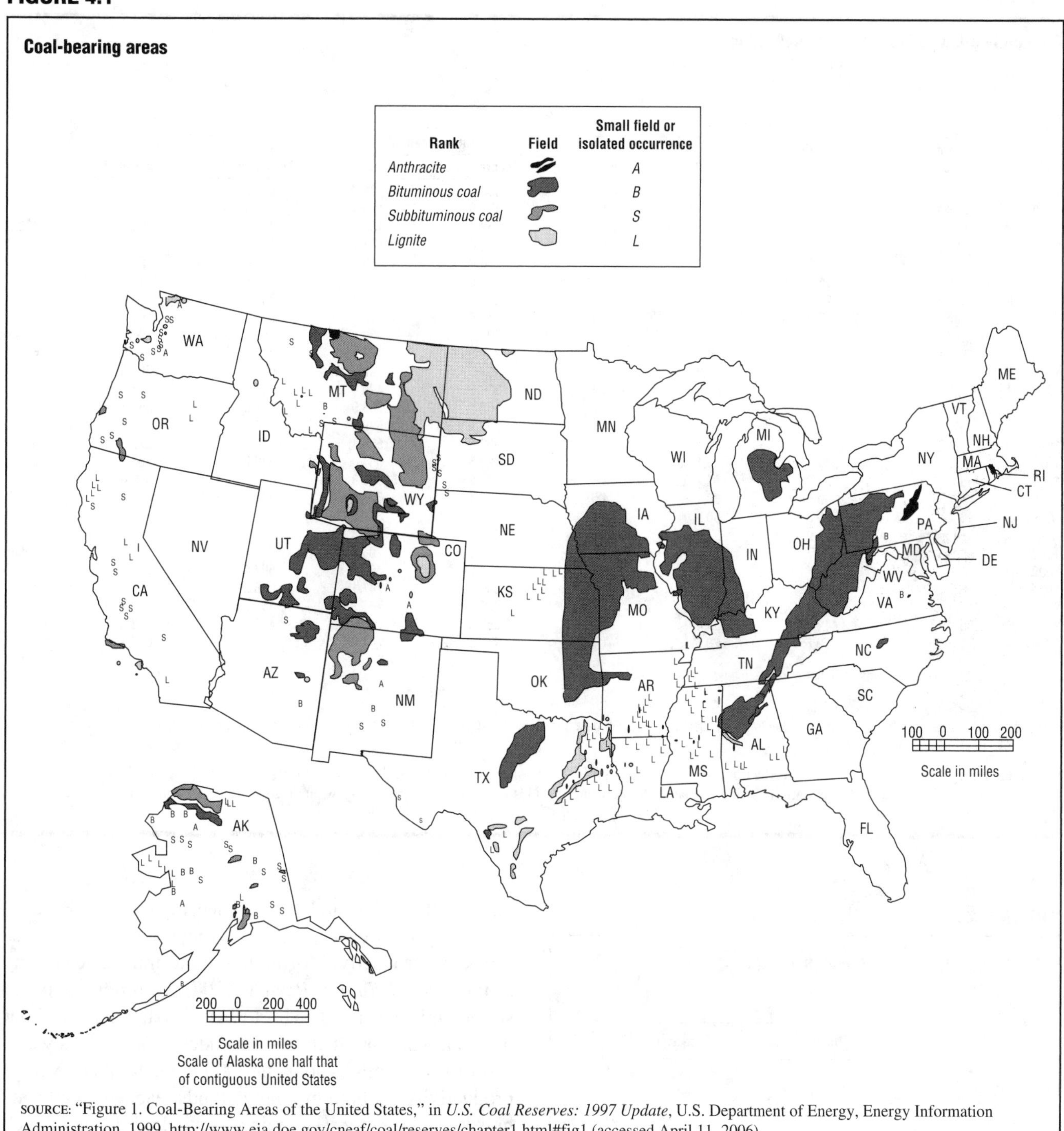

FIGURE 4.1

Coal-bearing areas

SOURCE: "Figure 1. Coal-Bearing Areas of the United States," in *U.S. Coal Reserves: 1997 Update*, U.S. Department of Energy, Energy Information Administration, 1999, http://www.eia.doe.gov/cneaf/coal/reserves/chapter1.html#fig1 (accessed April 11, 2006)

a vertical shaft to the coal deposit. Entrance to a slope mine, by contrast, is an incline from the aboveground opening. In a drift mine, the mineshaft runs horizontally from the opening in the hillside. As Figure 4.3 shows, workers and equipment are moved externally up the side of a hill or mountain to the entry.

Surface mines are usually less than two hundred feet deep and can be developed in flat or hilly terrain. On large plots of relatively flat ground workers use a technique known as area surface mining. Rock and soil that lie above the coal—called "overburden" or "spoil"—are loosened by drilling and blasting and then dug away. Another technique, contour surface mining, follows coal deposits along hillsides. (See Figure 4.3.) Open pit mining—a combination of area and contour mining—is used to mine thick, steeply inclined coal deposits.

Growth of surface mining and the closure of nonproductive mines led to increases in coal-mining productivity

TABLE 4.1

Coal production, selected years 1949–2004

[Million short tons]

	Rank				Mining method		Location		
Year	Bituminous coal[a]	Subbituminous coal	Lignite	Anthracite[a]	Underground	Surface[a]	East of the Mississippi[a]	West of the Mississippi[a]	Total[a]
1949	437.9	[b]	[b]	42.7	358.9	121.7	444.2	36.4	480.6
1950	516.3	[b]	[b]	44.1	421.0	139.4	524.4	36.0	560.4
1955	464.6	[b]	[b]	26.2	358.0	132.9	464.2	26.6	490.8
1960	415.5	[b]	[b]	18.8	292.6	141.7	413.0	21.3	434.3
1965	512.1	[b]	[b]	14.9	338.0	189.0	499.5	27.4	527.0
1970	578.5	16.4	8.0	9.7	340.5	272.1	567.8	44.9	612.7
1972	556.8	27.5	11.0	7.1	305.0	297.4	538.2	64.3	602.5
1974	545.7	42.2	15.5	6.6	278.0	332.1	518.1	91.9	610.0
1976	588.4	64.8	25.5	6.2	295.5	389.4	548.8	136.1	684.9
1978	534.0	96.8	34.4	5.0	242.8	427.4	487.2	183.0	670.2
1980	628.8	147.7	47.2	6.1	337.5	492.2	578.7	251.0	829.7
1982	620.2	160.9	52.4	4.6	339.2	499.0	564.3	273.9	838.1
1984	649.5	179.2	63.1	4.2	352.1	543.9	587.6	308.3	895.9
1986	620.1	189.6	76.4	4.3	360.4	529.9	564.4	325.9	890.3
1988	638.1	223.5	85.1	3.6	382.2	568.1	579.6	370.7	950.3
1990	693.2	244.3	88.1	3.5	424.5	604.5	630.2	398.9	1,029.1
1992	651.8	252.2	90.1	3.5	407.2	590.3	588.6	409.0	997.5
1994	640.3	300.5	88.1	4.6	399.1	634.4	566.3	467.2	1,033.5
1996	630.7	340.3	88.1	4.8	409.8	654.0	563.7	500.2	1,063.9
1998	640.6	385.9	85.8	5.3	417.7	699.8	570.6	547.0	1,117.5
1999	601.7	406.7	87.2	4.8	391.8	708.6	529.6	570.8	1,100.4
2000	574.3	409.2	85.6	4.6	373.7	700.0	507.5	566.1	1,073.6
2001	611.3[a]	434.4	80.0	1.9[a]	380.6	747.1[a]	528.8[a]	598.9[a]	1,127.7[a]
2002	572.1	438.4	82.5	1.4	357.4	736.9	492.9	601.4	1,094.3
2003	541.5[R]	442.6[R]	86.4[R]	1.3	352.8[R]	719.0[R]	469.2[R]	602.5[R]	1,071.8[R]
2004	546.6[E]	479.6[E]	83.5[E]	1.7[E]	367.5[E]	744.0[E]	484.1[E]	627.3[E]	1,111.5[P]

[a]Beginning in 2001, includes a small amount of refuse recovery.
[b]Included in "bituminous coal."
R=Revised. P=Preliminary. E=Estimate.
Note: Totals may not equal sum of components due to independent rounding. For data not shown for 1951–1969, see http://www.eia.doe.gov/emeu/aer/coal.html. For related information, see http://www.eia.doe.gov/fuelcoal.html.

SOURCE: Adapted from "Table 7.2. Coal Production, Selected Years, 1949–2004 (Million Short Tons)," in *Annual Energy Review 2004*, U.S. Department of Energy, Energy Information Administration, Office of Energy Markets and End Use, August, 2005, http://www.eia.doe.gov/emeu/aer/pdf/aer.pdf (accessed April 5, 2006)

FIGURE 4.2

Coal production by mining method, 1949–2004

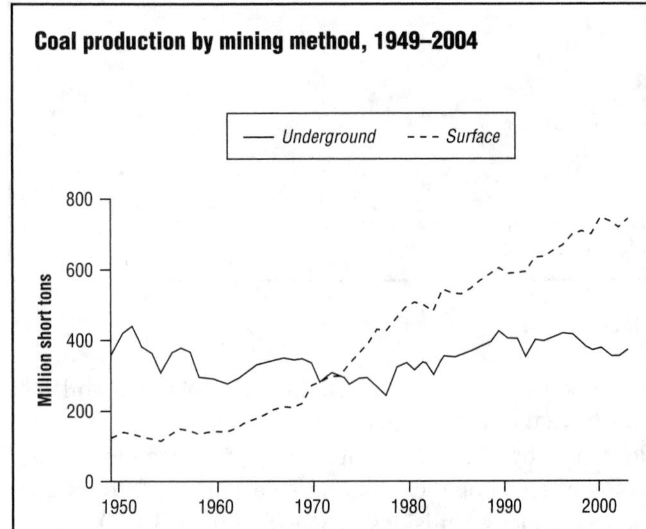

SOURCE: Adapted from "Figure 7.2. Coal Production, 1949–2004: by Mining Method," in *Annual Energy Review 2004*, U.S. Department of Energy, Energy Information Administration, Office of Energy Markets and End Use, August, 2005, http://www.eia.doe.gov/emeu/aer/pdf/aer.pdf (accessed April 5, 2006)

through the 1980s and 1990s. (See Figure 4.4.) Because surface mines are easier to work, they average up to three times the productivity of underground mines. According to the *Annual Energy Review 2004*, the productivity for surface mines was 10.5 short tons of coal per miner hour in 2004, while productivity for underground mines was 4 short tons per miner hour. In 2000 the combined average productivity for both mining methods reached an all-time high of 7 short tons per miner hour. In 2004 combined average productivity was 6.8 short tons per miner hour.

COAL-MINING SAFETY AND HEALTH RISKS

Mining safety in the United States is overseen by the Mine Safety & Health Administration (MSHA), which was formed in 1978 after Congress passed the Federal Mine Safety & Health Act of 1977 (PL 91-173; amended by PL 95-164). The law established mandatory health and safety standards for mines and required that mine operators and miners comply with them. It also provided assistance to the states to develop and enforce effective state mine health and safety programs and expanded

FIGURE 4.3

Coal mining methods

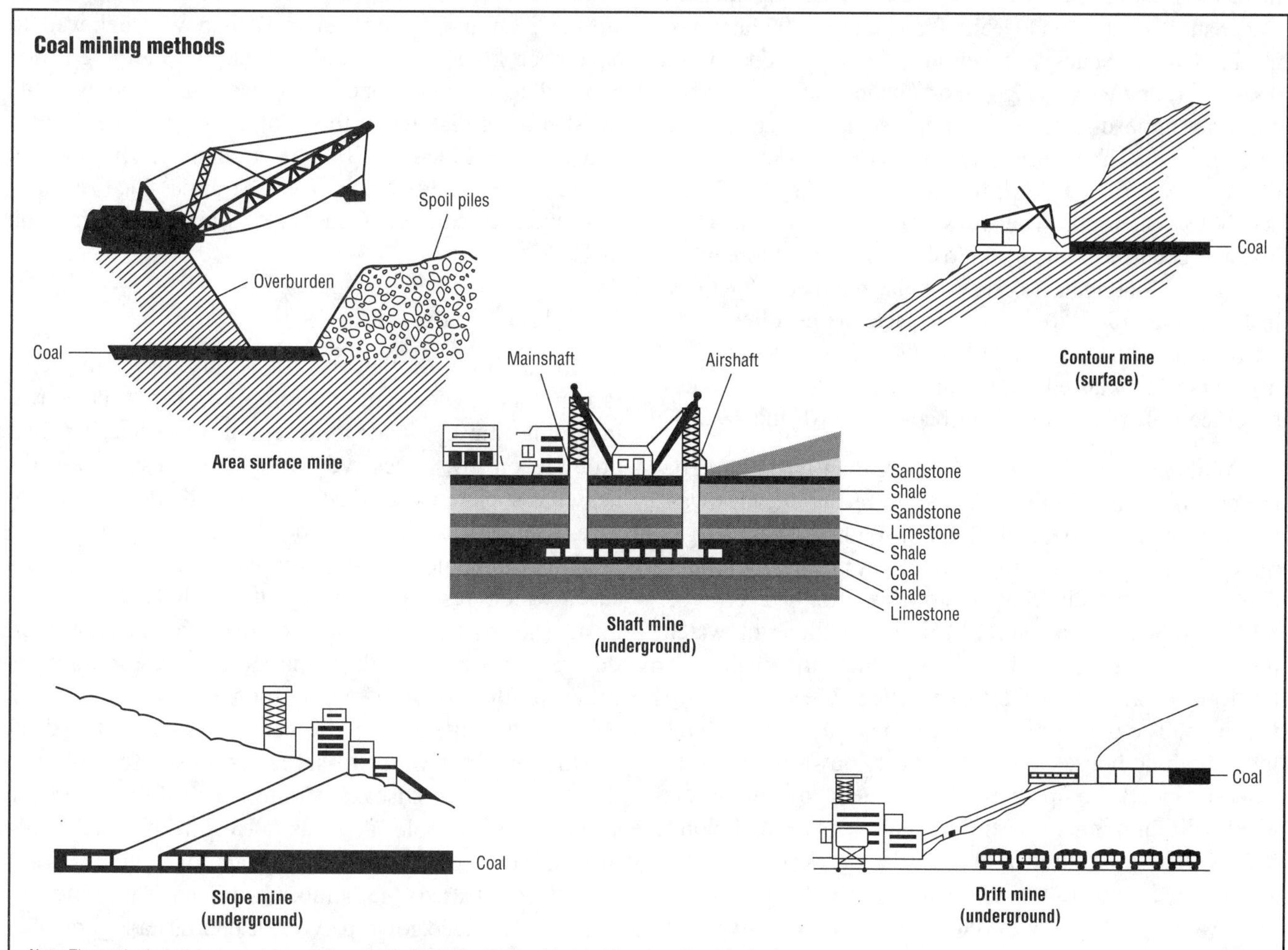

Note: The method of mining a coal deposit depends on the depth of the coal bed and the character of the land.

SOURCE: Adapted from "Figure 5. Coal Mining Methods," in *Coal Data: A Reference*, U.S. Department of Energy, Energy Information Administration, February 1995, http://tonto.eia.doe.gov/FTPROOT/coal/006493.pdf (accessed April 11, 2006)

FIGURE 4.4

Coal mining productivity, 1949–2004

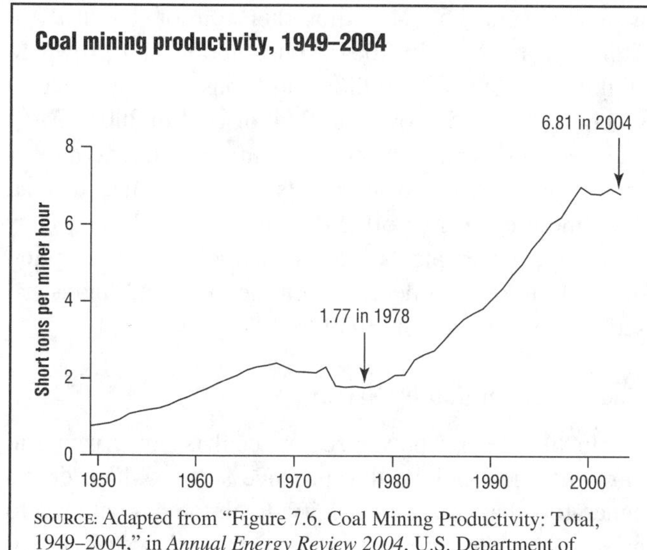

SOURCE: Adapted from "Figure 7.6. Coal Mining Productivity: Total, 1949–2004," in *Annual Energy Review 2004*, U.S. Department of Energy, Energy Information Administration, Office of Energy Markets and End Use, August, 2005, http://www.eia.doe.gov/emeu/aer/pdf/aer.pdf (accessed April 5, 2006)

research and development aimed at preventing accidents and diseases associated with mining occupations.

Throughout its history coal mining has been a physically challenging and dangerous occupation, with a recognized risk for injury or disease. As early as 1822 the term "miner's asthma" was used to describe the breathing difficulties and coughing often experienced by mine workers (http://www.umwa.org/blacklung/blacklung.-shtml). In addition, mine accidents can occur without warning, including cave-ins, fires, underground floods, equipment failures, and gas explosions. (Flammable gases, notably methane, are found naturally in coal mines.) In underground mines these accidents carry the additional risk of trapping miners in the mine without air, water, or food.

Fatalities

According to the MSHA, a total of 104,631 people were killed in coal-mining accidents between 1900 and July 2006 (http://www.msha.gov/stats/centurystats/coalstats.htm).

In records dating back to 1839, the National Institute for Occupational Safety and Health found that 13,805 fatalities in the United States had resulted from 614 coal-mine disasters (http://www.cdc.gov/niosh/mining/statistics/discoal.htm); a "mine disaster" is a mine accident that claims five or more lives. The number of disasters peaked during the period 1901 through 1925, when 297 large accidents occurred. The most deadly event in U.S. history occurred when explosions in the Monongah mines in Monongah, West Virginia, claimed 363 lives in December 1907; 1907 was, in fact, the deadliest year on record, with 3,242 fatalities. Of the twenty-six U.S. disasters that caused one hundred or more fatalities, seventeen of them took place between 1901 and 1925.

With a death toll of 125 and more than 1,100 injured, the deadliest coal-mining accident in recent decades was the massive Buffalo Creek flood in southern West Virginia in February 1972. After several days of heavy rain, a dam burst that was holding mine wastewater in a series of hillside pools. More than 132 million gallons of water then poured out and rushed through the valley below in the form of a black wave that reached fifteen to twenty feet high. The power of the water smashed structures and moved whole houses and railroad cars downstream. Terrified residents ran up nearby hills to get above the water level. Within minutes several communities located along a seventeen-mile stretch of Buffalo Creek were devastated, and the town of Saunders was completely destroyed (http://www.wvculture.org/hiStory/buffcreek/bctitle.html).

Coal mining has dramatically increased its safety record since the early and mid-twentieth century. Tighter regulations, improvements in technology, and preventive programs are credited with lowering—though not eliminating—many of the risks undertaken by miners. From 1981 to 2006 there were eleven coal-mine disasters that killed five or more people. The worst accident during this period was a mine fire that claimed twenty-seven lives at the Wilberg Mine in Orangeville, Utah, in December 1984. The year 2005 saw the fewest fatalities ever in the coal-mining industry: twenty-two.

Nonetheless, mine disasters and deaths can still occur. In January 2006 a mine disaster at the Sago Mine in Tallmansville, West Virginia, claimed the lives of twelve miners. One miner survived. The miners were trapped following an explosion of methane gas that may have been triggered by a lightning strike; investigators have theorized that a buildup of gas over the holidays might have contributed to the disaster, as the blast occurred shortly after the first shift returned to work on January 2. The explosion disabled the mine's internal communication system, which interfered with rescue operations. Rescue was also delayed because the air in the mine contained high concentrations of carbon monoxide and methane, which made it unsafe for rescue workers. Of those who perished, one was believed to have been killed by the initial blast, and the others succumbed to carbon monoxide poisoning. Sago was the worst mining disaster in the United States since thirteen miners were killed in 2001 in a mine in Brookwood, Alabama, and the worst in West Virginia since seventy-eight were killed at the Consol No. 9 mine in Farmington in 1968.

Long-Term Health Risks

In addition to the risk of injury or death, coal miners face a wide range of long-term health concerns, including muscle and joint conditions, hearing loss brought on by excessive noise, and respiratory illnesses associated with dust, fumes, and chemical exposure. One serious illness associated with mining is coal workers' pneumoconiosis, known as black lung disease, which results from repeated inhalation of coal dust. This risk has been drastically reduced, however, through the use of dust masks and respirators; by covering the walls of mine tunnels and shafts with pulverized white rock to lower the level of the dust; and by spraying water to settle the dust. According to the Centers for Disease Control and Prevention, in 1993, 1,631 people died as a result of black lung disease; by 2002 (the latest year for which figures have been released) the number had been reduced to 858 (http://www2a.cdc.gov/drds/WorldReportData/FigureTableDetails.asp?FigureTableID=24).

COAL IN THE DOMESTIC MARKET

Overall Production and Consumption

According to the *Annual Energy Review 2004*, the nation consumed 558.4 million short tons of coal in 1974. Thirty years later, in 2004, consumption had grown to slightly more than 1.1 billion short tons—nearly twice as much. (Figure 4.5 shows the flow of coal in 2004.) Most increases in consumption were in the electric power sector, as existing power plants were switched to coal from more expensive oil and natural gas and many new coal-fired power plants were constructed. Consumption of coal in the residential, commercial, and industrial sectors decreased from 1949 to 2004.

Coal Consumption by Sector

In the electric power sector, coal is pulverized and burned to heat boilers that produce steam, which drives generators that create electricity. Each ton of coal used to drive a generator produces about 2,000 kilowatt-hours of electricity. In household terms, each pound of coal produces enough electricity to light ten 100-watt light bulbs for one hour.

FIGURE 4.5

Coal flow, 2004

[Million short tons]

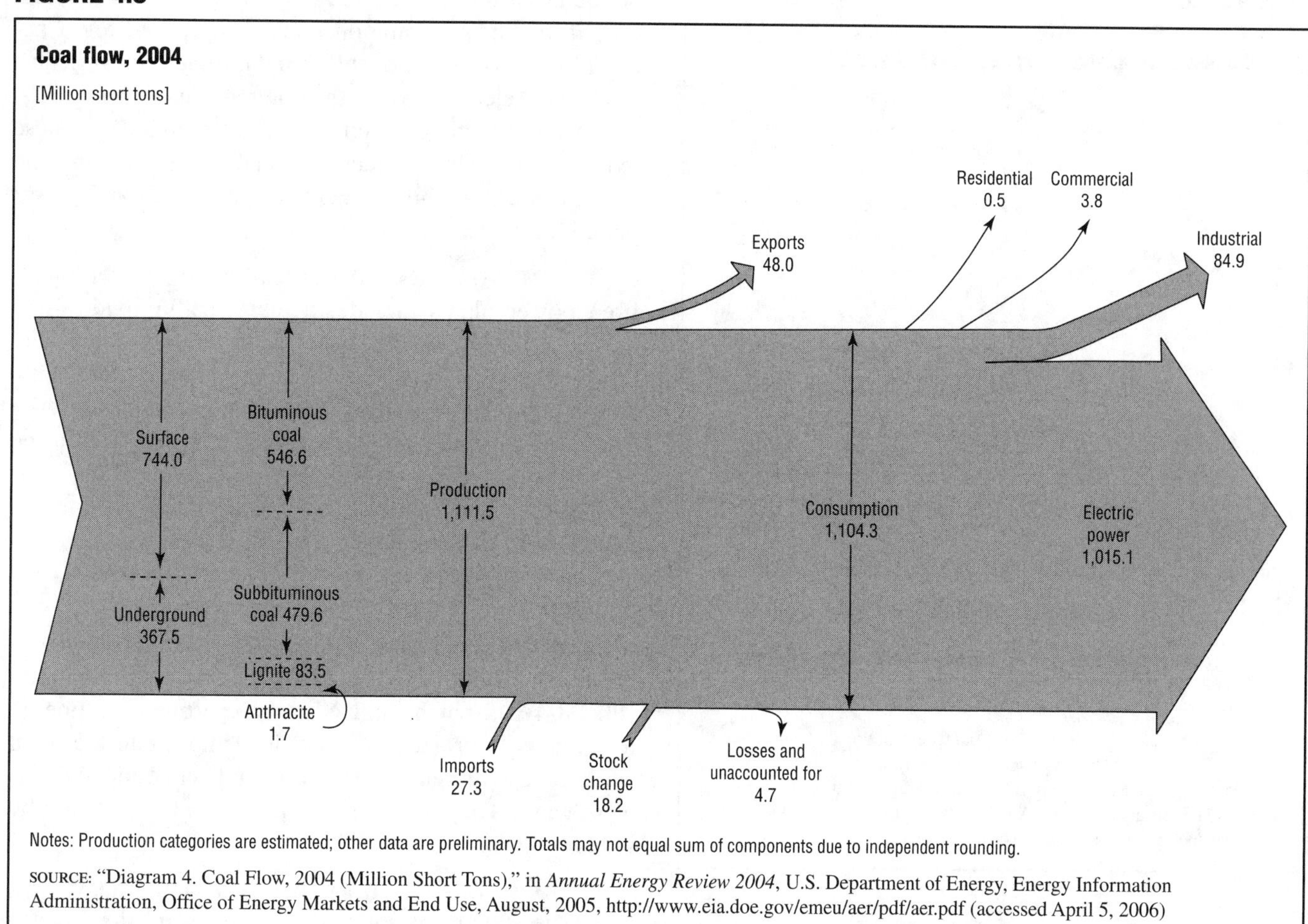

Notes: Production categories are estimated; other data are preliminary. Totals may not equal sum of components due to independent rounding.

SOURCE: "Diagram 4. Coal Flow, 2004 (Million Short Tons)," in *Annual Energy Review 2004*, U.S. Department of Energy, Energy Information Administration, Office of Energy Markets and End Use, August, 2005, http://www.eia.doe.gov/emeu/aer/pdf/aer.pdf (accessed April 5, 2006)

Electric power companies are by far the largest consumers of coal today. (See Figure 4.5 and Figure 4.6.) They accounted for 92% of domestic coal consumption, or slightly more than 1 billion short tons, in 2004. According to the *Annual Energy Review 2004*, coal-fired plants produced 2 trillion kilowatt-hours of electricity, or 50% of U.S. electricity net generation, in 2004. (Net generation refers to the power available to the system—it does not include power used at the generating plant, but it does include power that may be lost during transmission and distribution.)

The industrial sector was the second-largest consumer of coal in 2004, accounting for 8% of coal use (see Figure 4.6), or 84.9 million short tons (see Figure 4.5). Coal is used in many industrial applications, including the chemical, cement, paper, synthetic fuels, metals, and food-processing industries.

Coal was once a significant fuel source in the residential and commercial sectors. (See Figure 4.6.) In 1949 these sectors used 116.5 million short tons of coal. After the 1940s, however, coal was replaced by oil, natural gas, and electricity, which are cleaner and more convenient. By 1970, notes the *Annual Energy Review 2004*, only 16.1 million short tons of coal were used in the residential and commercial sectors. Since then residential and commercial coal use has continued to decline, falling to 4.3 million short tons in 2004, or far less than 1% of total coal use.

The Price of Coal

In 2004 the average price of a short ton of coal had fallen to $18.34 in real dollars—that is, adjusted for inflation—slightly higher than the all-time low of $16.78 in 2000 and only 36% of the 1975 price. (See Table 4.2.) On a per-Btu basis, coal remains the least expensive fossil fuel. In 2001 the average cost of coal was $1.29 per million Btu, compared with $6.87 per million Btu for natural gas and $3.99 per million Btu for residential fuel oil, according to the *Annual Energy Review 2004.*

COAL AND THE ENVIRONMENT

Problems

In 1306 King Edward I of England so objected to the noxious smoke from London's coal-burning fires that he banned the use of coal by everyone except blacksmiths. Since then, the potential for pollution has multiplied

FIGURE 4.6

Coal consumption, shares by sector, 1949 and 2004

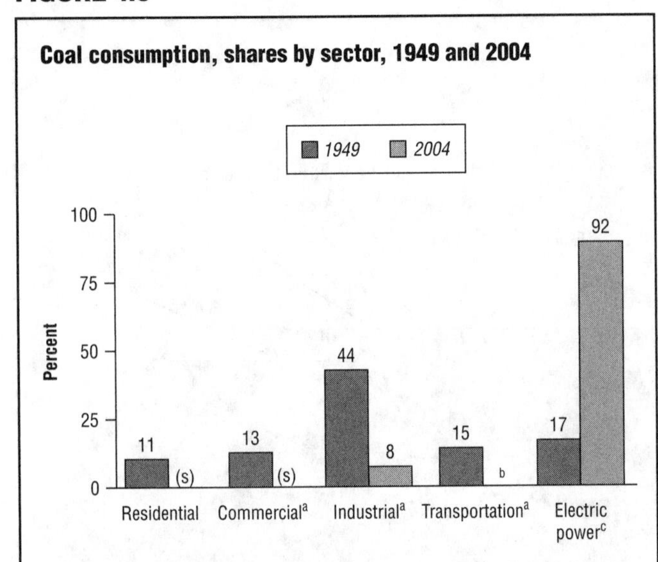

[a]Includes combined-heat-and-power plants and a small number of electricity-only plants.
[b]For 1978 forward, small amounts of transportation sector use are included in "industrial."
[c]Electricity-only and combined-heat-and-power plants whose primary business is to sell electricity, or electricity and heat, to the public.
(s)=Less than 0.5 million short tons or less than 0.5 percent, as appropriate.

SOURCE: Adapted from "Figure 7.3. Coal Consumption by Sector: Sector Shares, 1949 and 2004," in *Annual Energy Review 2004*, U.S. Department of Energy, Energy Information Administration, Office of Energy Markets and End Use, August, 2005, http://www.eia.doe.gov/emeu/aer/pdf/aer.pdf (accessed April 5, 2006)

exponentially, given the amount of fossil fuels, such as coal, that are burned worldwide.

THE "GREENHOUSE EFFECT." Coal-fired electric power plants emit gases that are considered harmful to the environment. Scientists believe that burning huge quantities of fossil fuels causes a "greenhouse effect," in which gases released from the fuels trap heat in Earth's atmosphere, raising temperatures. According to the Energy Information Administration, Earth has warmed about 1° Fahrenheit in the past 100 years. Some of the effects of the added heat are immediate, while others happen over long periods of time.

Much of the gas that causes the greenhouse effect is carbon dioxide. In 2003 the combustion of coal in the United States produced 2.1 billion metric tons of carbon dioxide, or 36% of total carbon dioxide emissions from all fossil fuels used in the United States. (See Figure 4.7.)

ACID RAIN. Acid rain is any form of precipitation that contains a greater-than-normal amount of acid. In many parts of the world it has caused significant damage to forests and lakes.

Even nonpolluted rain is slightly acidic: rainwater combines with the carbon dioxide normally found in the air to produce a weak acid called carbonic acid. However, pollutants in the air can increase the acidity of rain and other forms of precipitation, such as snow and fog. Chemicals such as oxides of sulfur and nitrogen, for instance, which are released during the combustion of fossil fuels, create highly acidic precipitation. The amount of these oxides in the air is directly related to the amount and content of automobile exhaust and industrial and power plant emissions.

COMMUNITY HEALTH ISSUES. Emissions from coal-fired power plants include mercury, sulfur oxides, and nitrogen oxides. Mercury can reach humans when they eat fish contaminated by mercury that, after being emitted into the air, settles in lakes and streams. Mercury can cause birth defects in newborns exposed to it in the womb. Sulfur oxides and nitrogen oxides contribute to air pollution, which is known to cause respiratory impairments. (See Table 4.3.)

Solutions

THE CLEAN COAL TECHNOLOGY LAW. In 1984 Congress established the Clean Coal Technology program (PL 98-473), which directed the Department of Energy to administer projects that would demonstrate that coal could be used in environmentally and economically efficient ways. The cost of the projects was to be shared by industry and government.

CLEAN COAL TECHNOLOGY AND THE CLEAN AIR ACT. The stated goal of both Congress and the Department of Energy has been to develop cost-effective ways to burn coal more cleanly, both to control acid rain and air pollution and to reduce the nation's dependence on imported fuels. One strategy is a slow, phased-in approach that allows utility companies and states to reduce their emissions in stages.

The burning of coal can be made cleaner by using physical or chemical methods. Scrubbers, a common physical method used to reduce sulfur dioxide emissions, filter coal emissions by spraying lime or a calcium compound and water across the emission stream before it leaves the smokestack. The sulfur dioxide bonds to the spray and settles as a mudlike substance that can be pumped out for disposal. However, scrubbers are expensive to operate, so mechanical and fabric particulate collectors are the most common emissions cleaners. While they are cheaper to operate than scrubbers, they are less effective. Some utilities use cooling towers to reduce the heat in emissions before they are released into the atmosphere and to reduce some pollutants. Chemical cleaning, a relatively new technology, uses biological or chemical agents to clean emissions.

The *Annual Energy Review 2004* noted that in 2003 coal-fired electricity plants that had environmental equipment installed had a production capacity of 328.6

TABLE 4.2

Coal prices, selected years 1949–2004

[Dollars per short ton]

Year	Bituminous coal		Subbituminous coal		Lignite[a]		Anthracite		Total	
	Nominal	Real[b]	Nominal	Real[b]	Nominal	Real[b]	Nominal	Real[b]	Nominal	Real[b]
1949	4.90[c]	29.97[c]	c	c	2.37	14.49	8.90	54.43	5.24	32.05
1950	4.86[c]	29.40[c]	c	c	2.41	14.58	9.34	56.50	5.19	31.40
1955	4.51[c]	24.06[c]	c	c	2.38	12.70	8.00	42.68	4.69	25.02
1960	4.71[c]	22.38[c]	c	c	2.29	10.88	8.01	38.07	4.83	22.96
1965	4.45[c]	19.75[c]	c	c	2.13	9.45	8.51	37.76	4.55	20.19
1970	6.30[c]	22.88[c]	c	c	1.86	6.76	11.03	40.06	6.34	23.03
1972	7.78[c]	25.79[c]	c	c	2.04	6.76	12.40	41.11	7.72	25.59
1974	16.01[c]	46.11[c]	c	c	2.19	6.31	22.19	63.90	15.82	45.56
1975	19.79[c]	52.08[c]	c	c	3.17	8.34	32.26	84.89	19.35	50.92
1976	20.11[c]	50.03[c]	c	c	3.74	9.30	33.92	84.39	19.56	48.66
1978	22.64[c]	49.48[c]	c	c	5.68	12.41	35.25	77.04	21.86	47.77
1980	29.17	53.98	11.08	20.50	7.60	14.06	42.51	78.66	24.65	45.61
1982	32.15	51.25	13.37	21.31	9.79	15.61	49.85	79.47	27.25	43.44
1984	30.63	45.27	12.41	18.34	10.45	15.45	48.22	71.27	25.61	37.85
1986	28.84	40.48	12.26	17.21	10.64	14.93	44.12	61.92	23.79	33.39
1988	27.66	36.54	10.45	13.81	10.06	13.29	44.16	58.34	22.07	29.16
1990	27.43	33.62	9.70	11.89	10.13	12.42	39.40	48.29	21.76	26.67
1992	26.78	31.00	9.68	11.21	10.81	12.51	34.24	39.64	21.03	24.34
1994	25.68	28.45	8.37	9.27	10.77	11.93	36.07	39.96	19.41	21.50
1996	25.17	26.82	7.87	8.39	10.92	11.64	36.78	39.19	18.50	19.71
1998	24.87	25.78	6.96	7.21	11.08	11.49	42.91	44.48	17.67	18.32
1999	23.92	24.44	6.87	7.02	11.04	11.28	35.13	35.90	16.63	16.99
2000	24.15	24.15	7.12	7.12	11.41	11.41	40.90	40.90	16.78	16.78
2001	25.36	24.77	6.67	6.51[R]	11.52	11.25	47.67	46.55[R]	17.38	16.97[R]
2002	26.57	25.53[R]	7.34	7.05[R]	11.07	10.63[R]	47.78	45.90[R]	17.98	17.27[R]
2003	26.73[R]	25.22[R]	7.73[R]	7.29[R]	11.20[R]	10.57[R]	49.55[R]	46.75[R]	17.85[R]	16.84[R]
2004[E]	30.47	28.15	8.51	7.86	12.35	11.41	60.16	55.58	19.85	18.34

[a]Because of withholding to protect company confidentiality, lignite prices exclude Texas for 1955–1977 and Montana for 1974–1978. As a result, lignite prices for 1974–1977 are for North Dakota only.
[b]In chained (2000) dollars, calculated by using gross domestic product implicit price deflators.
[c]Through 1978, subbituminous coal is included in "bituminous coal."
R=Revised. E=Estimate.
Notes: Prices are free-on-board (f.o.b.) rail/barge prices, which are the f.o.b. prices of coal at the point of first sale, excluding freight or shipping and insurance costs. For data not shown for 1951–1969, see http://www.eia.doe.gov/emeu/aer/coal.html. For related information, see http://www.eia.doe.gov/fuelcoal.html.

SOURCE: Adapted from "Table 7.8. Coal Prices, Selected Years, 1949–2004 (Dollars per Short Ton)," in *Annual Energy Review 2004*,U.S. Department of Energy, Energy Information Administration, Office of Energy Markets and End Use, August, 2005, http://www.eia.doe.gov/emeu/aer/pdf/aer.pdf (accessed April 5, 2006)

gigawatts (1 gigawatt equals 1,000 megawatts). Statistics were collected on three types of environmental equipment: particulate collectors, cooling towers, and scrubbers. Some plants had all three types of equipment; some had two, and some had only one. All the plants had particulate collectors, nearly half had cooling towers, but less than one-third had scrubbers.

The Clean Air Act of 1990 (PL 101–549) placed restrictions on sulfur dioxide and nitrogen oxide emissions, which first took effect in 1995 and were tightened in 2000. Each round of regulation requires utilities to find lower-sulfur coal or to install cleaner technology, such as scrubbers. The first Clean Air Act, passed in 1970, sought to change air-quality standards at new generating stations while exempting existing coal-fired plants. Under the 1990 act, older plants are also bound by the 1970 regulations. In addition, plants with coal-fired boilers must be built to reduce sulfur emissions by 70% to 90%. When new plants that burn high-sulfur coal are built, about 30% of their construction costs are spent on pollution control equipment. Up to 5% of the plants' power output is used to operate this equipment. Research is under way to develop technology to lower these costs.

When President George W. Bush took office in 2001, he promised to ease the regulations for older coal-burning plants to keep them up and running. His energy policy would have allowed the plants to modify their equipment and structures without adding pollution control measures, as long as they met certain conditions. Fifteen state governments and environmental and public health groups brought suit to stop this interpretation of the Clean Air Act. In April 2006 the Bush policy was rejected by the courts.

THE CLEAN AIR INTERSTATE RULE AND THE CLEAN AIR MERCURY RULE. In March 2005 the Environmental Protection Agency announced the Clean Air Interstate Rule. It focused on twenty-eight eastern states and the District of Columbia where sulfur dioxide and nitrogen oxide emissions contributed significantly to fine particle and ozone pollution in downwind states. Under the

FIGURE 4.7

Carbon dioxide emissions from energy consumption, by fuel, 2003

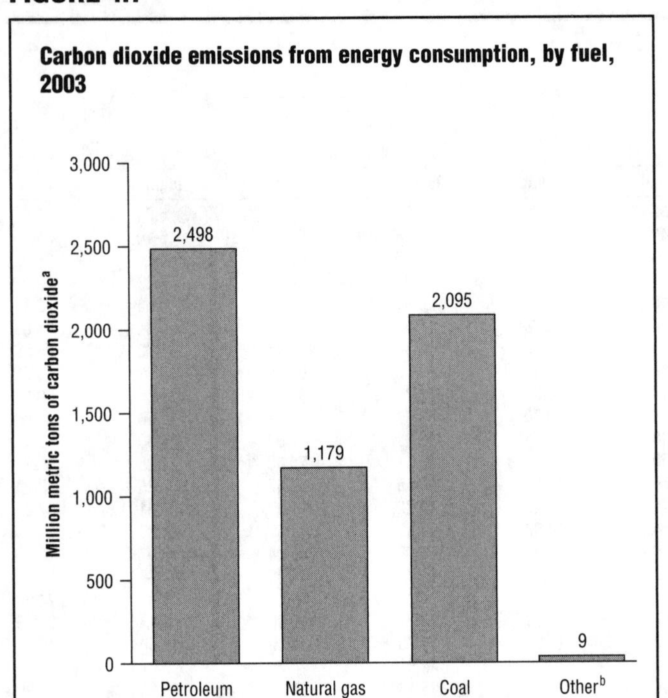

[a]Metric tons of carbon dioxide can be converted to metric tons of carbon equivalent by multiplying by 12/44.
[b]Coal coke net imports, municipal solid waste, and geothermal.

SOURCE: Adapted from "Figure 12.3. Carbon Dioxide Emissions From Energy Consumption by Sector by Energy Source, 2003: Total by Fuel," in *Annual Energy Review 2004*, U.S. Department of Energy, Energy Information Administration, Office of Energy Markets and End Use, August, 2005, http://www.eia.doe.gov/emeu/aer/pdf/aer.pdf (accessed April 5, 2006)

TABLE 4.3

Air pollutants, health risks, and contributing sources

Pollutants	Health risks	Contributing sources
Ozone* (O_3)	Asthma, reduced respiratory function, eye irritation	Cars, refineries, dry cleaners
Particulate matter (PM-IO)	Bronchitis, cancer, lung damage	Dust, pesticides
Carbon monoxide (CO)	Blood oxygen carrying capacity reduction, cardiovascular and nervous system impairments	Cars, power plants, wood stoves
Sulphur dioxide (SO_2)	Respiratory tract impairment, destruction of lung tissue	Power plants, paper mills
Lead (Pb)	Retardation and brain damage, esp. children	Cars, nonferrous smelters, battery plants
Nitrogen dioxide (NO_2)	Lung damage and respiratory illness	Power plants, cars, trucks

*Ozone refers to tropospheric ozone which is hazardous to human health.

SOURCE: Fred Seitz and Christine Plepys, "Table 1. Criteria Air Pollutants, Health Risks and Sources," in *Healthy People 2000: Statistical Notes, Number 9*, Centers for Disease Control and Prevention, National Center for Health Statistics, September 1995, http://www.cdc.gov/nchs/data/statnt/statnt09.pdf (accessed April 11, 2006)

regulations, coal-burning power plants would have to install advanced pollution-control technologies, use coal that burned cleaner, or make other changes to reduce emissions of sulfur dioxide and nitrogen oxide. When fully implemented, the rule is expected to reduce sulfur dioxide emissions by 70% and nitrogen oxide emissions by 60% from 2003 levels. The agency also announced the Clean Air Mercury Rule, which made the United States the first country to regulate mercury emissions from power plants. Mercury, when ingested by children, can have serious health consequences. When fully implemented, the two rules are expected to reduce mercury emissions from electric power plants by nearly 70% from 1999 levels.

COAL EXPORTS

Since 1950 the United States has produced more coal than it has consumed, allowing it to become a significant exporter. However, exports of this energy source have declined dramatically since 1991, when the U.S. exported 109 million short tons of coal. In 2004 the U.S. exported 48 million short tons. (See Table 4.4.)

In 2004 coal made up 28% of all U.S. energy exports. (See Figure 1.7 in Chapter 1.) The countries that bought the most U.S. coal were Canada, Brazil, Japan, the Netherlands, Italy, and the United Kingdom. (See Table 4.4.)

INTERNATIONAL COAL PRODUCTION AND CONSUMPTION

The *Annual Energy Review 2004* reported that slightly more than 5.4 billion short tons of coal were produced worldwide in 2003, accounting for 24% of world energy production. China produced the most coal—more than 1.6 billion short tons—followed by the United States, which mined almost 1.1 billion short tons. (See Figure 4.8.) Other major coal producers were India, Australia, Russia, South Africa, Germany, and Poland.

World consumption of coal in 2003 totaled slightly more than 5.4 billion short tons. China consumed the most, using more than 1.5 billion short tons, followed by the United States, which used slightly less than 1.1 billion short tons. (See Figure 4.9.) Other major consumers included India, Germany, Russia, South Africa, and Japan.

FUTURE TRENDS IN THE COAL INDUSTRY

The *Annual Energy Outlook 2006* (Energy Information Administration, 2006) provides forecasts for domestic coal production. Output is expected to increase to nearly 1.3 billion short tons by 2015, to more than 1.5 billion short tons by 2025, and to 1.7 billion short tons by 2030. (See Table 4.5.) Domestic coal consumption is projected to match (2015) and surpass (2025 and 2030)

TABLE 4.4

Coal exports by country of destination, selected years 1960–2004

[Million short tons]

Year	Canada	Brazil	Europe: Belgium and Luxembourg	Denmark	France	Germany*	Italy	Netherlands	Spain	United Kingdom	Other	Total	Japan	Other	Total
1960	12.8	1.1	1.1	0.1	0.8	4.6	4.9	2.8	0.3	0.0	2.4	17.1	5.6	1.3	38.0
1961	12.1	1.0	1.0	0.1	0.7	4.3	4.8	2.6	0.2	0.0	2.0	15.7	6.6	1.0	36.4
1962	12.3	1.3	1.3	s	0.9	5.1	6.0	3.3	0.8	s	1.8	19.1	6.5	1.0	40.2
1964	14.8	1.1	2.3	s	2.2	5.2	8.1	4.2	1.4	0.0	2.6	26.0	6.5	1.1	49.5
1966	16.5	1.7	1.8	s	1.6	4.9	7.8	3.2	1.2	s	2.5	23.1	7.8	1.0	50.1
1968	17.1	1.8	1.1	0.0	1.5	3.8	4.3	1.5	1.5	0.0	1.9	15.5	15.8	0.9	51.2
1970	19.1	2.0	1.9	0.0	3.6	5.0	4.3	2.1	3.2	s	1.8	21.8	27.6	1.2	71.7
1972	18.7	1.9	1.1	0.0	1.7	2.4	3.7	2.3	2.1	2.4	1.1	16.9	18.0	1.2	56.7
1974	14.2	1.3	1.1	0.0	2.7	1.5	3.9	2.6	2.0	1.4	0.9	16.1	27.3	1.8	60.7
1976	16.9	2.2	2.2	s	3.5	1.0	4.2	3.5	2.5	0.8	2.1	19.9	18.8	2.1	60.0
1978	15.7	1.5	1.1	0.0	1.7	0.6	3.2	1.1	0.8	0.4	2.2	11.0	10.1	2.5	40.7
1980	17.5	3.3	4.6	1.7	7.8	2.5	7.1	4.7	3.4	4.1	6.0	41.9	23.1	6.0	91.7
1982	18.6	3.1	4.8	2.8	9.0	2.3	11.3	5.9	5.6	2.0	7.6	51.3	25.8	7.5	106.3
1984	20.4	4.7	3.9	0.6	3.8	0.9	7.6	5.5	2.3	2.9	5.3	32.8	16.3	7.2	81.5
1986	14.5	5.7	4.4	2.1	5.4	0.8	10.4	5.6	2.6	2.9	8.4	42.6	11.4	11.4	85.5
1988	19.2	5.3	6.5	2.8	4.3	0.7	11.1	5.1	2.5	3.7	8.5	45.1	14.1	11.3	95.0
1990	15.5	5.8	8.5	3.2	6.9	1.1	11.9	8.4	3.8	5.2	9.5	58.4	13.3	12.7	105.8
1991	11.2	7.1	7.5	4.7	9.5	1.7	11.3	9.6	4.7	6.2	10.4	65.5	12.3	13.0	109.0
1992	15.1	6.4	7.2	3.8	8.1	1.0	9.3	9.1	4.5	5.6	8.5	57.3	12.3	11.4	102.5
1994	9.2	5.5	4.9	0.5	2.9	0.3	7.5	4.9	4.1	3.4	7.3	35.8	10.2	10.7	71.4
1996	12.0	6.5	4.6	1.3	3.9	1.1	9.2	7.1	4.1	6.2	9.8	47.2	10.5	14.2	90.5
1998	20.7	6.5	3.2	0.3	3.2	1.2	5.3	4.5	3.2	5.9	6.9	33.8	7.7	9.4	78.0
1999	19.8	4.4	2.1	0.0	2.5	0.6	4.0	3.4	2.5	3.2	4.3	22.5	5.0	6.7	58.5
2000	18.8	4.5	2.9	0.1	3.0	1.0	3.7	2.6	2.7	3.3	5.7	25.0	4.4	5.8	58.5
2001	17.6	4.6	2.8	0.0	2.2	0.9	5.4	2.1	1.6	2.5	3.3	20.8	2.1	3.6	48.7
2002	16.7	3.5	2.4	0.0	1.3	1.0	3.1	1.7	1.9	1.9	2.4	15.6	1.3	2.6	39.6
2003	20.8	3.5	1.8	0.3	1.3	0.5	2.8	2.0	1.8	1.5	3.2	15.1	s	3.6	43.0
2004	17.8	4.4	1.7	0.1	1.1	0.6	2.1	2.5	1.5	2.0	3.6	15.2	4.4	6.2	48.0

*Through 1990, data for Germany are for the former West Germany only. Beginning in 1991, data for Germany are for the unified Germany, i.e., the former East Germany and West Germany.
s=Less than 0.05 million short tons.
Note: Totals may not equal sum of components due to independent rounding.

SOURCE: Adapted from "Table 7.4. Coal Exports by Country of Destination, 1960–2004 (Million Short Tons)," in *Annual Energy Review 2004*, U.S. Department of Energy, Energy Information Administration, Office of Energy Markets and End Use, August, 2005, http://www.eia.doe.gov/emeu/aer/pdf/aer.pdf (accessed April 5, 2006)

production, reaching nearly 1.3 billion short tons in 2015, almost 1.6 billion short tons by 2025, and nearly 1.8 billion short tons by 2030. Electric power generation is expected to use the majority of the coal produced in 2015 (approximately 1.2 billion short tons), 2025 (nearly 1.4 billion short tons), and 2030 (1.5 billion short tons). The projections assume the provisions of the Clean Air Act, the Clean Air Interstate Rule, and the Clean Air Mercury Rule are being enforced.

Coal minemouth prices—the price of coal at the mouth of the mine before transportation and other costs are added—are projected to rise slightly from $20.07 per short ton in 2004 (see Table 4.5) to $20.39 in 2015, to $20.63 in 2025, and to $21.73 in 2030. Prices are expected to go up because of increased demand for coal from the electric power sector.

The *Annual Energy Outlook 2006* also predicts that U.S. coal exports—about 4% of the coal mined in 2004—will decline by 2030 to only 1% of coal mined. All projections expect the United States to become a net importer of coal over the 2004–2030 period.

All of the projections could be affected by new environmental legislation—more stringent acid rain regulations, for example—as well as by conservation initiatives and additional clean-coal technologies. Furthermore, because pollutants travel as winds circle the globe, the U.S. environment and the domestic coal industry may be affected by coal use in other countries. For instance, China, with nearly five times the population of the United States and a growing economy, may surpass the United States in carbon emissions by 2020.

FIGURE 4.8

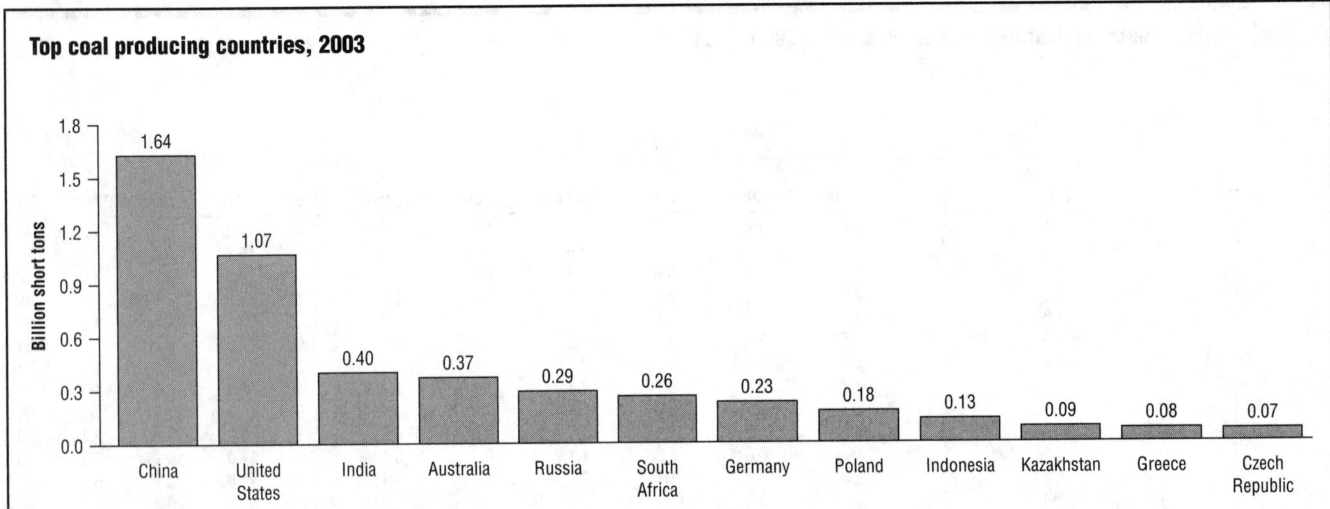

SOURCE: Adapted from "Figure 11.14. World Coal Production: Top Producing Countries, 2003," in *Annual Energy Review 2004*, U.S. Department of Energy, Energy Information Administration, Office of Energy Markets and End Use, August, 2005, http://www.eia.doe.gov/emeu/aer/pdf/aer.pdf (accessed April 5, 2006)

FIGURE 4.9

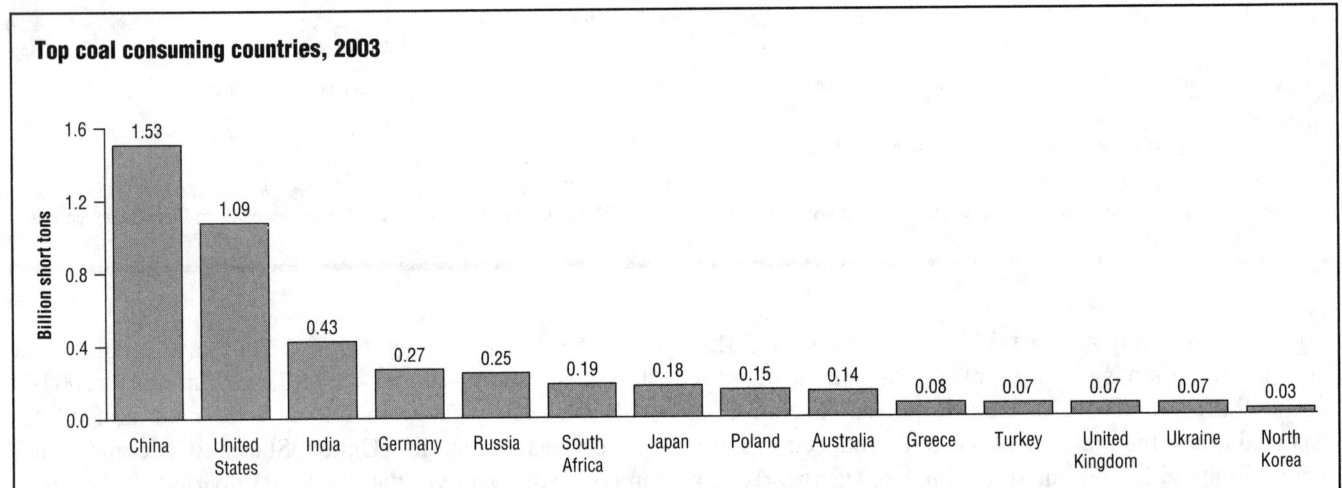

SOURCE: Adapted from "Figure 11.15. World Coal Consumption: Top Consuming Countries, 2003," in *Annual Energy Review 2004*, U.S. Department of Energy, Energy Information Administration, Office of Energy Markets and End Use, August, 2005, http://www.eia.doe.gov/emeu/aer/pdf/aer.pdf (accessed April 5, 2006)

TABLE 4.5

Comparison of coal forecasts, 2015, 2025, and 2030

[Million short tons, except where noted]

Projection	2004	AEO2006 forecast: Reference	AEO2006 forecast: Low economic growth	AEO2006 forecast: High economic growth
			2015	
Production	1,125	1,272	1,251	1,318
Consumption by sector				
Electric power	1,015	1,161	1,145	1,199
Coke plants	24	22	21	23
Coal-to-liquids	0	22	19	27
Industrial/other	65	71	69	72
Total	**1,104**	**1,276**	**1,254**	**1,321**
Net coal exports	20.7	−4.8	−4.8	−4.8
Exports	48.0	22.0	22.0	22.0
Imports	27.3	26.7	26.7	26.8
Minemouth price				
(2004 dollars per short ton)	20.07	20.39	20.04	20.67
(2004 dollars per million Btu)	0.98	1.01	0.99	1.02
Average delivered price to electricity generators				
(2004 dollars per short ton)	27.43	28.12	27.74	28.50
(2004 dollars per million Btu)	1.36	1.40	1.39	1.42
			2025	
Production	1,125	1,530	1,394	1,710
Consumption by sector				
Electric power	1,015	1,354	1,248	1,486
Coke plants	24	21	19	23
Coal-to-liquids	0	146	115	192
Industrial/other	65	71	68	73
Total	**1,104**	**1,592**	**1,450**	**1,774**
Net coal exports	20.7	−62.8	−57.9	−65.5
Exports	48.0	19.6	19.6	18.4
Imports	27.3	82.4	77.4	84.0
Minemouth price				
(2004 dollars per short ton)	20.07	20.63	19.40	21.73
(2004 dollars per million Btu)	0.98	1.03	0.98	1.09
Average delivered price to electricity generators				
(2004 dollars per short ton)	27.43	29.02	27.48	30.87
(2004 dollars per million Btu)	1.36	1.44	1.37	1.52
			2030	
Production	1,125	1,703	1,497	1,936
Consumption by sector				
Electric power	1,015	1,502	1,331	1,680
Coke plants	24	21	19	23
Coal-to-liquids	0	190	153	247
Industrial/other	65	72	68	75
Total	**1,104**	**1,784**	**1,571**	**2,025**
Net coal exports	20.7	−82.7	−69.3	−89.0
Exports	48.0	16.7	16.4	16.8
Imports	27.3	99.4	85.7	105.8
Minemouth price				
(2004 dollars per short ton)	20.07	21.73	19.91	23.05
(2004 dollars per million Btu)	0.98	1.09	1.00	1.15
Average delivered price to electricity generators				
(2004 dollars per short ton)	27.43	30.58	28.28	32.79
(2004 dollars per million Btu)	1.36	1.51	1.41	1.61

SOURCE: Adapted from "Table 25. Comparison of Coal Forecasts, 2015, 2025, and 2030 (Million Short Tons, Except where Noted)," in *Annual Energy Outlook 2006*, U.S. Department of Energy, Energy Information Administration, Office of Integrated Analysis and Forecasting, February 2006, http://www.eia.doe.gov/oiaf/aeo/pdf/0383(2006).pdf (accessed April 5, 2006)

CHAPTER 5
NUCLEAR ENERGY

In the early 1970s many Americans favored the use of nuclear power to generate electricity. Reactors had been in operation since 1956, providing what appeared to be cleaner, more efficient energy than power plants that burned fossil fuels. In addition, the nation had plenty of uranium, the fuel used in U.S. reactors, so nuclear plants could help reduce U.S. dependence on foreign energy sources.

By the beginning of this century, however, opinion had turned: Voices were being raised in the United States—as well as around the world—in opposition to building additional nuclear power plants. In fact, some people wanted existing plants shut down. Two incidents—an accident in 1979 at the Three Mile Island nuclear power plant in Pennsylvania, in which the reactor core lost coolant and partially melted, and the 1986 catastrophe at Chernobyl in the Soviet Union, in which a reactor exploded and released huge amounts of radioactivity into the atmosphere—greatly increased concerns about the safety of nuclear power. Reports of design flaws, cracks, and leaks in other reactors fueled public fears. Furthermore, the safe disposal of radioactive waste, which is a by-product of nuclear energy, had become a scientific and political headache.

Supporters of nuclear power say that it is as safe as any other form of energy production, as long as it is monitored properly. They point to the growing problems related to the use of fossil fuels: global warming, acid rain, and the damage caused by mining and transporting fossil fuels. This concern over fossil fuels has led a small number of environmentalists who had previously opposed nuclear power to reconsider their position. Nonetheless, environmental, safety, and economic concerns have restrained growth in the nuclear industry since the mid-1970s. Figure 5.1 shows that from 1988 through 2004 nuclear energy's share of electricity production leveled off at about 20%.

HOW NUCLEAR ENERGY WORKS

In a nuclear power plant, fuel (uranium in the United States) in the core of the reactor generates a nuclear reaction (fission) that produces heat. In a pressurized water reactor (see Figure 5.2), the heat from the reaction is carried away by water under high pressure, which heats a second water stream, producing steam. The steam runs through a turbine, making the attached generator spin, which produces electricity. The large cooling towers associated with nuclear plants cool the steam after it has run through the turbines. A boiling water reactor works much the same way, except that the water surrounding the core boils and directly produces the steam, which is then piped to the turbine generator.

The challenges faced by those who operate a nuclear power reactor include finding fuel (uranium 235) that will sustain a chain reaction; maintaining the reaction at a level that yields heat but does not escalate out of control and explode; and coping with the radiation produced by the chain reaction.

Radioactivity

Radioactivity is the spontaneous emission of energy and/or high-energy particles from the nucleus of an atom. One type of radioactivity is produced naturally and is emitted by radioactive isotopes (or radioisotopes), such as radioactive carbon (carbon 14, or C-14) and radioactive hydrogen (H-3, or tritium). The energy and high-energy particles that radioactive isotopes emit include alpha rays, beta rays, and gamma rays.

Isotopes are atoms of an element that have the usual number of protons but different numbers of neutrons in their nuclei. For example, twelve protons and twelve neutrons make up the nucleus of the element carbon. One isotope of carbon, C-14, has twelve protons and fourteen neutrons in its nucleus.

FIGURE 5.1

Nuclear share of electricity net generation, 1957–2004

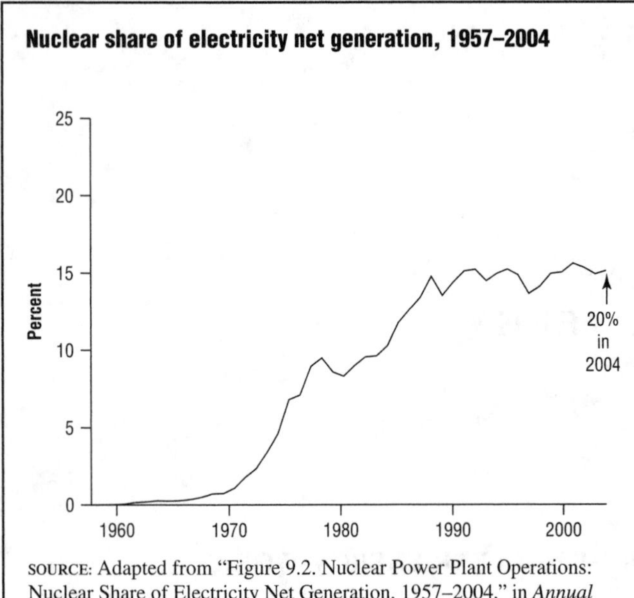

SOURCE: Adapted from "Figure 9.2. Nuclear Power Plant Operations: Nuclear Share of Electricity Net Generation, 1957–2004," in *Annual Energy Review 2004*, U.S. Department of Energy, Energy Information Administration, Office of Energy Markets and End Use, August, 2005, http://www.eia.doe.gov/emeu/aer/pdf/aer.pdf (accessed April 5, 2006)

FIGURE 5.2

Pressurized water reactor

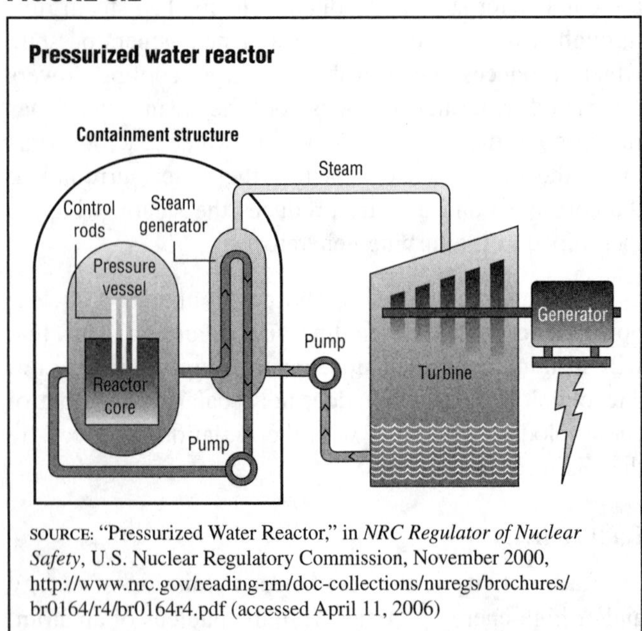

SOURCE: "Pressurized Water Reactor," in *NRC Regulator of Nuclear Safety*, U.S. Nuclear Regulatory Commission, November 2000, http://www.nrc.gov/reading-rm/doc-collections/nuregs/brochures/br0164/r4/br0164r4.pdf (accessed April 11, 2006)

Radioisotopes (such as C-14) are unstable isotopes, and their nuclei decay, or break apart, at a steady rate. Decaying radioisotopes produce other isotopes as they emit energy and/or high-energy particles. If the newly formed nuclei are radioactive as well, they emit radiation and change into other nuclei. The final products in this chain are stable, nonradioactive nuclei.

Radiation and radioisotopes reach our bodies daily, emitted from sources in outer space and from rocks and soil on Earth. Figure 5.3 shows that radon is the largest

FIGURE 5.3

Sources of radiation

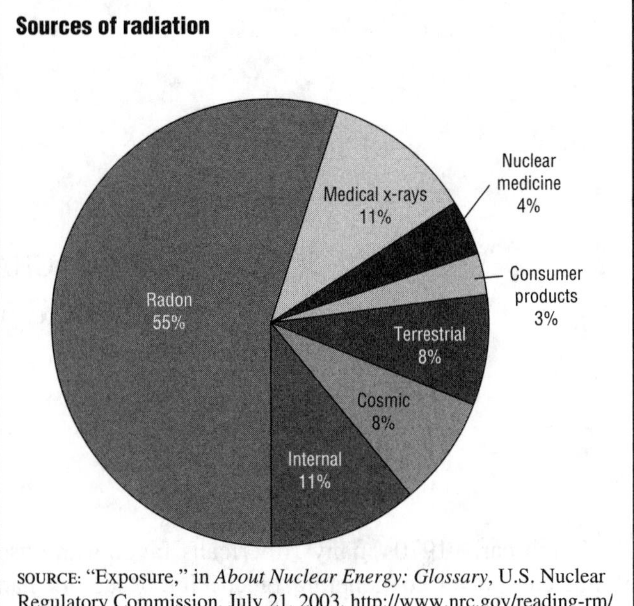

SOURCE: "Exposure," in *About Nuclear Energy: Glossary*, U.S. Nuclear Regulatory Commission, July 21, 2003, http://www.nrc.gov/reading-rm/basic-ref/glossary/exposure.html (accessed April 11, 2006)

source of radiation to which humans are exposed. This gas is formed in rocks and soil from the radioactive decay of radium. Most prevalent in the northern half of the United States, radon can enter cracks in basement walls and remain trapped there. Prolonged exposure to high levels of radioactive radon is thought to lead to lung cancer. Other radioisotopes, in minute quantities, are used in medicine as diagnostic tools.

Radiation was discovered at the beginning of the twentieth century by Antoine Henri Becquerel, Marie Curie, and Pierre Curie. Decades later other scientists determined that they could unleash energy by artificially breaking apart atomic nuclei. Such a process is called nuclear fission. Scientists discovered that they could produce the most energy by bombarding the nuclei of an isotope of uranium called uranium 235, or U-235. The fission of a U-235 atom releases several neutrons, which have the potential to penetrate other U-235 atomic nuclei and cause them to fission. In this way the fission of a single U-235 atom begins a cascading chain of nuclear reactions, as shown in Figure 5.4. If this series of reactions is regulated to occur slowly, as it is in nuclear power plants, the energy emitted can be captured for a variety of uses, such as generating electricity. If this series of reactions is allowed to occur all at once, as in an atomic bomb, the energy emitted is explosive. (Plutonium 239 can also be used to generate a chain reaction similar to that of U-235.)

Mining Nuclear Fuel

In the United States, U-235 is used as nuclear fuel. Most of it is found in the form of ore in Wyoming and

FIGURE 5.4

Nuclear chain reaction

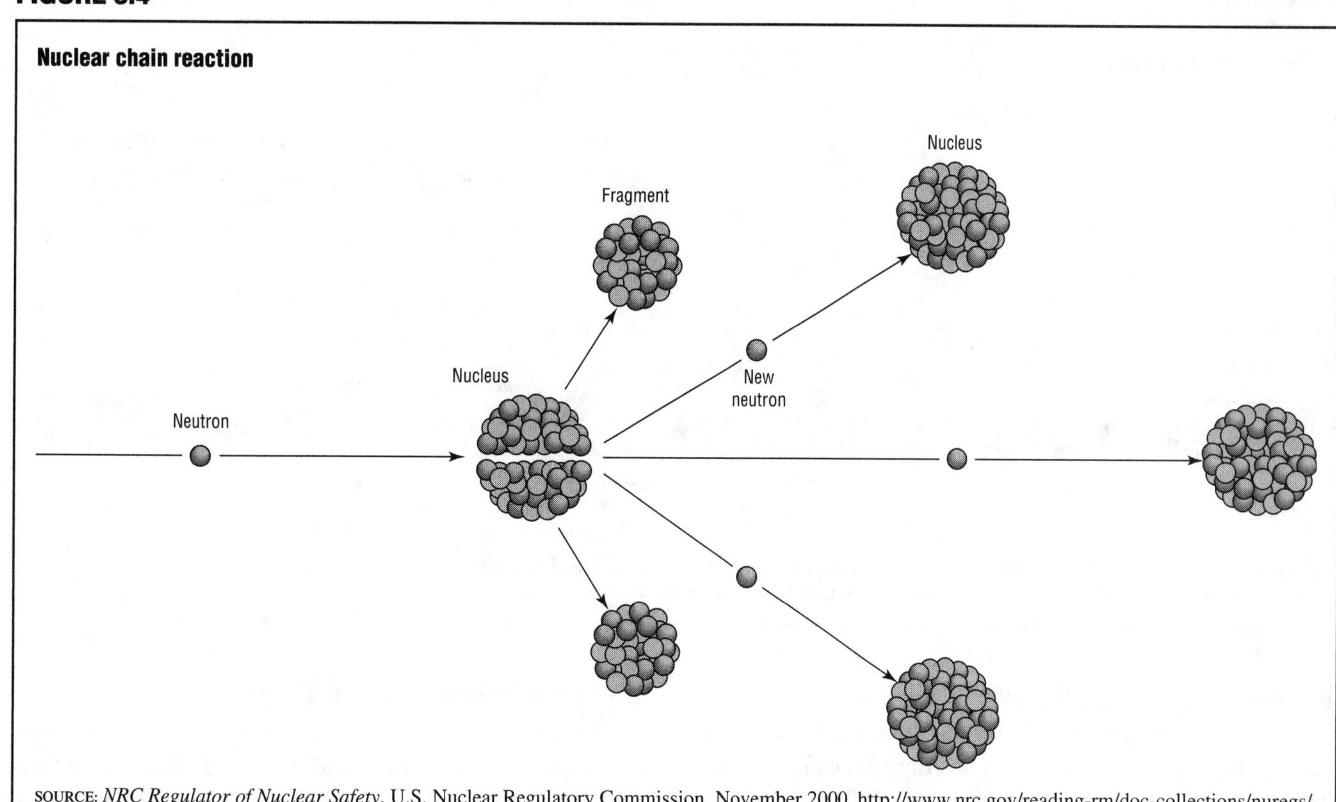

SOURCE: *NRC Regulator of Nuclear Safety*, U.S. Nuclear Regulatory Commission, November 2000, http://www.nrc.gov/reading-rm/doc-collections/nuregs/brochures/br0164/r4/br0164r4.pdf (accessed April 11, 2006)

New Mexico, where it is mined using methods similar to those for other metal ores. The one difference is that uranium mining can expose workers to radioactivity. Uranium atoms split by themselves at a slow rate, causing radioactive substances such as radon to accumulate slowly in the deposits.

After it is mined, uranium must be concentrated, because uranium ore generally contains only 0.1% uranium metal by weight. To concentrate the uranium, it goes through a process called milling: The ore is first crushed, and then various chemicals are poured slowly through the crushed ore to dissolve the uranium. When the uranium—called yellowcake because of its color—is extracted from this chemical solution, it is 85% pure by weight. However, this uranium is 99.3% nonfissionable U-238 and only 0.7% fissionable U-235. Other processes are necessary to create enriched uranium, which has a higher percentage of fissionable uranium.

To enrich uranium, yellowcake is first converted into uranium hexafluoride (UF_6). Cylinders of this gas are sent to a gaseous diffusion plant, shown in Figure 5.5, where the uranium is heated in a furnace (UF_6 vaporization). The enriched uranium is then converted into oxide powder (UO_2), which is made into fingertip-sized fuel pellets. The pellets are less than one-half inch in diameter, but each one can produce as much energy as 120 gallons of oil. The pellets are stacked in rods, which are tubes about twelve feet long. Many rods are bundled together in assemblies, and hundreds of these assemblies make up the core of a nuclear reactor.

DOMESTIC NUCLEAR ENERGY PRODUCTION

The percentage of U.S. electricity supplied by nuclear power grew considerably during the 1970s and early to mid-1980s and then leveled off. (See Figure 5.1.) According to the *Annual Energy Review 2004* (Energy Information Administration, 2005, http://tonto.eia.doe.gov/FTPROOT/multifuel/038404.pdf), nuclear power supplied only 4.5% of the total electricity generated in the United States in 1973. In 2004 nuclear electricity net generation reached 788.6 billion kilowatt-hours, or about 20% of the nation's electricity. Thirty-one states had 104 nuclear reactors in operation. Figure 5.6 shows that most of these reactors were located east of the Mississippi River.

No new nuclear power plants have been ordered since 1978, and many have closed. (See Table 5.1.) As shown in Figure 5.7, the number of operable nuclear generating units peaked in 1990 with 112 units; as of 2004, 104 units were operating. The decline in nuclear power plants stems from several related issues. Financing is difficult to find, and construction has become more expensive, partly because of longer delays for licensing, but also because of regulations instituted

FIGURE 5.5

The nuclear fuel cycle

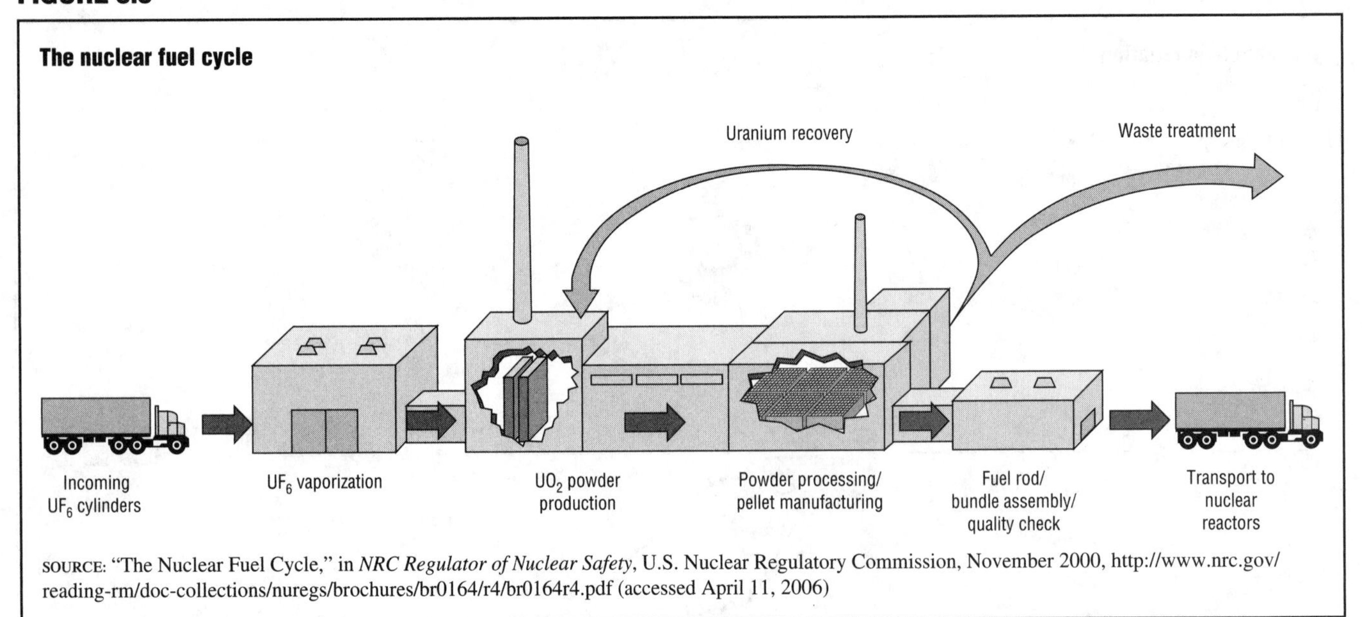

SOURCE: "The Nuclear Fuel Cycle," in *NRC Regulator of Nuclear Safety*, U.S. Nuclear Regulatory Commission, November 2000, http://www.nrc.gov/reading-rm/doc-collections/nuregs/brochures/br0164/r4/br0164r4.pdf (accessed April 11, 2006)

following the Three Mile Island accident. Still, output of electricity at existing plants has increased, achieved largely through an increase in average capacity factor. The capacity factor is the proportion of electricity produced to what could have been produced at full-power operation. The *Annual Energy Review 2004* reveals that in 2004 the average capacity factor for U.S. nuclear power plants was 90.5%, an all-time high. Better training for operators, longer operating cycles between refueling, and control-system improvements contributed to increased plant performance.

OUTLOOK FOR DOMESTIC NUCLEAR ENERGY

In its *Annual Energy Outlook 2006* (2006, http://www.eia.doe.gov/oiaf/aeo/index.html), the Energy Information Administration predicted that capacity of nuclear power plants would increase from 99.6 gigawatts in 2004 to 108.8 gigawatts in 2030. About 6.0 gigawatts of capacity would come from new plants and 3.2 gigawatts from power uprates—an increase in the power output of a nuclear power plant. A power uprate is accomplished by adding a more highly enriched uranium fuel to the existing fuel. The plant must be able to operate safely at the higher power level.

The forecasts are based on continued operation of all existing nuclear plants through 2030. Nuclear generation is predicted to grow from 780 billion kilowatt-hours in 2004 to 871 billion kilowatt-hours in 2030. However, the agency notes that even with such an increase in production, the share that nuclear power contributes to electricity generation in the United States will decrease from 20% in 2004 to about 15% in 2030.

INTERNATIONAL PRODUCTION

At 2,523.1 billion kilowatt-hours, nuclear power provided about 16% of the total 15,843.9 billion kilowatt-hours of electricity produced in the world during 2003. (See Table 5.2.) This amounted to 6.4% of the world's total energy, according to the *Annual Energy Review 2004*. As of December 31, 2004, electricity was being produced by 440 nuclear plants worldwide. Fifty-two other plants were under construction, on order, or had had their construction halted (Nuclear Regulatory Commission, *Information Digest 2005–2006*, 2005, http://www.nrc.gov/reading-rm/doc-collections/nuregs/staff/sr1350/).

In 2003 the United States led the world in nuclear power electricity generation with 763.7 billion kilowatt-hours, followed by France (419 billion) and Japan (237.2 billion). (See Table 5.2.) Together the three countries generated 56% of the world's nuclear electric power. About 78% of France's electrical power is produced by nuclear energy, followed by Belgium (57%), Sweden (49%), and Switzerland (41%). Japan produced 23% of its electricity by nuclear power generation, and the United States 20%. Canada produced less than 13% of its electricity with nuclear power.

Worldwide, the Energy Information Administration projected nuclear generating capacity would increase slightly from 361 gigawatts in 2002 to 422 gigawatts in 2025 (*International Energy Outlook 2005*, 2005, http://www.eia.doe.gov/oiaf/archive/ieo05/index.html). Some countries whose capacities are projected to increase are:

- Canada, from 10.6 gigawatts in 2002 to 14.5 gigawatts in 2020, then decreasing to 13.8 gigawatts by 2025
- Japan, from 45.9 gigawatts to 54.8 gigawatts by 2025

FIGURE 5.6

Operating nuclear reactors, June 2005

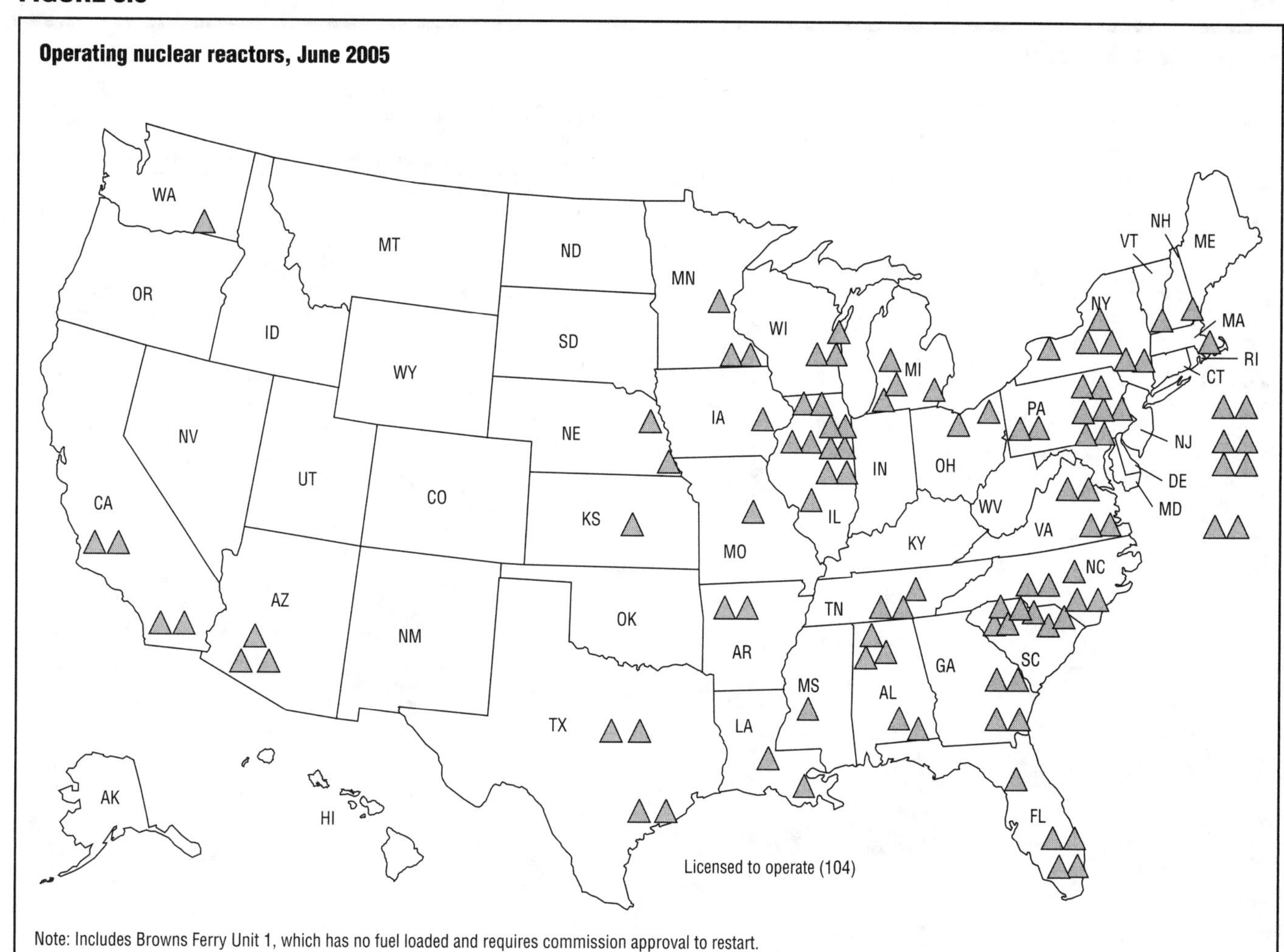

Note: Includes Browns Ferry Unit 1, which has no fuel loaded and requires commission approval to restart.

SOURCE: Adapted from "Figure 18. U.S. Operating Commercial Nuclear Power Reactors," in *Information Digest 2005–2006 Edition*, NUREG-1350, vol. 17, July 2005, U.S. Nuclear Regulatory Commission, Division of Planning, Budget, and Analysis, Office of the Chief Financial Officer, http://www.nrc.gov/reading-rm/doc-collections/nuregs/staff/sr1350/v17/sr1350v17.pdf (accessed April 11, 2006)

- China, from 2.2 gigawatts in 2002 to 26.0 gigawatts by 2025
- India, from 2.9 gigawatts in 2002 to 15.3 gigawatts by 2025
- United States, from 98.9 gigawatts in 2002 to 102.7 gigawatts in 2025

Some countries were planning to cut nuclear capacity, largely for environmental reasons. Sweden, for example, had legislated a nuclear phase-out by 2010 after a referendum in 1980. However, because of economic considerations, only two of its nuclear reactors had been shut down by early 2006.

AGING NUCLEAR POWER PLANTS

At some point, the nuclear plants now operating worldwide may need to be retired. Most of them were designed to last about thirty years. Some plants showed serious levels of deterioration after as few as fifteen years; some have lasted well beyond their thirty-year expectations. There are three methods of retiring, or decommissioning, a reactor. *Safe enclosure*, or mothballing, involves removing the fuel from the plant, monitoring any radioactive contamination (which is usually very low or nonexistent), and guarding the structure to prevent anyone from entering until eventual dismantling and decontamination activities occur. *Entombment*, which was used at Chernobyl, involves permanently encasing the structure in a long-lived material such as concrete. This procedure allows the radioactive material to remain safely on-site. *Immediate dismantling* involves decontaminating and tearing down the facility within a few months or years. This method is initially more expensive than the other options but removes the long-term costs of monitoring both the structure and the radiation levels. It also frees the site for other uses, including the construction of another nuclear power plant. Nuclear power companies are also seeking alternative uses for nuclear shells, including the conversion of old nuclear plants to gas-fired plants.

TABLE 5.1

Nuclear generating units, 1953–2004

Year	Orders[a]	Cancelled orders[b]	Construction permits[c]	Low-power operating licenses[d]	Full-power operating licenses[e]	Shutdowns[f]	Operable units[g]
1953	1	0	0	0	0	0	0
1954	0	0	0	0	0	0	0
1955	3	0	1	0	0	0	0
1956	1	0	3	0	0	0	0
1957	2	0	1	1	1	0	1
1958	4	0	0	0	0	0	1
1959	4	0	3	1	1	0	2
1960	1	0	7	1	1	0	3
1961	0	0	0	0	0	0	3
1962	2	0	1	7	6	0	9
1963	4	0	1	3	2	0	11
1964	0	0	3	2	3	1	13
1965	7	0	1	0	0	0	13
1966	20	0	5	1	2	1	14
1967	29	0	14	3	3	2	15
1968	16	0	23	0	0	2	13
1969	9	0	7	4	4	0	17
1970	14	0	10	4	3	0	20
1971	21	0	4	5	2	0	22
1972	38	7	8	6	6	1	27
1973	42	0	14	12	15	0	42
1974	28	9	23	14	15	2	55
1975	4	13	9	3	2	0	57
1976	3	1	9	7	7	1	63
1977	4	10	15	4	4	0	67
1978	2	13	13	3	4	1	70
1979	0	6	2	0	0	1	69
1980	0	15	0	5	2	0	71
1981	0	9	0	3	4	0	75
1982	0	18	0	6	4	1	78
1983	0	6	0	3	3	0	81
1984	0	6	0	7	6	0	87
1985	0	2	0	7	9	0	96
1986	0	2	0	7	5	0	101
1987	0	0	0	6	8	2	107
1988	0	3	0	1	2	0	109
1989	0	0	0	3	4	2	111
1990	0	1	0	1	2	1	112
1991	0	0	0	0	0	1	111
1992	0	0	0	0	0	2	109
1993	0	0	0	1	1	0	110
1994	0	1	0	0	0	1	109
1995	0	2	0	1	0	0	109
1996	0	0	0	0	1	1	109
1997	0	0	0	0[h]	0[h]	2	107
1998	0	0	0	0	0	3	104
1999–2004	0	0	0	0	0	0	104
Total	**259**	**124**	**177**	**132**	**132**	**28**	—

[a]Placement of an order by a utility or government agency for a nuclear steam supply system.
[b]Cancellation by utilities of ordered units. Includes WNP 1 (Washington state); the licensee intends to request that the construction permit be cancelled. Does not include three units (Bellefonte 1 and 2 and Watts Bar 2) where construction has been stopped indefinitely.
[c]Issuance by regulatory authority of a permit, or equivalent permission, to begin construction. Numbers reflect permits issued in a given year, not extant permits.
[d]Issuance by regulatory authority of license, or equivalent permission, to conduct testing but not to operate at full power.
[e]Issuance by regulatory authority of full-power operating license, or equivalent permission. Units generally did not begin immediate operation.
[f]Ceased operation permanently.
[g]Total of nuclear generating units holding full-power licenses, or equivalent permission to operate, at the end of the year. Although Browns Ferry 1 was shut down in 1985, the unit has remained fully licensed and thus has continued to be counted as operable during the shutdown; in May 2002, the Tennessee Valley Authority announced its intention to have the unit resume operation in 2007.
[h]Under new regulations beginning in 1997, the terms "low-power operating licenses" and "full-power operating licenses" are no longer applicable; while no new licenses have been granted under the new regulations, applications were made in 2003 for three "early site permits."
—=Not applicable.
Note: For related information, see http://www.eia.doe.gov/fuelnuclear.html.

SOURCE: Adapted from "Table 9.1. Nuclear Generating Units, 1953–2004," in *Annual Energy Review 2004*, U.S. Department of Energy, Energy Information Administration, Office of Energy Markets and End Use, August 2005, http://www.eia.doe.gov/emeu/aer/pdf/aer.pdf (accessed April 5, 2006)

Paying for the closing, decontamination, or dismantling of nuclear plants has become an issue of intense public debate. The industry contends that the cost of decommissioning retired plants and handling radioactive wastes will continue to escalate, causing serious financial problems for electric utilities.

FIGURE 5.7

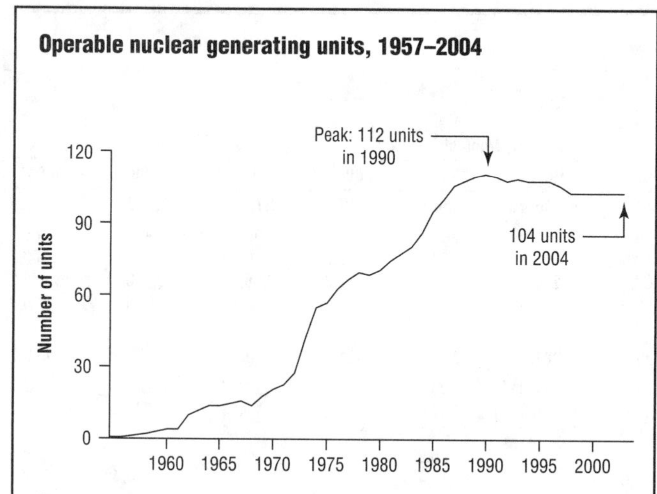

Notes: Operable units are units holding full-power operating license, or equivalent permission to operate. Data are at end of year.

SOURCE: Adapted from "Figure 9.1. Nuclear Generating Units: Operable Units, 1957–2004," in *Annual Energy Review 2004*, U.S. Department of Energy, Energy Information Administration, Office of Energy Markets and End Use, August 2005, http://www.eia.doe.gov/emeu/aer/pdf/aer.pdf (accessed April 5, 2006)

A NEW GENERATION OF NUCLEAR PLANTS

In 2001 the Generation IV International Forum, a group of nuclear nations, agreed to work together to create a new generation of nuclear reactors. Their intent was to develop new systems by 2030 that "present significant improvements in economics, safety and reliability and sustainability over currently operating reactor technologies" (Generation IV International Forum, http://gif.inel.gov/roadmap/). As of 2006 the members included Argentina, Brazil, Canada, the European Union, France, Japan, the Republic of Korea, South Africa, Switzerland, the United Kingdom, and the United States.

The group chose six nuclear power technologies for development:

- *Gas-Cooled Fast Reactor (GFR)*—This technology would provide a closed system, helium-cooled reactor that minimizes long-lived radioactive waste by recycling it.
- *Very-High-Temperature Reactor (VHTR)*—With core outlet temperatures reaching 1,000° Celsius (1,832° Fahrenheit), the VHTR would be used to produce hydrogen and to meet specific high-temperature heating needs of the petrochemical industry and others.
- *Supercritical-Water-Cooled Reactor (SWCR)*—SWCR is a high-temperature, high-pressure water-cooled reactor that increases efficiency by operating above the thermodynamic critical point of water. Water under high pressure does not boil and turn to steam; the thermodynamic critical point of water refers to the temperature and pressure at which the liquid state ceases to exist and the liquid and gaseous forms become indistinguishable. For water, this point is 374° Celsius (705° Fahrenheit).
- *Sodium-Cooled Fast Reactor (SFR)*—Intended for electricity generation, this system uses a sodium-cooled reactor and a closed fuel cycle to increase safety and reduce high-level radioactive waste.
- *Lead-Cooled Fast Reactor (LFR)*—Designed with small electricity grids and developing countries in mind, the LFR proposes a factory-built system that would need refueling only every fifteen to twenty years. It features a fast-spectrum lead or liquid metal-cooled reactor and a closed cycle for reducing waste.
- *Molten Salt Reactor (MSR)*—In this system, the fuel is a liquid mixture of sodium, zirconium, and uranium fluorides, which eliminates the need for fuel production. By-products can be recycled back into the fuel mixture.

Each of the new technologies builds on the knowledge gained during the three other periods of nuclear reactor development: Generation I—the experimental reactors developed in the 1950s and 1960s; Generation II—large, central-station nuclear power reactors, such as the 104 plants still operating in the United States, built in the 1970s and 1980s; and Generation III—the advanced light-water reactors built in the 1990s, primarily in East Asia, to meet that region's expanding electricity needs. In February 2005 the United States joined Canada, France, Japan, and the United Kingdom in an agreement to develop these six nuclear energy systems. In February 2006 the United States also signed an agreement with France and Japan to work together to develop sodium-cooled fast reactor systems.

NUCLEAR SAFETY

Safety has been an issue from the beginning of the industry. For example, some plant sites, especially those near earthquake fault lines, have raised serious questions for governments, industry leaders, and environmentalists. Plants can be shut down for a variety of reasons. For example, in 1987 the Nuclear Regulatory Commission shut down the Peach Bottom nuclear plant in Delta, Pennsylvania, because control-room operators were found sleeping on duty. In February 2005 the Kewaunee power station in Carlton, Wisconsin, was shut down after it was discovered that emergency shutdown systems could be compromised by flooding from water storage tanks located in an adjacent part of the plant. Although no flood actually occurred, operations at the plant were halted until the shutdown systems could be protected from a potential water threat. Most safety questions, however, have been focused by major accidents since the late 1970s.

TABLE 5.2

World net electric power generation, 1980, 1990, and 2003

[Billion kilowatt-hours]

	Fossil fuels			Nuclear electric power			Hydroelectric power[a]			Total[b]		
Region and country	**1980**	**1990**	**2003[P]**	**1980**	**1990**	**2003[P]**	**1980**	**1990**	**2003[P]**	**1980**	**1990**	**2003[P]**
North America	**1,880.1**	**2,292.0**	**3,087.4**	**287.0**	**648.9**	**844.5**	**546.9**	**606.5**	**619.4**	**2,721.6**	**3,623.9**	**4,659.6**
Canada	79.8	101.9	154.5	35.9	69.2	70.8	251.0	293.9	332.5	367.9	468.6	566.3
Mexico	46.0	85.7	173.3	0.0	2.8	10.0	16.7	23.2	19.7	63.6	116.6	209.2
United States	1,753.8	2,103.8	2,758.6	251.1	576.9	763.7	279.2	289.4	267.3	2,289.6	3,038.0	3,883.2
Other	0.5	0.7	0.9	0.0	0.0	0.0	0.0	0.0	0.0	0.5	0.7	0.9
Central and South America	**99.8**	**114.8**	**221.1**	**2.2**	**9.0**	**20.4**	**201.5**	**365.1[R]**	**561.4**	**308.2**	**497.2[R]**	**828.7**
Argentina	22.2	20.9	40.3	2.2	7.0	7.0	17.3	20.2	33.4	41.8	48.3	83.3
Brazil	7.5	8.1	26.7	0.0	1.9	13.4	128.4	204.6	302.9	138.3	219.6	359.2
Paraguay	R, s	s	s	0.0	0.0	0.0	0.7	27.2[R]	51.2	0.8	27.2[R]	51.3
Venezuela	17.6	21.0	27.9	0.0	0.0	0.0	14.4	36.6	59.6	32.0	57.6	87.4
Other	52.4	64.8	126.3	0.0	0.0	0.0	40.6	76.4[R]	114.3	95.3	**144.4[R]**	247.4
Western Europe	**1,180.1**	**1,171.8[R]**	**1,496.6**	**219.2**	**711.3[R]**	**883.0**	**431.7**	**453.4**	**483.7**	**1,844.5**	**2,355.9[R]**	**2,973.7**
Belgium	38.3	25.0	31.9	11.9	40.6	45.0	0.3	0.3	0.2	50.8	66.5	78.8
Finland	22.0	22.8	38.6	6.6	18.3	21.6	10.1	10.8	9.3	38.7	51.8	79.6
France	118.0	44.3	55.2	63.4	298.4	419.0	68.3	52.8	58.6	250.8	397.6	536.9
Germany	390.3	358.9	349.0	55.6	145.1	157.0	18.8	17.2	20.8	469.9	526.0	558.1
Italy	125.5	167.5	222.4	2.1	0.0	0.0	45.0	31.3	36.3	176.4	202.1	270.1
Netherlands	58.0	63.2	82.0	3.9	3.3	3.8	0.0	0.1	0.1	62.9	67.7[R]	91.0
Norway	0.1	0.2	0.5	0.0	0.0	0.0	82.7	119.9	104.4	82.9	120.4	105.6
Spain	74.5	66.5	131.7	5.2	51.6	58.8	29.2	25.2	40.6	109.2	143.9	247.3
Sweden	10.1	3.2	8.7	25.3	64.8	62.2	58.1	71.8	52.5	94.3	141.5	127.9
Switzerland	0.9	0.6	1.1	12.9	22.4	26.1	32.5	29.5	34.4	46.4	53.0	63.4
Turkey	12.0	32.3	98.5	0.0	0.0	0.0	11.2	22.9	35.0	23.3	55.2	133.6
United Kingdom	228.9	230.0	273.7	32.3	62.5[R]	84.5	3.9	5.1	4.5	265.1	299.0[R]	369.9
Other	101.4	157.2[R]	203.5	0.0	4.4	5.0	71.7	66.6	87.0	173.8	231.1[R]	311.5
Eastern Europe and former U.S.S.R.	**1,309.3**	**1,471.5**	**1,090.7**	**83.2**	**251.3**	**305.0**	**210.4[R]**	**251.7[R]**	**267.8**	**1,603.2[R]**	**1,974.8[R]**	**1,668.2**
Czech Republic	—	—	51.6	—	—	24.6	—	—	1.4	—	—	78.2
Kazakhstan	—	—	50.9	—	—	0.0	—	—	9.5	—	—	60.3
Poland	111.1	125.0	138.6	0.0	0.0	0.0	2.3[R]	1.4[R]	1.7	113.8[R]	126.7[R]	141.2
Romania	51.4	49.7	31.3	0.0	0.0	4.5	12.5	10.9	15.9	63.9	60.6	51.7
Russia	—	—	571.5	—	—	138.4	—	—	170.6	—	—	883.3
Ukraine	—	—	82.9	—	—	76.7	—	—	10.3	—	—	169.9
Other	1,146.8	1,296.7	164.0	83.2	251.3	60.8	195.5	239.4	58.6	1,425.6	1,787.5	283.5
Middle East	**82.8**	**217.3[R]**	**483.6**	**0.0**	**0.0**	**0.0**	**9.6**	**12.5**	**22.5**	**92.4**	**229.9[R]**	**506.2**
Iran	15.7	49.8	132.3	0.0	0.0	0.0	5.6	6.0	10.0	21.3	55.9	142.3
Saudi Arabia	20.5	64.9	145.1	0.0	0.0	0.0	0.0	0.0	0.0	20.5	64.9	145.1
Other	46.6	102.6[R]	206.2	0.0	0.0	0.0	4.1	6.5	12.5	50.7	109.1[R]	218.7
Africa	**129.1**	**243.7[R]**	**372.4**	**0.0**	**8.4**	**12.7**	**60.1**	**54.8[R]**	**85.0**	**189.2**	**307.4[R]**	**471.1**
Egypt	8.6	31.5	71.4	0.0	0.0	0.0	9.7	9.9	12.6	18.3	41.4	84.3
South Africa	92.1	146.6	202.2	0.0	8.4	12.7	1.0	1.0	0.8	93.1	156.0	215.9
Other	28.4	65.6	98.8	0.0	0.0	0.0	49.4	43.9[R]	71.6	77.8	110.0[R]	170.9
Asia and Oceania	**907.7**	**1,626.8**	**3,613.0**	**92.7**	**279.9**	**457.6**	**262.7[R]**	**404.1[R]**	**606.0**	**1,268.0[R]**	**2,337.6[R]**	**4,736.6**
Australia	74.5	131.8	197.1	0.0	0.0	0.0	12.8	14.0	15.9	87.7	146.4	215.8
China	227.9	465.2	1,484.2	0.0	0.0	41.7	57.6	125.1	278.5	285.5	590.3	1,806.8
India	69.7	198.9	467.7	3.0	5.6	16.4	46.5	70.9	68.5	119.3	275.5	556.8
Indonesia	10.6	35.3	94.8	0.0	0.0	0.0	2.2[R]	6.7[R]	8.4	12.8[R]	43.0[R]	109.5
Japan	381.6	524.0	648.3	78.6	192.2	237.2	87.8	88.4	104.1	549.1	821.8[R]	1,017.5
South Korea	29.8	45.5	197.4	3.3	50.2	123.2	1.5	4.6	4.8	34.6	100.4	326.2
Taiwan	31.3	43.6	121.8	7.8	31.6	37.4	2.9	8.2	6.8	42.0	83.3	166.0
Thailand	12.3	38.7	105.0	0.0	0.0	0.0	1.3	4.9	7.2	13.6	43.7	114.7
Other	70.1	143.8	296.7	s	0.4	1.8	50.0[R]	81.2[R]	111.6	123.5[R]	233.2[R]	423.4
World	**5,588.8**	**7,138.0[R]**	**10,364.8**	**684.4**	**1,908.8[R]**	**2,523.1**	**1,722.8[R]**	**2,148.2[R]**	**2,645.8**	**8,027.1[R]**	**11,326.6[R]**	**15,843.9**

[a]Excludes pumped storage, except for the United States.
[b]Wood, waste, geothermal, solar, wind, batteries, chemicals, hydrogen, pitch, purchased steam, sulfur, and miscellaneous technologies are included in total.
R=Revised. P=Preliminary. —=Not applicable. s=Less than 0.05 billion kilowatthours.
Note: Totals may not equal sum of components due to independent rounding. Web page: For related information, see http://www.eia.doe.gov/international.

SOURCE: "Table 11.16. World Net Generation of Electricity by Type, 1980, 1990, and 2003 (Billion Kilowatthours)," in *Annual Energy Review 2004*, U.S. Department of Energy, Energy Information Administration, Office of Energy Markets and End Use, August 2005, http://www.eia.doe.gov/emeu/aer/pdf/aer.pdf (accessed April 5, 2006)

Three Mile Island

On March 28, 1979, the Three Mile Island nuclear facility near Harrisburg, Pennsylvania, was the site of the worst nuclear accident in U.S. history. Information released several years after the accident revealed that Unit 2, one of the reactors operating at the site, came much closer to meltdown than either the Nuclear Regulatory Commission or the industry had previously indicated. Temperatures inside the reactor, which were first said to have reached 3,500° Fahrenheit, are now known to have reached at least 4,800° Fahrenheit. The temperature needed to melt uranium dioxide fuel is 5,080° Fahrenheit. When meltdown occurs, an uncontrolled explosion may result, unlike the controlled nuclear reaction of normal operation.

The emergency system at Three Mile Island was designed to dump water on the hot core of the reactor and spray water into the reactor building to stop the production of steam. During the accident, however, the valves leading to the emergency water pumps closed. Another valve was stuck in the open position, drawing water away from the core, which then became partially uncovered and began to melt.

The accident resulted in no deaths or injuries to plant workers or the nearby community. On average, area residents were exposed to less radiation than that of a chest X-ray. Nevertheless, the incident raised concerns about nuclear safety, which resulted in more rigorous safety standards in the nuclear power industry and at the Nuclear Regulatory Commission. Antinuclear sentiment was fueled as well, heightening Americans' wariness of nuclear power as an energy source.

The nuclear reactor in Unit 2 at Three Mile Island has been in monitored storage since it underwent cleanup. Operation of the reactor in Unit 1 resumed in 1985.

Chernobyl

On April 26, 1986, the most serious nuclear accident ever occurred at Chernobyl, a nuclear plant in what is now Ukraine (then part of the Soviet Union). At least thirty-one people died and hundreds were injured when one of the four reactors exploded during a badly run test. Millions of people were exposed to some levels of radiation when radioactive particles were released into the atmosphere. About 350,000 people were eventually evacuated from the area, including residents of what are now the neighboring nations of Belarus and the Russian Federation, who were endangered by fallout carried by prevailing winds.

The cleanup was a huge project. Helicopters dropped tons of limestone, sand, clay, lead, and boron on the smoldering reactor to stop the radiation leakage and reduce the heat. Workers built a giant steel and cement sarcophagus to entomb the remains of the reactor and contain the radioactive waste.

According to the International Atomic Energy Agency (*Chernobyl's Legacy: Health, Environmental, and Socio-Economic Impacts*, 2005, http://www.iaea.org/Publications/Booklets/Chernobyl/chernobyl.pdf), about 1,000 people involved in the initial cleanup, including emergency workers and the military, received high doses of radiation. Eventually more than 600,000 people would be involved in decontamination and containment activities. The long-term effects of whatever exposure they received is being monitored. The most measurable effect of the radiation, according to the agency, has been the incidence of thyroid cancer in children. At least 4,000 children—considered a very large number, given the size of the population—have been treated for the cancer, almost all successfully.

While some of the evacuated land has been declared fit for habitation again, several areas that received heavy concentrations of radiation are expected to be closed for decades. The three remaining nuclear reactors at Chernobyl continue to operate. The site itself is off-limits to all but official personnel.

An International Agreement on Safety

In September 1994 forty nations, including the United States, signed the International Convention on Nuclear Safety, an agreement that requires them to shut down nuclear power plants if necessary safety measures cannot be guaranteed. The agreement applies to land-based civil nuclear power plants and seeks to avert accidents like the 1986 explosion at Chernobyl. Ukraine, which inherited the Chernobyl plant after the collapse of the Soviet Union, signed the agreement. Signers must submit reports on atomic installations and, if necessary, make improvements to upgrade safety at the sites. Neighboring countries may call for an urgent study if they are concerned about a reactor's safety and the potential fallout that could affect their own population or crops.

RADIOACTIVE WASTE

Working in a laboratory in Chicago, Illinois, in 1942, physicist Enrico Fermi (1901–54) assembled enough uranium to cause a nuclear fission reaction. His discovery transformed both warfare and energy production. The experiment also produced a small packet of radioactive waste material that may remain dangerous for 100,000 years. That first radioactive waste lies buried under a foot of concrete and two feet of dirt on a hillside in Illinois.

Radioactive waste is produced at all stages of the nuclear fuel cycle, from the initial mining of the uranium to the final disposal of the spent fuel from the reactor. *Radioactive waste* is a term that encompasses a broad range of material with widely varying characteristics. Some is barely radioactive and safe to handle, while other types are intensely hot and highly radioactive. Some waste decays to safe levels of radioactivity in a matter of days or weeks, while other types will remain dangerous for thousands of years. The U.S. Department of Energy and the Nuclear Regulatory Commission have defined the major types of radioactive waste.

Uranium Mill Tailings

Uranium mill tailings are sandlike wastes produced in uranium refining operations. Although they emit low levels of radiation, their large volumes (10 million to 15 million tons annually) pose a hazard, particularly from radon emissions and groundwater contamination. The dangers of uranium mill tailings were not realized until the early 1970s, so many miners and residents of the western United States were exposed to them. Cancer incidences are high among miners who worked prior to the 1970s. In addition, the World Information Source on Energy reports that those residing near uranium mills have shown increased risks of leukemia, lung and renal cancer, and birth defects (http://www.wise-uranium.org/uhr.html).

Low-Level Waste

Low-level waste, which contains lesser levels of radioactivity, includes trash (such as wiping rags, swabs, and syringes), contaminated clothing (such as shoe covers and protective gloves), and hardware (such as luminous dials, filters, and tools). This waste comes from nuclear reactors, industrial users, government users (but not nuclear weapons sites), research universities, and medical facilities. In general, low-level waste decays relatively quickly (in ten to one hundred years).

High-Level Waste

Spent nuclear fuel (used reactor fuel) is high-level radioactive waste. Uranium fuel can be used for twelve to eighteen months, after which it is no longer as efficient in splitting its atoms and producing the heat needed to generate electricity. It must be removed from the reactor and replaced with fresh fuel. Some of the spent fuel is reprocessed to recover the usable uranium and plutonium, but the radioactive material that remains is dangerous for thousands of years.

Transuranic (TRU) Wastes

Transuranic wastes are eleven man-made radioactive elements with atomic numbers greater than that of uranium (ninety-two) and therefore beyond (*trans-*) uranium (*-uranic*) on the periodic table of the elements. Their half-lives—the time it takes for half the radioisotopes present in a sample to decay to nonradioactive elements—are thousands of years. They are found in trash produced mainly by nuclear weapons plants and are therefore part of the nuclear waste problem but not directly the concern of nuclear power utilities.

Mixed Waste

Mixed waste is high-level, low-level, or transuranic waste that also contains hazardous nonradioactive waste. Such waste poses serious institutional problems, because the radioactive portion is regulated by the Department of Energy or the Nuclear Regulatory Commission under the Atomic Energy Act (PL 83-703), while the Environmental Protection Agency regulates the nonradioactive elements under the Resource Conservation and Recovery Act (PL 95-510).

RADIOACTIVE WASTE DISPOSAL

Disposing of radioactive waste is unquestionably one of the major problems associated with the development of nuclear power; radioactive waste is also a by-product of nuclear weapons plants, hospitals, and scientific research. Although federal policy is based on the assumption that radioactive waste can be disposed of safely, new storage and disposal facilities for all types of radioactive waste have frequently been delayed or blocked by concerns about safety, health, and the environment.

The highly toxic wastes must be isolated from the environment until the radioactivity decays to a safe level. In the case of plutonium, for example, the half-life (or, the point at which half of the radioactive nuclei have decayed) is 26,000 years. At that rate, it will take at least 100,000 years before radioactive plutonium is no longer dangerous. Any facilities built to store such materials must last at least that long.

Regulation of Radioactive Waste Disposal

As of January 2006 the Nuclear Regulatory Commission had entered into agreements with thirty-three states to regulate the management, storage, and disposal of certain nuclear waste within their borders.

Disposal of Uranium Mill Tailings

Mill tailings are usually deposited in large piles next to the mill that processed the ore. In 1978 Congress passed the Uranium Mill Tailing Radiation Control Act (PL 95-604), which requires mill owners to follow Environmental Protection Agency standards for cleanup of uranium and thorium after milling operations have permanently closed. The companies must cover the mill tailings to control the release of radon gas for 1,000 years. Figure 5.8 shows the locations of uranium mill tailings disposal sites, which are close to deposits of uranium ore.

FIGURE 5.8

Uranium mill tailings disposal sites

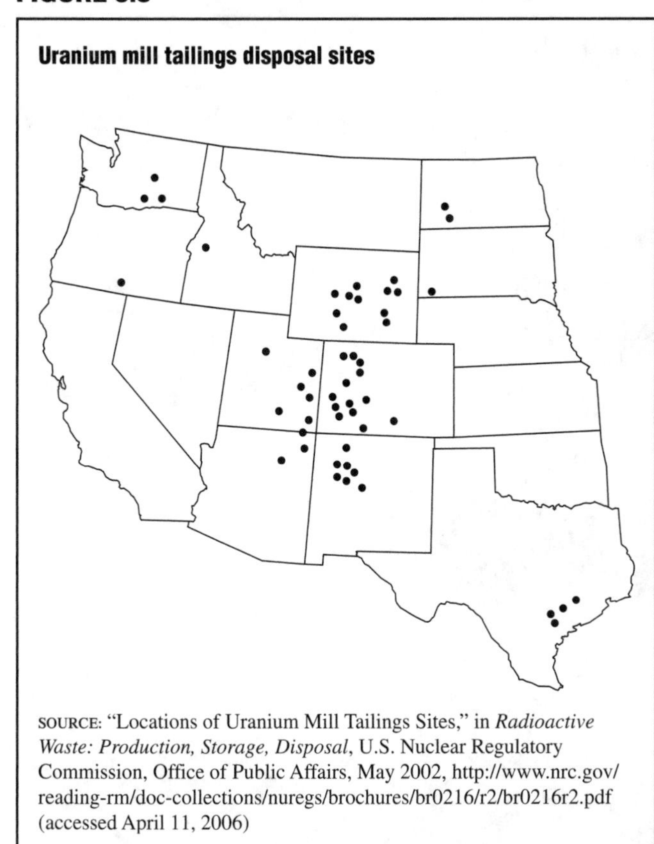

SOURCE: "Locations of Uranium Mill Tailings Sites," in *Radioactive Waste: Production, Storage, Disposal*, U.S. Nuclear Regulatory Commission, Office of Public Affairs, May 2002, http://www.nrc.gov/reading-rm/doc-collections/nuregs/brochures/br0216/r2/br0216r2.pdf (accessed April 11, 2006)

Disposal of Low-level Waste

According to the Nuclear Regulatory Commission, as of March 2005 three low-level licensed nuclear waste facilities were operating in the United States (*Information Digest, 2005–2006 Edition*): in Barnwell, South Carolina, in Richland, Washington, and in a remote part of Utah, where Envirocare of Utah accepts large amounts of mill tailings and low-level waste, such as contaminated soil or debris from demolished buildings. Four low-level radioactive waste facilities have been closed: in West Valley, New York (closed 1975); in Morehead/Maxey Flats, Kentucky (closed 1977); in Sheffield, Illinois (closed 1978); and in Beatty, Nevada (closed 1993).

The design of a low-level waste facility is shown in Figure 5.9. Wastes are buried in shallow underground sites in specially designed canisters. Underground storage may or may not include protection by concrete vaults.

The Low-Level Radioactive Waste Policy Amendments Act of 1985 (PL 99-240) encouraged states to enter into compacts, which are legal agreements among states for low-level radioactive waste disposal. Figure 5.10 shows these compacts. Although each compact is responsible for the development of disposal capacity for the low-level waste generated within the compact, new disposal sites have yet to be built. As Figure 5.10 shows, two of the three operational low-level sites are located in the Northwest Compact and one is in the Atlantic Compact. Nuclear power facilities located in compacts without low-level waste disposal sites must petition the compact to export their low-level radioactive waste to one of the three operating disposal sites.

Disposal of High-Level Waste

A major step toward shifting the responsibility for disposal of high-level radioactive wastes (spent fuel) from the nuclear power industry to the federal government was taken in 1982, when Congress passed the Nuclear Waste Policy Act (PL 97-425). It established national policy; set a detailed timetable for the disposal and management of high-level nuclear waste; and authorized construction of the first high-level nuclear waste repository. A 1987 amendment to the Nuclear Waste Policy Act directed investigation of Yucca Mountain in Nevada as a potential site. In early 2006 no long-term, permanent disposal repository for high-level waste existed. While waiting for the development of the Yucca Mountain site, spent fuel was being stored at away-from-reactor storage facilities, such as the General Electric Company facility in Morris, Illinois, or at the nuclear power plants that generated the waste.

In July 2002 the Congress approved the Yucca Mountain site. The State of Nevada challenged the decision, but in 2004 federal courts dismissed the challenge. In April 2006 the Secretary of Energy announced proposals to facilitate licensing and construction of the repository, which was expected to take about five years.

Figure 5.11 shows the location of the proposed disposal facility at Yucca Mountain. Aboveground structures for handling and packaging nuclear waste would cover approximately 400 acres and be surrounded by a three-mile buffer zone. Underground about 1,400 acres would be mined, with tunnels leading to areas where sealed metal containers would be placed. This type of deep geologic disposal is widely considered by governments, scientists, and engineers to be the best option for isolating highly radioactive waste.

The repository would be designed to contain radioactive material by using layers of man-made and natural barriers. Regulations require that a repository isolate waste until the radiation decays to a level that is about the same as that from a natural underground uranium deposit. This decay time was originally estimated at be about 10,000 years, but the National Academy of Sciences has more recently recommended a higher standard of about 300,000 years.

After the repository has been filled to capacity, regulations require the Department of Energy to keep the facility open and to monitor it for at least fifty years from the fill date. Eventually, the repository shafts would be filled with rock and earth and sealed. At ground level,

FIGURE 5.9

Low-level waste disposal site

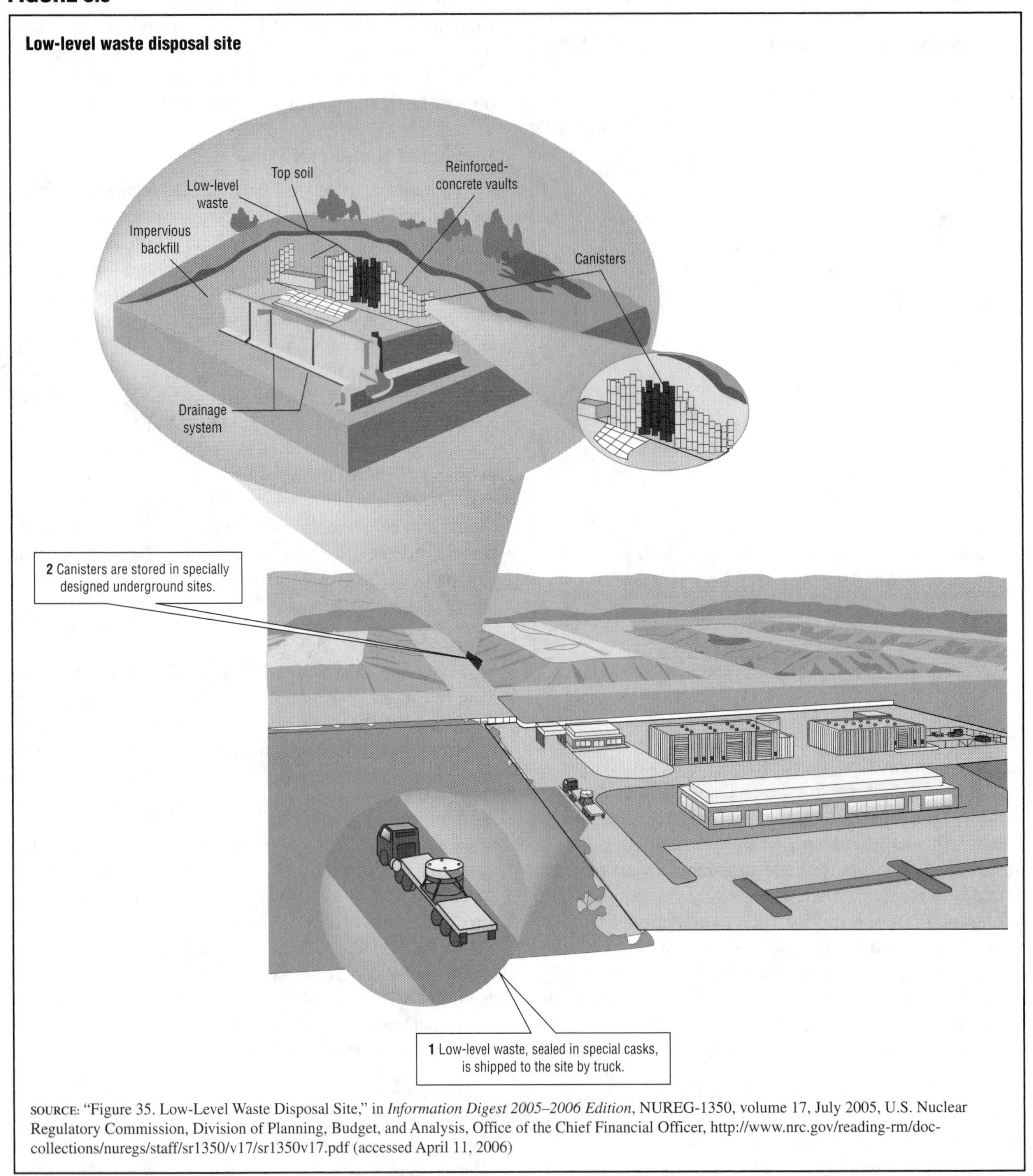

SOURCE: "Figure 35. Low-Level Waste Disposal Site," in *Information Digest 2005–2006 Edition*, NUREG-1350, volume 17, July 2005, U.S. Nuclear Regulatory Commission, Division of Planning, Budget, and Analysis, Office of the Chief Financial Officer, http://www.nrc.gov/reading-rm/doc-collections/nuregs/staff/sr1350/v17/sr1350v17.pdf (accessed April 11, 2006)

facilities would be removed and, as much as possible, the site returned to its original condition.

Scientists expect some of the man-made barriers in a repository to break down over thousands of years. Once that happens, natural barriers will be counted on to stop or slow the movement of radiation particles. The most likely way for particles to reach humans and the environment would be through water, which is why the low water tables at Yucca Mountain are so crucial. In addition, Yucca Mountain contains minerals called zeolites that would stick to the particles and slow their movement throughout the environment.

The long delay in providing disposal sites for spent nuclear fuel has been expensive for the industry. Several aging power plants are being maintained at a cost of $20

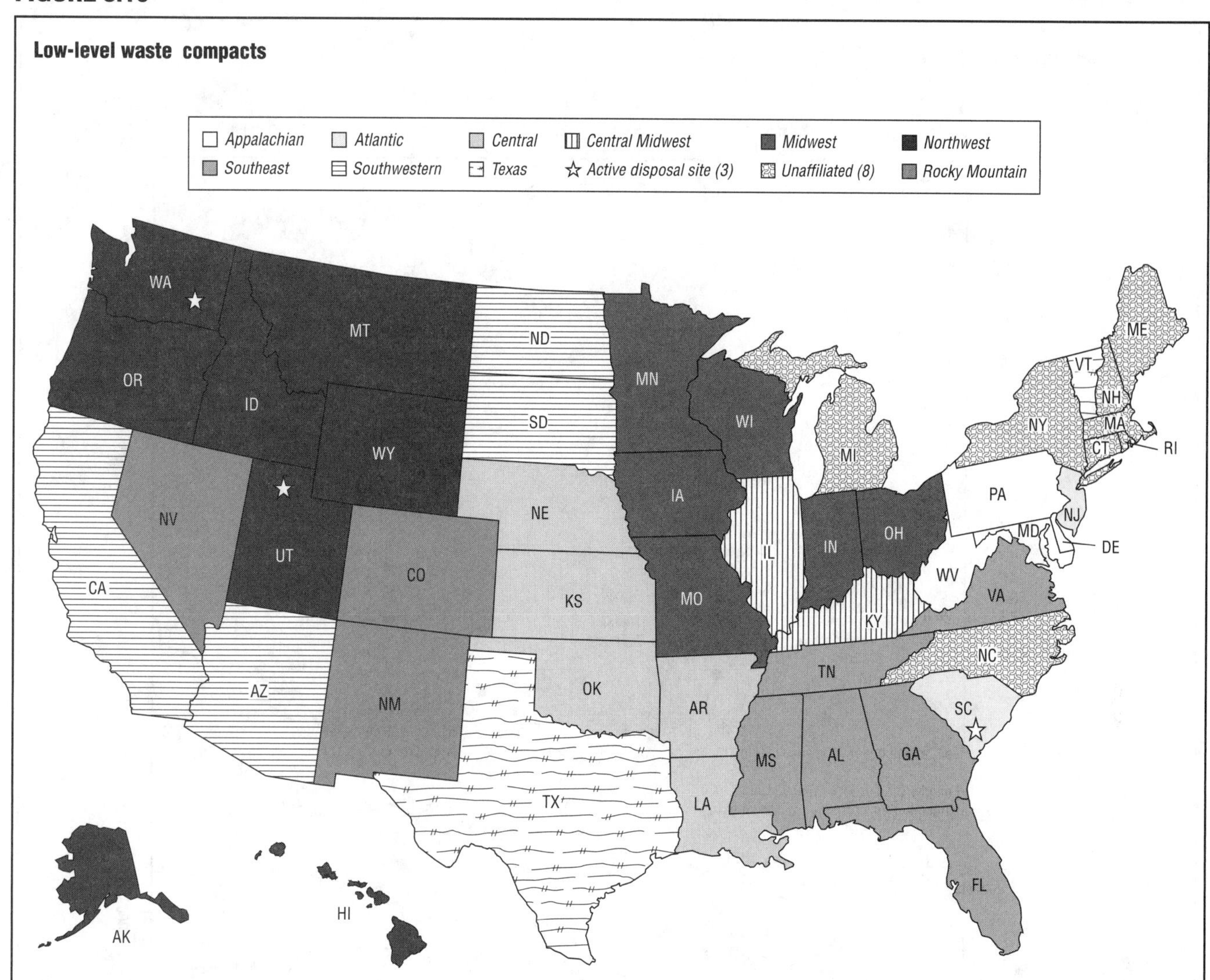

FIGURE 5.10

Low-level waste compacts

Note: Data as of March 2005. Puerto Rico is unaffiliated. There are three active, licensed low-level waste disposal facilities located in agreement states.
Barnwell, located in Barnwell, South Carolina—Currently, Barnwell accepts waste from all U.S. generators except those in Rocky Mountain and northwest compacts. Beginning in 2008, Barnwell will only accept waste from the Atlantic compact states (Connecticut, New Jersey, and South Carolina). Barnwell is licensed by the state of South Carolina to receive waste in classes A–C.
Hanford, located in Hanford, Washington—Hanford accepts waste from the northwest and Rocky Mountain compacts. Hanford is licensed by the state of Washington to receive waste in classes A–C.
Envirocare, located in Clive, Utah—Envirocare accepts waste from all regions of the United States. Envirocare is licensed by the state of Utah for class A waste only.

SOURCE: "Figure 36. U.S. Low-Level Waste Compacts," in *Information Digest 2005–2006 Edition*, NUREG-1350, volume 17, July 2005, U.S. Nuclear Regulatory Commission, Division of Planning, Budget, and Analysis, Office of the Chief Financial Officer, http://www.nrc.gov/reading-rm/doc-collections/nuregs/staff/sr1350/v17/sr1350v17.pdf (accessed April 11, 2006)

million per reactor per year simply because there is no place to send the waste once the plants are decommissioned. Table 5.3 shows that as of 2002 more than 47,000 metric tonnes (tons) of nuclear uranium waste were sitting in spent fuel pools at the 104 operating and nineteen permanently closed nuclear power plants.

Disposal of Transuranic Waste

The Waste Isolation Pilot Plant is the first disposal facility licensed to dispose of transuranic waste, and does not accept waste from commercial sources or electrical power plants. It opened in March 1999 in the desert in southeastern New Mexico. Its disposal rooms were mined 2,150 feet underground in a 2,000-foot-thick salt formation that has been stable for more than 200 million years. Beginning with one or two shipments per week, operations grew over the years, expanding to an all-time high of 33 shipments received during a single week in February 2006. In September 2006 the plant received its 5,000th shipment of transuranic waste and announced that thirteen sites around the country, most notably the Rocky Flats Environmental Technology site in Colorado, had been completely cleaned-up by that time.

FIGURE 5.11

Yucca mountain radioactive waste site

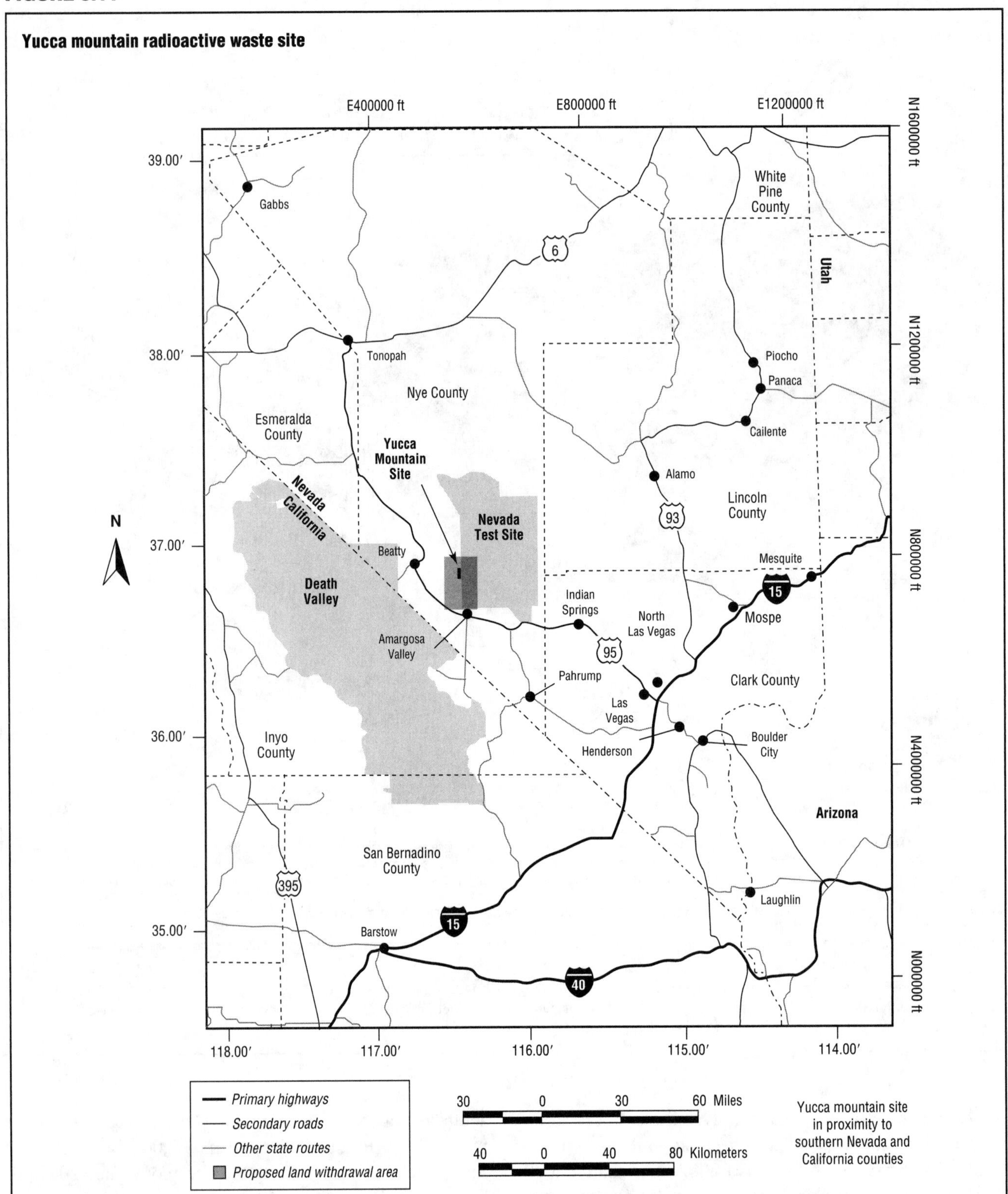

SOURCE: "Figure 1-5. Map Showing the Location of Yucca Mountain in Relation to Major Highways; Surrounding Counties, Cities, and Towns in Nevada and California; the Nevada Test Site; and Death Valley National Park," in *Yucca Mountain Science and Engineering Report: Technical Information Supporting Site Recommendation Consideration Revision 1*, U.S. Department of Energy, February 2002, http://www.ocrwm.doe.gov/documents/ser_b/figures/chap1/f01-05.htm (accessed April 11, 2006)

TABLE 5.3

Total commercial spent nuclear fuel discharges, 1968–2002

Reactor type	Stored at reactor sites	Stored at away-from-reactor facilities	Total
	Number of assemblies		
Boiling-water reactor	90,398	2,957	93,355
Pressurized-water reactor	69,800	491	70,291
High-temperature gas cooled reactor	1,464	744	2,208
Total	**161,662**	**4,192**	**165,854**
	Metric tonnes of uranium (MTU)		
Boiling-water reactor	16,153.60	554	16,707.60
Pressurized-water reactor	30,099.00	192.6	30,291.60
High-temperature gas cooled reactor	15.4	8.8	24.2
Total	**46,268.00**	**755.4**	**47,023.40**

Notes: A number of assemblies discharged prior to 1972, which were reprocessed, are not included in this table (no data is available for assemblies reprocessed before 1972). Totals may not equal sum of components because of independent rounding.

SOURCE: "Table 1. Total U.S. Commercial Spent Nuclear Fuel Discharges, 1968–2002," in *Energy Information Administration Spent Nuclear Fuel*, U.S. Department of Energy, Energy Information Administration, October 1, 2004, http://www.eia.doe.gov/cneaf/nuclear/spent_fuel/ussnfdata.html (accessed April 11, 2006)

Disposal of Mixed Waste

Several commercial facilities in the United States accept mixed waste. However, only the Envirocare facility in Clive, Utah, is permitted to accept solid mixed waste. It stores the waste in aboveground, capped embankments designed to last at least 1,000 years. The storage facility is located in an isolated area twenty miles from the nearest water supply and more than forty miles from the nearest populated area. Containment efforts are supported by area soil and weather conditions, as well, with little permeability in the clay soil and fewer than six inches of rain per year.

Sites in Florida, Tennessee, and Texas offer storage for liquid or sludge materials or use thermal or chemical processes to neutralize them. Among them, Diversified Scientific Services, Inc. in Kingston, Tennessee, generates electrical power from liquid mixed waste through an industrial boiler system, and Texas-based NSSI provides drum storage.

CHAPTER 6

RENEWABLE ENERGY

WHAT IS RENEWABLE ENERGY?

Imagine energy sources that use no oil, produce no pollution, create no radioactive waste, cannot be affected by political events and cartels, and yet are economical. Although it sounds impossible, some experts claim that technological advances could make wide use of renewable energy sources possible within a few decades. They may become substantially better energy sources than fossil fuels and nuclear power.

Renewable energy is naturally regenerated. Sources include the sun (solar), wind, water (hydropower), vegetation (biomass), and the heat of the earth (geothermal).

A HISTORICAL PERSPECTIVE

Before the eighteenth century, most energy came from renewable sources. People burned wood for heat, used sails to harness the wind and propel boats, and installed waterwheels on streams to run mills that ground grain. The large-scale shift to nonrenewable energy sources began in the 1700s with the Industrial Revolution, a period marked by the rise of factories, first in Europe and then in North America. As demand for energy grew, coal replaced wood as the main fuel. Coal was the most efficient fuel for the steam engine, one of the most important inventions of the Industrial Revolution.

Until the early 1970s most Americans were unconcerned about the sources of the nation's energy. Supplies of coal and oil, which together provided more than 90% of U.S. energy, were believed to be plentiful. The decades preceding the 1970s were characterized by cheap gasoline and little public discussion of energy conservation.

That carefree approach to energy consumption ended in the 1970s. An oil crisis, caused in part by the devaluation of the dollar, but largely by an oil embargo by the Organization of Petroleum Exporting Countries, made Americans acutely aware of their dependence on foreign energy. Throughout the United States, people waited in line to fill their gas tanks—in some places gasoline was rationed—and lower heat settings for offices and homes were encouraged. In a country where mobility and convenience were highly valued, the oil crisis was a shock to the system. Developing alternative sources of energy to supplement and perhaps eventually replace fossil fuels became suddenly important. As a result, the administration of President Jimmy Carter (1977–1981) encouraged federal funding for research into alternative energy sources.

In 1978 Congress passed the Public Utilities Regulatory Policies Act (PL 95-617), which was designed to help the struggling alternative energy industry. The act exempted small alternative producers from state and federal utility regulations and required existing local utilities to buy electricity from them. The renewable energy industries grew rapidly, gaining experience, improving technologies, and lowering costs. This law was the single most important factor in the development of the commercial renewable energy market.

In the 1980s President Ronald Reagan favored private-sector financing, so he proposed the reduction or elimination of federal expenditures for alternative energy sources. Although federal funds were severely cut, the U.S. Department of Energy continued to support some research and development. President Bill Clinton's administration reemphasized the importance of renewable energy and increased funding in several areas. The administration of President George W. Bush supported funding for research and development of renewable technologies and tax credits for the purchase of hybrid and alternative-fuel cars.

DOMESTIC RENEWABLE ENERGY USAGE

In 2004 the United States consumed approximately 6.1 quadrillion Btu of renewable energy, about 6% of the nation's total energy consumption. (See Table 6.1 and

TABLE 6.1

Energy consumption by source, selected years 1949–2004

[Quadrillion Btu]

	Fossil fuels						Renewable energy[a]							
Year	Coal	Coal coke net imports	Natural gas[b]	Petroleum[c, d]	Total	Nuclear electric power	Conventional hydroelectric power	Wood, waste, alcohol[d, e]	Geothermal	Solar	Wind	Total	Electricity net imports	Total[d]
1949	11.981	−0.007	5.145	11.883	29.002	0.000	1.425	1.549	NA	NA	NA	2.974	0.005	31.982
1950	12.347	0.001	5.968	13.315	31.632	0.000	1.415	1.562	NA	NA	NA	2.978	0.006	34.616
1955	11.167	−0.010	8.998	17.255	37.410	0.000	1.360	1.424	NA	NA	NA	2.784	0.014	40.208
1960	9.838	−0.006	12.385	19.919	42.137	0.006	1.608	1.320	0.001	NA	NA	2.929	0.015	45.087
1965	11.581	−0.018	15.769	23.246	50.577	0.043	2.059	1.335	0.004	NA	NA	3.398	s	54.017
1970	12.265	−0.058	21.795	29.521	63.522	0.239	2.634	1.431	0.011	NA	NA	4.076	0.007	67.844
1972	12.077	−0.026	22.698	32.947	67.696	0.584	2.864	1.503	0.031	NA	NA	4.398	0.026	72.704
1974	12.663	0.056	21.732	33.455	67.906	1.272	3.177	1.540	0.053	NA	NA	4.769	0.043	73.991
1976	13.584	s	20.345	35.175	69.104	2.111	2.976	1.713	0.078	NA	NA	4.768	0.029	76.012
1978	13.766	0.125	20.000	37.965	71.856	3.024	2.937	2.038	0.064	NA	NA	5.039	0.067	79.986
1980	15.423	−0.035	20.394	34.202	69.984	2.739	2.900	2.485	0.110	NA	NA	5.494	0.071	78.289
1982	15.322	−0.022	18.505	30.232	64.037	3.131	3.266	2.615	0.105	NA	NA	5.985	0.100	73.253
1984	17.071	−0.011	18.507	31.051	66.617	3.553	3.386	2.880	0.165	s	s	6.431	0.135	76.736
1986	17.260	−0.017	16.708	32.196	66.148	4.380	3.071	2.841	0.219	s	s	6.132	0.122	76.782
1988	18.846	0.040	18.552	34.222	71.660	5.587	2.334	2.937	0.217	s	s	5.489	0.108	82.844
1990	19.173	0.005	19.730	33.553	72.460	6.104	3.046	2.662	0.336	0.060	0.029	6.133	0.008	84.704[R]
1992	19.122	0.035	20.835	33.527	73.519	6.479	2.617	2.847	0.349	0.064	0.030	5.907	0.087	85.992[R]
1994	19.909	0.058	21.842	34.670	76.480	6.694	2.683	2.939	0.338	0.069	0.036	6.065	0.153	89.283[R]
1996	21.002	0.023	23.197	35.757	79.978	7.087	3.590	3.127	0.316	0.071	0.033	7.137	0.137	94.256[R]
1998	21.656	0.067	22.936	36.934	81.592	7.068	3.297	2.835	0.328	0.070	0.031	6.561	0.088	95.192[R]
1999	21.623	0.058	23.010	37.960	82.650	7.610	3.268	2.885	0.331	0.069	0.046	6.599	0.099	96.836[R]
2000	22.580	0.065	23.916	38.404	84.965	7.862	2.811	2.907	0.317	0.066	0.057	6.158	0.115	98.961[R]
2001	21.914[R]	0.029	22.906	38.333	83.182[R]	8.033	2.242[R]	2.640	0.311	0.065	0.070[R]	5.328[R]	0.075	96.472[R]
2002	21.904[R]	0.061	23.628[R]	38.401	83.994[R]	8.143	2.689[R]	2.648[R]	0.328	0.064	0.105	5.835[R]	0.078	97.877[R]
2003	22.321[R]	0.051	23.069[R]	39.047[R]	84.487[R]	7.959[R]	2.825[R]	2.740[R]	0.339[R]	0.064[R]	0.115[R]	6.082[R]	0.022	98.311[R]
2004[P]	22.390	0.138	22.991	40.130	85.649	8.232	2.725	2.845	0.340	0.063	0.143	6.116	0.039	99.740

[a]Electricity net generation from conventional hydroelectric power, geothermal, solar, and wind; consumption of wood, waste, and alcohol fuels; geothermal heat pump and direct use energy; and solar thermal direct use energy.
[b]Natural gas, plus a small amount of supplemental gaseous fuels that cannot be identified separately.
[c]Petroleum products supplied, including natural gas plant liquids and crude oil burned as fuel. Beginning in 1993, also includes ethanol blended into motor gasoline.
[d]Beginning in 1993, ethanol blended into motor gasoline is included in both "petroleum" and "wood, waste, alcohol," but is counted only once in total consumption.
[e]"Alcohol" is ethanol blended into motor gasoline.
R=Revised. P=Preliminary. NA=Not available. s=Less than 0.0005 and greater than −0.0005 quadrillion Btu.
Notes: Totals may not equal sum of components due to independent rounding. For data not shown for 1951–1969, see http://www.eia.doe.gov/emeu/aer/overview.html.

SOURCE: Adapted from "Table 1.3. Energy Consumption by Source, Selected Years, 1949–2004 (Quadrillion Btu)," in *Annual Energy Review 2004*, U.S. Department of Energy, Energy Information Administration, Office of Energy Markets and End Use, August 2005, http://www.eia.doe.gov/emeu/aer/pdf/aer.pdf (accessed April 5, 2006)

Figure 6.1.) Biomass sources (wood, waste, and alcohol) contributed 2.8 quadrillion Btu, while hydroelectric power provided 2.7 quadrillion Btu. Together, biomass and hydroelectric power provided more than 90% of renewable energy in 2004. Geothermal energy was the third-largest source, with about 0.3 quadrillion Btu. Solar power contributed 0.06 quadrillion Btu, and wind provided 0.1 quadrillion Btu.

BIOMASS ENERGY

Biomass refers to organic material such as plant and animal waste, wood, seaweed and algae, and garbage. The use of biomass is not without environmental problems. Deforestation can occur from widespread use of wood, especially if forests are clear-cut, which can result in soil erosion and mudslides. Burning wood, like burning fossil fuels, also pollutes the environment. Biomass can be burned directly or converted to biofuel by thermochemical conversion and biochemical conversion.

Direct Burning

Direct combustion is the easiest and most commonly used method of using biomass as fuel. Materials such as dry wood or agricultural wastes are chopped and burned to produce steam, electricity, or heat for industries, utilities, and homes. Industrial-size wood boilers are operating throughout the country. The burning of agricultural wastes is also becoming more widespread. In Florida, sugarcane producers use the residue from harvested cane to generate much of their energy. Residential use of wood as fuel generated 332 trillion Btu in 2004, considerably less than the 860 trillion–940 trillion Btu generated in homes in the 1980s (Energy Information Administration, *Annual Energy Review 2004*, 2005).

FIGURE 6.1

Renewable energy consumption as a share of total energy, 2004

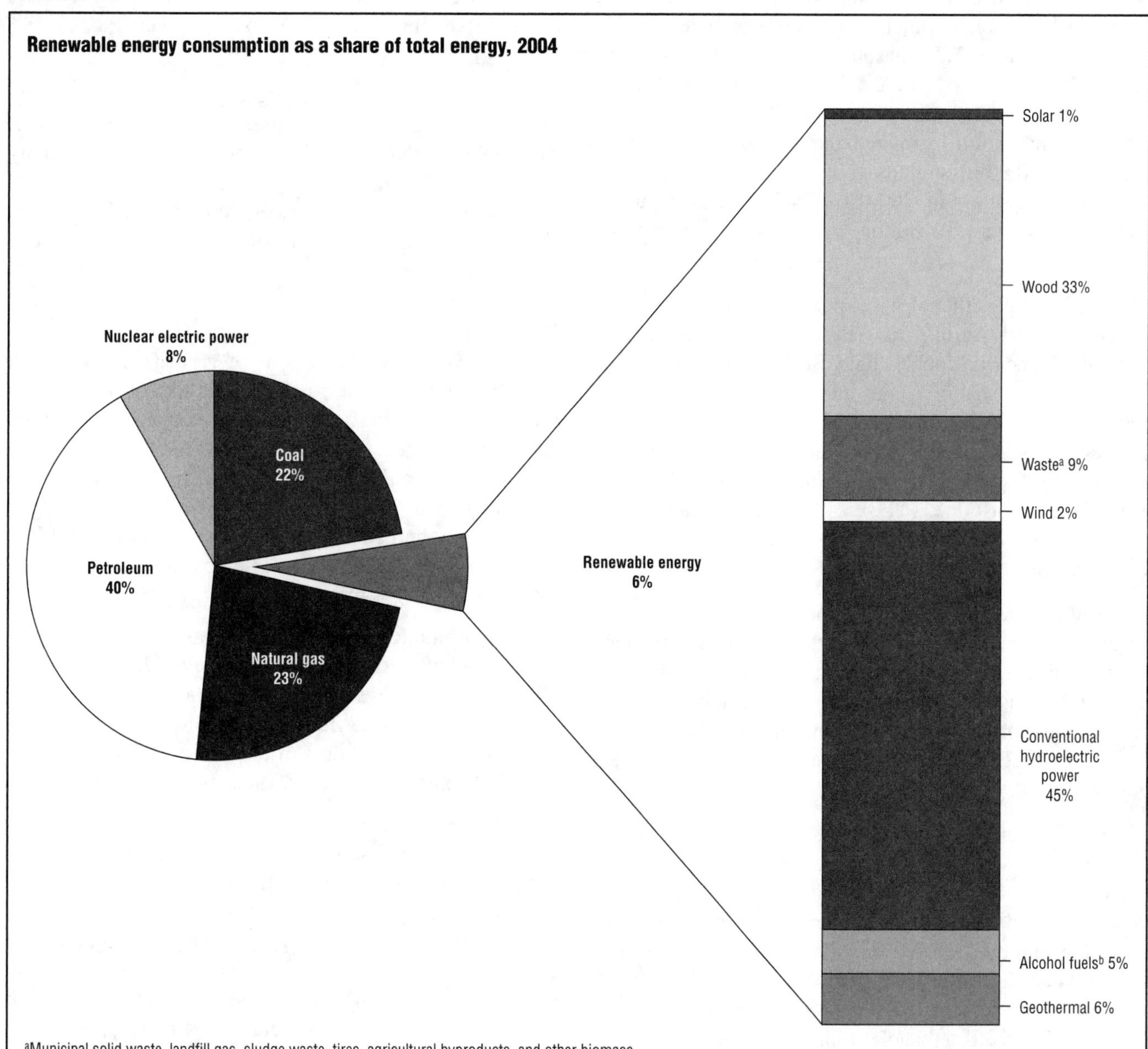

[a]Municipal solid waste, landfill gas, sludge waste, tires, agricultural byproducts, and other biomass.
[b]Ethanol blended into motor gasoline.

SOURCE: Adapted from "Figure 10.1. Renewable Energy Consumption by Major Sources: Renewable Energy as Share of Total Energy, 2004," in *Annual Energy Review 2004*, U.S. Department of Energy, Energy Information Administration, Office of Energy Markets and End Use, August 2005, http://www.eia.doe.gov/emeu/aer/pdf/aer.pdf (accessed April 5, 2006)

Thermochemical Conversion

Thermochemical conversion involves heating biomass in an oxygen-free or low-oxygen atmosphere, which transforms the material into simpler substances that can be used as fuels. Products such as charcoal and methanol are produced this way.

Biochemical Conversion

Biochemical conversion uses enzymes, fungi, or other microorganisms to convert high-moisture biomass into either liquid or gaseous fuels. Bacteria convert manure, agricultural wastes, paper, and algae into methane, which is used as fuel. Sewage treatment plants have used anaerobic (oxygen-free) digestion for many years to generate methane gas. Small-scale digesters have been used on farms, primarily in Europe and Asia, for hundreds of years. Biogas pits (a biomass-based technology) are a significant source of energy in China.

Another type of biochemical conversion, fermentation, uses yeast to decompose carbohydrates, yielding ethyl alcohol (ethanol), a colorless, nearly odorless, flammable liquid, and carbon dioxide. Most of the ethanol manufactured for use as fuel in the United States is derived from corn, wood, and sugar. Ethanol is mixed with gasoline to create gasohol, which is sold in three blends: 10% gasohol, which is a mixture of 10% ethanol and 90% gasoline; 7.7% gasohol, which is at least 7.7% ethanol but less than 10%; and 5.7% gasohol, which is at least 5.7% ethanol but less than 7.7%. The Federal Highway Administration estimated that in 2003 Americans used about 20.5 billion gallons of 10% gasohol, up from 16.3 billion gallons in 2000. In 2003 Americans used slightly more than 12 billion gallons of less-than-10% gasohol.

The use of ethanol is expected to increase. The Energy Policy Act of 2005 (PL 109-58) required fuel suppliers to nearly double their use of ethanol by 2012 to reduce the nation's dependence on foreign fuel sources. Fuel distributors are also being required by several major cities and more than twenty-five states to remove the additive methyl tertiary butyl ether, or MTBE, from gasoline. MTBE is known to contaminate groundwater. Many suppliers are replacing the additive with ethanol.

Automobiles can be built to run directly on ethanol or on any mixture of gasoline and ethanol. However, ethanol is difficult and expensive to produce in bulk. Development of this fuel source may depend more on the political support of legislators from farming states and a desire for energy independence than on savings at the gas pump.

Some scientists have suggested that ethanol made from refuse—for example, corncobs and rice hulls—could liberate the alcohol fuel industry from its dependence on food crops, such as corn and sugarcane, and make the fuel cheaper. Worldwide enough corncobs and rice hulls are left after crop production to produce more than 40 billion gallons of ethanol. Other scientists argue that wood-derived ethanol could eventually create a sustainable liquid fuel industry. If new trees were planted to replace those that were cut for fuel, they say, those trees would not only be available for harvesting years later but, in the meantime, would also alleviate global warming by processing carbon dioxide. Other scientists counter with the warning that an increased demand for wood for transportation fuels might accelerate the destruction of old-growth forests and endanger ecosystems.

Another alternative fuel is ethanol-85 (E-85), which is a blend of 85% ethanol and 15% unleaded gasoline. In 2006 there were approximately 1.5 million automobiles on U.S. roads capable of using E-85 as a fuel. Such vehicles are called "flex-fuel" vehicles because they can run on E-85, gasohol, or gasoline.

Methanol (methyl alcohol) fuels have also been tested successfully. Using methanol instead of diesel fuel virtually eliminates sulfur emissions and reduces other environmental pollutants usually emitted from trucks and buses. Producing methanol is costly, however.

Burning biofuels in vehicle engines is part of the "carbon cycle," in which vegetation makes use of the products of automobile combustion. (See Figure 6.2.) Automobile exhaust generated from fossil fuels, however, contains pollutants. In addition, generating excessive amounts of carbon dioxide from either fossil fuels or biofuels is thought to add to global warming because it traps heat in the atmosphere.

Municipal Waste Recovery

Each year millions of tons of garbage are buried in landfills and city dumps. This method of disposal is becoming increasingly costly, and many landfills across the nation are near capacity. Some communities discovered that they could solve both problems—cost and capacity—by constructing waste-to-energy plants. Not only is the garbage burned and reduced in volume by 90%, energy in the form of steam or electricity is also generated in a cost-effective way.

Use of municipal waste as fuel has increased steadily since the 1980s. According to the *Annual Energy Review 2004*, municipal waste (including landfill gas, sludge waste, tires, and agricultural by-products) generated 22.7 billion kilowatt-hours of electricity in 2004, up from 0.1 billion kilowatt-hours in 1982. (See Table 6.2.)

MASS BURN SYSTEMS. Most waste-to-energy plants in the United States use the mass burn system (also called direct combustion). Because the waste does not have to be sorted or prepared before burning—except for removing obviously noncombustible, oversized objects—the system eliminates expensive sorting, shredding, and transportation machinery that may be prone to break down. The waste is simply carried to the plant in trash trucks and dropped into a storage pit. Overhead cranes lift the garbage into a hopper that controls the amount of waste that is fed into the furnace. The burning waste produces heat, which is used to produce steam. The steam can be used directly for industrial needs or can be sent through a turbine to power a generator to produce electricity.

REFUSE-DERIVED FUEL SYSTEMS. At refuse-derived fuel plants, waste is first processed to remove noncombustible objects and to create homogeneous and uniformly sized fuel. Large items such as bedsprings, dangerous materials, and flammable liquids are removed by hand. The trash is then shredded and screened to remove glass, rocks, and other material that cannot be burned. The remaining material is usually sifted a second time with an air separator to yield fluff, which is placed in storage bins. It can also be compressed into pellets or briquettes for long-term storage. This fuel can be used by itself or with other fuels, such as coal or wood.

FIGURE 6.2

The carbon cycle

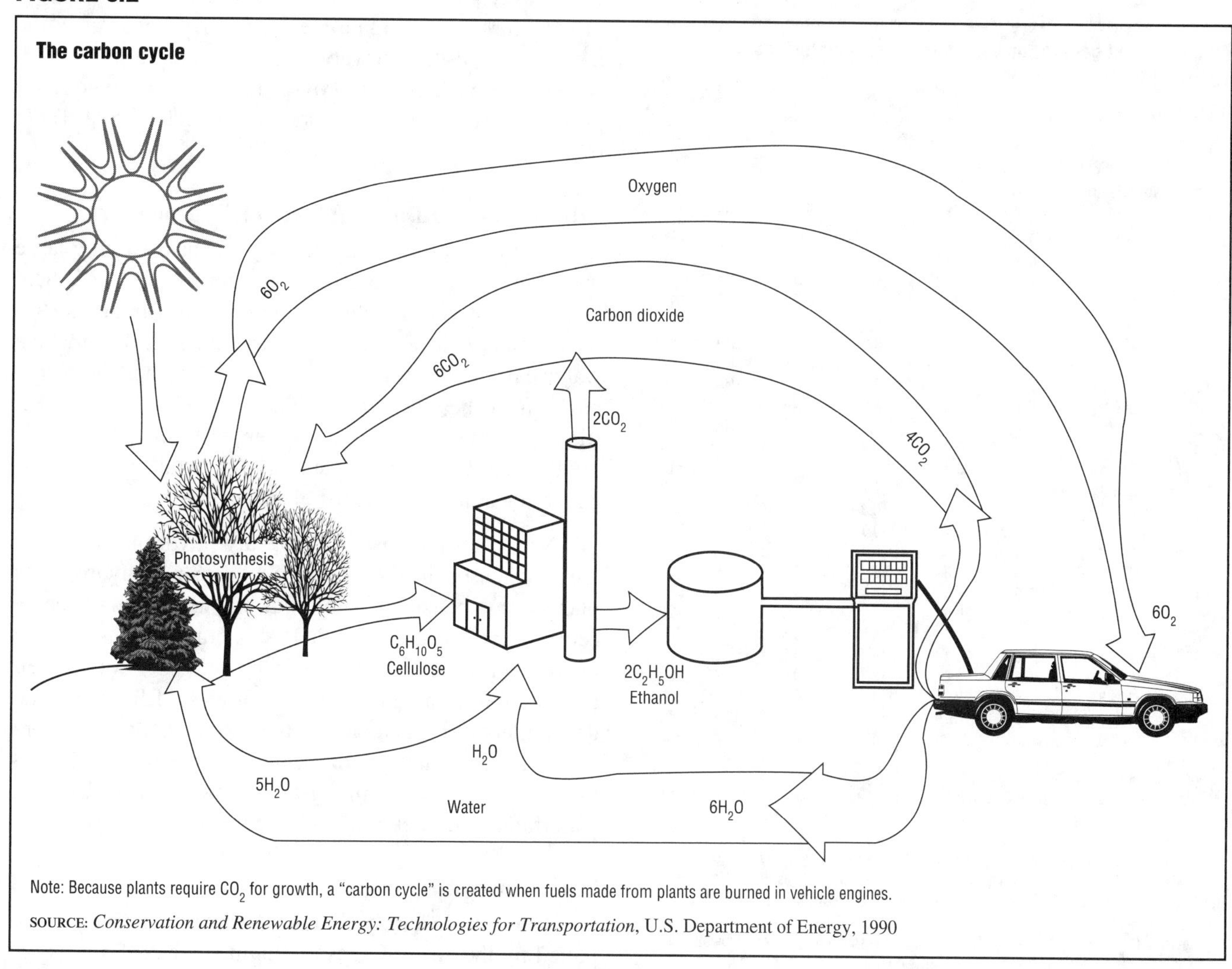

Note: Because plants require CO_2 for growth, a "carbon cycle" is created when fuels made from plants are burned in vehicle engines.

SOURCE: *Conservation and Renewable Energy: Technologies for Transportation*, U.S. Department of Energy, 1990

PERFORMANCE OF WASTE-TO-ENERGY SYSTEMS. Most waste-to-energy systems can produce two to four pounds of steam for every pound of garbage burned. A 1,000-ton-per-day mass burn system usually converts an average of 310,250 tons of trash each year and recovers 2 trillion Btu of energy. In addition, the plant will emit 96,000 tons of ash (32% of waste input) for landfill disposal. A refuse-derived fuel plant produces less ash but sends almost the same amount of waste to the landfill because of the noncombustibles that are removed from the trash before it is burned.

DISADVANTAGES OF WASTE-TO-ENERGY PLANTS. The major problem with increasing the use of municipal waste-to-energy plants is their effect on the environment. The emission of particles into the air is partially controlled by electrostatic precipitators, and many gases can be eliminated by proper combustion techniques. However, large amounts of dioxin (a dangerous air pollutant) and other toxins are often emitted from these plants. Noise from trucks, fans, and processing equipment can also be unpleasant for nearby residents.

Landfill Gas Recovery

Landfills contain a large amount of biodegradable matter that is compacted and covered with soil. Methanogens, which are anaerobic microorganisms, thrive in this oxygen-depleted environment. They metabolize the biodegradable matter in the landfill, producing methane gas and carbon dioxide as by-products. In the past, as landfills aged, these gases built up and leaked out, which prompted some communities to drill holes in landfills and burn off the methane to prevent dangerously large amounts from exploding.

The energy crisis of the 1970s made landfill methane gas an energy resource too valuable to waste. The first landfill gas recovery site was built in 1975 at the Palos Verdes Landfill in Rolling Hills Estates, California. By December 2005, according to the Environmental Protection Agency (EPA), 395 landfill gas energy sites were operating in the United States and another 600 landfill sites had been identified for potential development (http://www.epa.gov/lmop/proj/index.htm). The EPA estimated

TABLE 6.2

Electricity net generation by renewables, selected years 1949–2004

[Billion kilowatt-hours]

Year	Conventional hydroelectric power	Wood[a]	Waste[b]	Geothermal	Solar[c]	Wind	Total
1949	94.8	0.4	NA	NA	NA	NA	95.2
1950	100.9	0.4	NA	NA	NA	NA	101.3
1955	116.2	0.3	NA	NA	NA	NA	116.5
1960	149.4	0.1	NA	s	NA	NA	149.6
1965	197.0	0.3	NA	0.2	NA	NA	197.4
1970	251.0	0.1	0.2	0.5	NA	NA	251.8
1972	275.9	0.1	0.2	1.5	NA	NA	277.7
1974	304.2	0.1	0.2	2.5	NA	NA	306.9
1976	286.9	0.1	0.2	3.6	NA	NA	290.8
1978	283.5	0.2	0.1	3.0	NA	NA	286.8
1980	279.2	0.3	0.2	5.1	NA	NA	284.7
1982	312.4	0.2	0.1	4.8	NA	NA	317.5
1984	324.3	0.5	0.4	7.7	s	s	332.9
1986	294.0	0.5	0.7	10.3	s	s	305.5
1988[d]	226.1	0.9	0.7	10.3	s	s	238.1
1990	292.9	32.5	13.3	15.4	0.4	2.8	357.2
1992	253.1	36.5	17.8	16.1	0.4	2.9	326.9
1994	260.1	37.9	19.1	15.5	0.5	3.4	336.7
1996	347.2	36.8	20.9	14.3	0.5	3.2	423.0
1998	323.3	36.3	22.4	14.8	0.5	3.0	400.4
2000	275.6	37.6	23.1	14.1	0.5	5.6	356.5
2001	217.0	35.2	21.8	13.7	0.5	6.7	294.9
2002	264.3	38.7	22.9	14.5	0.6	10.4	351.3
2003	275.8[R]	37.5[R]	23.7[R]	14.4[R]	0.5	11.2[R]	363.2[R]
2004[P]	269.6	37.3	22.7	14.4	0.6	14.2	358.8

[a]Wood, black liquor, and other wood waste.
[b]Municipal solid waste, landfill gas, sludge waste, tires, agricultural byproducts, and other biomass.
[c]Solar thermal and photovoltaic energy.
[d]Through 1988, all data except hydroelectric are for electric utilities only; hydroelectric data through 1988 include industrial plants as well as electric utilities. Beginning in 1989, data are for electric utilities, independent power producers, commercial plants, and industrial plants.
R=Revised. P=Preliminary. NA=Not available. s=Less than 0.05 billion killowatt-hours.
Notes: Totals may not equal sum of components due to independent rounding. For data not shown for 1951–1969, see http://www.eia.doe.gov/emeu/aer/elect.html. For related information, see http://www.eia.doe.gov/fuelelectric.html.

SOURCE: Adapted from "Table 8.2a. Electricity Net Generation: Total (All Sectors), Selected Years, 1949–2004 (Sum of Tables 8.2b and 8.2d; Billion Kilowatt-hours)," in *Annual Energy Review 2004*, U.S. Department of Energy, Energy Information Administration, Office of Energy Markets and End Use, August 2005, http://www.eia.doe.gov/emeu/aer/pdf/aer.pdf (accessed April 5, 2006)

that landfill gas generated about 9 billion kilowatt-hours of electricity per year.

In a typical operation, garbage is allowed to decompose for several months. When a sufficient amount of methane gas has developed, it is piped to a generating plant, where it is used to create electricity. In its purest form, methane gas can be used like natural gas. Depending on the extraction rates, most sites can produce gas for about twenty years. Besides the energy provided, tapping the methane reduces landfill odors and the chances of explosions.

HYDROPOWER

In the past, flowing water turned waterwheels of mills to grind grain; today hydropower plants convert the energy of flowing water into mechanical energy, turning turbines to create electricity. Hydropower is the most widely used renewable energy source in the world. In the United States it provided 75% of all electricity produced from renewable sources in 2004. (See Table 6.2.)

Advantages and Disadvantages of Hydropower

Today hydropower is the only means of storing large quantities of energy for almost instant use. Water is held in a large reservoir behind a dam, with a hydroelectric power plant below. The dam creates a height from which water can flow at a fast rate. When it reaches the power plant, it pushes the turbine blades attached to the electrical generator. Whenever power is needed, the valves are opened, the moving water spins the turbines, and the generator quickly produces electricity.

Nearly all the best sites for large hydropower plants are being used in the United States. Small hydropower plants are expensive to build but may eventually become economical because of their low operating costs. One of the disadvantages of small hydropower generators is their reliance on rain and melting snow to fill reservoirs; drought conditions can affect the water supply. Additionally, environmental groups strongly protest the construction of new dams, pointing to ruined streams, dried up waterfalls, and altered aquatic habitats.

The Future of Hydropower

The last federally funded hydropower dam, completed by the Army Corps of Engineers in 1986, was the Richard B. Russell Dam and Lake on the Savannah River, which forms the border between South Carolina and Georgia. Since then, local governments have been required to contribute half the cost of any new dam proposed in the United States. While expansion and efficiency improvements at existing dams offer significant potential for additional energy, hydropower's future contribution to U.S. energy generation should remain relatively constant. Additional supplies of hydroelectric power for the United States will likely come from Canada.

Most of the new development in hydropower is occurring in developing nations, which see it as an effective method of supplying power to growing populations. These massive public-works projects usually require huge amounts of money—most of it borrowed from the developed world. Hydroelectric dams are considered worth the cost and potential environmental threats because they bring cheap electric power to the citizenry.

In May 2006 the Chinese government announced that construction was completed on the world's largest dam, the Three Gorges Dam on the Yangtze River in Hubei province, China (http://english.gov.cn/2006-05/20/content_286525.htm). Five times the size of the Hoover

FIGURE 6.3

Cross-section of the Earth showing source of geothermal energy

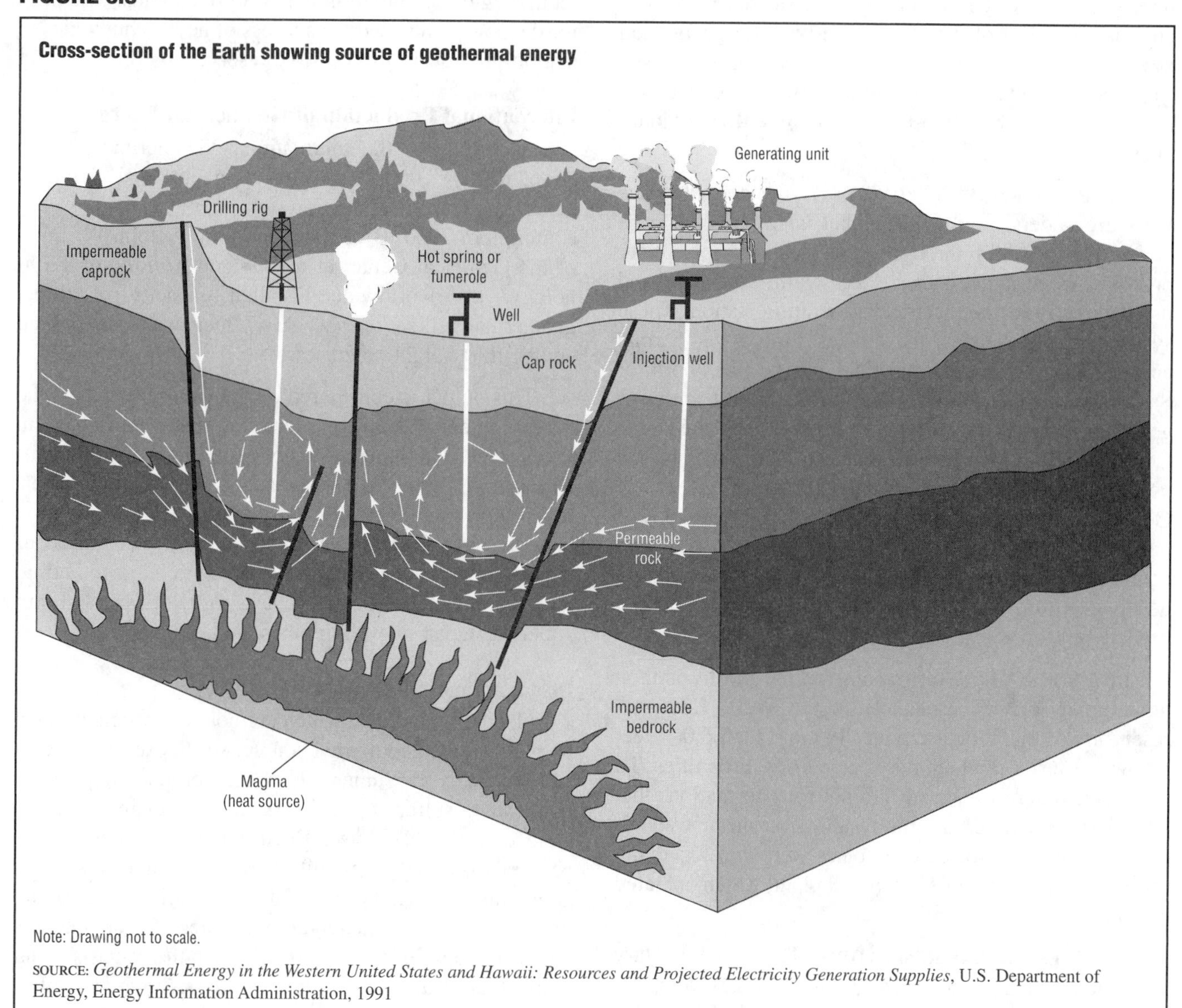

Note: Drawing not to scale.

SOURCE: *Geothermal Energy in the Western United States and Hawaii: Resources and Projected Electricity Generation Supplies*, U.S. Department of Energy, Energy Information Administration, 1991

Dam in the United States, the dam is 185 meters (607 feet) tall and 2,309 meters (7,575 feet, or 1.4 miles) in length. It is scheduled to begin generating power in 2009, after its twenty-six generators have been installed and water has filled the giant reservoir. Once operational, the dam is expected to produce 84.7 billion kilowatt-hours of electricity per year.

GEOTHERMAL ENERGY

Although bubbling hot springs became public baths as early as ancient Rome, using hot water and underground steam to produce power is a relatively recent development. Electricity was first generated from natural steam in Italy in 1904. The world's first natural steam power plant was built in 1958 in a volcanic region of New Zealand. A field of twenty-eight geothermal power plants covering thirty square miles in northern California was completed in 1960.

What Is Geothermal Energy?

Geothermal energy is the natural, internal heat of Earth trapped in rock formations deep underground. Only a fraction of it can be extracted, usually through large fractures in Earth's crust. Hot springs, geysers, and fumaroles (holes in or near volcanoes from which vapor escapes) are the most easily exploitable sources. (See Figure 6.3.) Geothermal reservoirs provide hot water or steam that can be used for heating buildings and processing food. Pressurized hot water or steam can also be directed toward turbines, which spin, generating electricity for residential and commercial customers.

Types of Geothermal Energy

Like most natural energy sources, geothermal energy is usable only when it is concentrated in one spot—in this case, in what is known as a "thermal reservoir." There are four types of reservoirs, including hydrothermal reservoirs, dry rock reservoirs, geopressurized reservoirs, and magma. Most of the known reservoirs for geothermal power in the United States are located west of the Mississippi River, and the highest-temperature geothermal resources occur for the most part west of the Rocky Mountains.

HYDROTHERMAL RESERVOIRS. Hydrothermal reservoirs are underground pools of hot water covered by a permeable formation through which steam escapes under pressure. Once at the surface the steam is purified and piped directly to the electrical generating station. These systems are the cheapest and simplest form of geothermal energy. The Geysers thermal field, ninety miles north of San Francisco, California, is the world's largest source of geothermal power. According to the Web site of Calpine Corporation, which operates nineteen of the twenty-one power generating plants at the site, as of 2006 The Geysers generated enough electricity to satisfy the power needs of a city the size of San Francisco and provided nearly 60% of the electricity used in the region extending northward from the Golden Gate Bridge to the Oregon border (http://www.geysers.com/).

DRY ROCK. These formations are the most common geothermal sources, especially in the West. However, reservoirs of this type are typically more than 6,000 feet below the surface, which poses numerous difficulties. To tap them, water is injected into hot rock formations that have been fractured, and the resulting steam or water is collected. Hot dry rock technologies were developed and tested at New Mexico's Fenton Hill plant, which operated between 1970 and 1996.

GEOPRESSURIZED RESERVOIRS. These sedimentary formations contain hot water and methane gas. Supplies of geopressurized energy remain uncertain, and drilling is expensive. Scientists are developing new technology to exploit the methane content in these reservoirs.

MAGMA. This molten or partially liquefied rock is found from 10,000 to 33,000 feet below Earth's surface. Because magma is so hot, ranging from 1,650 to 2,200° Fahrenheit, it is a good geothermal resource. Extracting energy from magma is still in the experimental stages.

Domestic Production of Geothermal Energy

In 2004 geothermal energy produced 14.4 billion kilowatt-hours of electricity, or 4% of the 358.8 billion kilowatt-hours of electricity produced by renewable energy sources in the United States. (See Table 6.2.) According to the International Geothermal Association, at the end of 2003 the United States had 24% of the installed geothermal generating capacity of the world and the largest installed generating capacity of any single country. However, most of the easily exploited geothermal reserves in the United States have already been developed. Continued growth in the U.S. market depends on the regulatory environment, oil price trends, who pays for the new plants, and the success of new technologies to exploit previously inaccessible reserves.

International Production of Geothermal Energy

According to the International Geothermal Association, nearly 8,000 megawatts of geothermal electrical generating capacity was present in more than twenty countries in 2000. By 2004, according to John W. Lund of the Geo-Heat Center at the Oregon Institute of Technology, the worldwide installed capacity of direct geothermal utilization was 9,047 megawatts distributed among thirty-eight countries.

This output is considered only a small fraction of the overall potential: Many countries are believed to have in excess of 100,000 megawatts of geothermal energy available. As with other fuel sources, however, world geothermal reserves are unevenly distributed. They occur mostly in seismically active areas at the margins or borders of the planet's nine tectonic plates. Areas rich in geothermal reserves include the west coasts of North and South America, Japan, the Philippines, and Indonesia.

Disadvantages of Geothermal Energy

Geothermal plants, which are not very efficient, must be built near a geothermal source, so they are not accessible to many consumers. They also produce unpleasant odors when sulfur is released during processing and generate considerable noise. Environmental concerns have been raised about potentially harmful pollutants, such as ammonia, arsenic, boron, hydrogen sulfide, and radon, that are often found in geothermal waters. Other concerns are the collapse of the land from which the water is being drained and water shortages from massive withdrawals.

WIND ENERGY

Winds are created by the uneven heating of the atmosphere by the sun, the irregularities of Earth's surface, and the rotation of the planet. They are strongly influenced by bodies of water, weather patterns, vegetation, and other factors. When "harvested" by turbines, wind can be used to generate electricity.

Early windmills produced mechanical energy to pump water and grind grain in mills. By the late 1890s, Americans had begun experimenting with wind power to generate electricity. Their early efforts produced enough electricity to light one or two modern light bulbs.

Beginning in the late twentieth century, industrial and developing countries alike started using wind power on a significant scale to complement existing power sources and to bring electricity to remote regions. Wind turbines cost less to install per unit of kilowatt capacity than either coal or nuclear facilities. After installing a windmill, there are few additional costs.

FIGURE 6.4

Wind capacity map, 2003

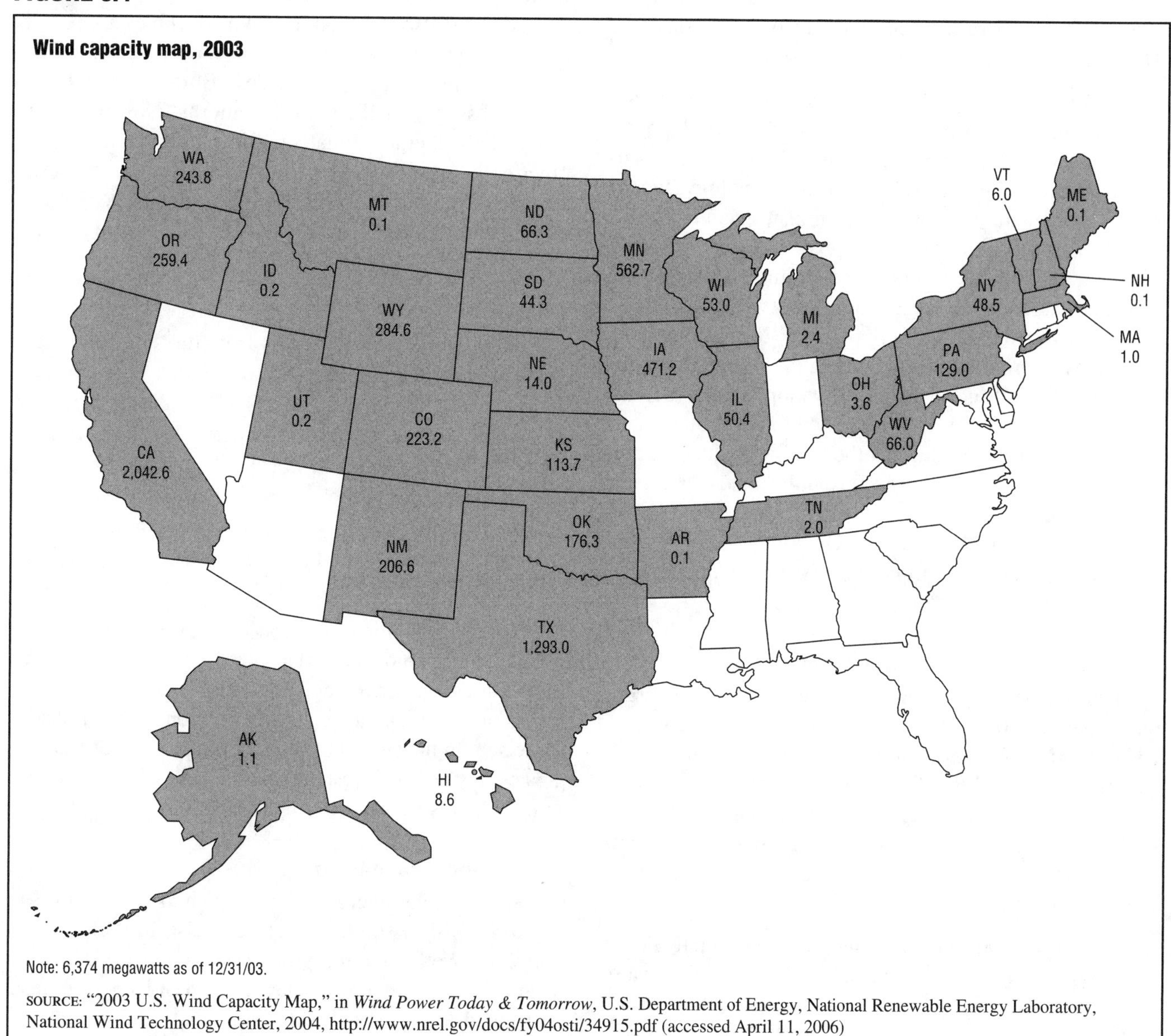

Note: 6,374 megawatts as of 12/31/03.

SOURCE: "2003 U.S. Wind Capacity Map," in *Wind Power Today & Tomorrow*, U.S. Department of Energy, National Renewable Energy Laboratory, National Wind Technology Center, 2004, http://www.nrel.gov/docs/fy04osti/34915.pdf (accessed April 11, 2006)

Compared with the pinwheel-shaped farm windmills that still dot rural America, today's state-of-the-art wind turbines look more like airplane propellers. Their sleek fiberglass design allows them to generate an abundance of mechanical energy, which can be converted to electricity.

The most favorable locations for wind turbines are in mountain passes and along coastlines, where wind speeds are generally highest and most consistent. Of all the places in the world, Europe has the greatest coastal wind resources. Western Europe and the United States accounted for nearly 90% of all new wind energy installations in 2003 (*International Energy Outlook 2005*, Energy Information Administration, 2005). As of 2004, electricity-producing wind turbines operated in ninety-five countries.

Domestic Energy Production by Wind Turbines

The U.S. wind energy industry began in California in 1981 with the installation of 144 relatively small turbines with a combined capacity of 7 megawatts of electricity. Within a year the number of turbines had increased ten times and by 1986 a hundred times.

During 1998 and 1999, wind farm activity expanded into other states, motivated by financial incentives (such as tax credits for wind energy production), regulatory incentives, and state mandates (in Iowa and Minnesota). In 1999 Iowa, Minnesota, and Texas added capacity exceeding 100 megawatts each. In 2003 the total installed generating capacity of the United States was 6,374 megawatts, and wind power plants operated in thirty-two states. (See Figure 6.4.) By 2004, according to the American Wind Energy Association, twelve states—California, Colorado, Iowa,

Kansas, Minnesota, New Mexico, Pennsylvania, Oklahoma, Oregon, Texas, Washington, and Wyoming—contained 94% of the U.S. wind energy capacity.

In *Annual Energy Outlook 2006* (2006), the Energy Information Administration projected that wind power capacity in the United States would triple from 2004 to 2030. Refinements in wind-turbine technology would drive some of the increase. Government encouragement would also be important, however. Wind Powering America, an initiative announced in June 1999 by the U.S. Department of Energy, sought to have 80,000 megawatts of generation capacity in place by 2020 and to have wind power provide 5% of the nation's electricity. In addition, the Wind Energy Development Program, announced in 2005 by the Bureau of Land Management (part of the Department of the Interior), focused on development of wind energy on public lands in eleven western states: Arizona, California, Colorado, Idaho, Montana, Nevada, New Mexico, Oregon, Utah, Washington, and Wyoming (http://windeis.anl.gov/documents/docs/WindPEISROD.pdf).

International Development of Wind Energy

During the decade following the 1973 oil embargo, more than 10,000 wind machines were installed worldwide, ranging in size from portable units to multimegawatt turbines. In China, for example, small wind turbines allow people to watch their favorite television shows, an activity that has increased wind energy demand. In fact, in 2001 China was the world's largest manufacturer of small wind turbines.

According to the Global Wind Energy Council, global wind-power-generating capacity was 47,000 megawatts in 2004, up from 3,531 megawatts in 1994 (http://www.gwec.net/fileadmin/documents/GWEC_Brochure.pdf). Europe accounted for 72% of the global wind power in 2004. The World Wind Energy Association projected in 2004 that global wind-power capacity would top 100,000 megawatts by 2008.

Interest in wind energy has been driven, in part, by the declining cost of capturing wind energy. For new turbines at sites with strong winds, prices declined from more than $0.38 per kilowatt-hour in 1980 to about $0.04 per kilowatt-hour in 2002 (Lester R. Brown, "Wind Power Set to Become World's Leading Energy Source," Earth Policy Institute, 2003). Decreasing costs could make wind power competitive with gas and coal power plants, even before considering wind's environmental advantages.

Advantages and Disadvantages of Wind Energy

The main problem with wind energy is that the wind does not always blow. In addition, some people find the whirring noise of wind turbines annoying and object to clusters of wind turbines in mountain passes and along shorelines, where they interfere with scenic views. Environmentalists also point out that wind turbines are responsible for the loss of thousands of endangered birds that inadvertently fly into the blades. Birds frequently use windy passages in their travel patterns. However, wind farms do not emit climate-altering carbon dioxide and other pollutants, respiratory irritants, or radioactive waste. Because wind farms do not require water to operate, they are especially well suited to semiarid and arid regions.

SOLAR ENERGY

Solar energy, which comes from the sun, is a renewable, widely available energy source that does not generate huge amounts of pollution or radioactive waste. Solar-powered cars have competed in long-distance races, and solar energy has been used for many years to power spacecraft. Although many people consider solar energy a product of the space age, architectural researchers at the Massachusetts Institute of Technology built the first solar-heated house in 1939.

Solar radiation is nearly constant outside Earth's atmosphere, but the amount of solar energy reaching any point on Earth varies with changing atmospheric conditions, such as clouds and dust, and the changing position of Earth relative to the sun. In the United States, exposure to the sun's rays is greatest in the Southwest, although almost all regions have some solar resources. (See Figure 6.5.)

Passive and Active Solar Systems

Passive solar energy systems, such as greenhouses or windows with a southern exposure, use heat flow, evaporation, or other natural processes to collect and transfer heat. (See Figure 6.6.) They are considered the least costly and least difficult solar systems to implement.

Active solar systems require collectors and storage devices as well as motors, pumps, and valves to operate the systems that transfer heat. (See Figure 6.6.) Collectors consist of an absorbing plate that transfers the sun's heat to a working fluid (liquid or gas), a translucent cover plate that prevents the heat from radiating back into the atmosphere, and insulation on the back of the collector panel to further reduce heat loss. Excess solar energy is transferred to a storage facility so it may provide power on cloudy days.

In both passive and active systems, the conversion of solar energy into a form of power is made at the site where it is used. The most common and least expensive active solar systems are used for heating water.

Solar Thermal Energy Systems

In a solar thermal energy system, mirrors or lenses constantly track the sun's position and focus its rays onto solar receivers that contain water or other fluids. The

FIGURE 6.5

Solar resources

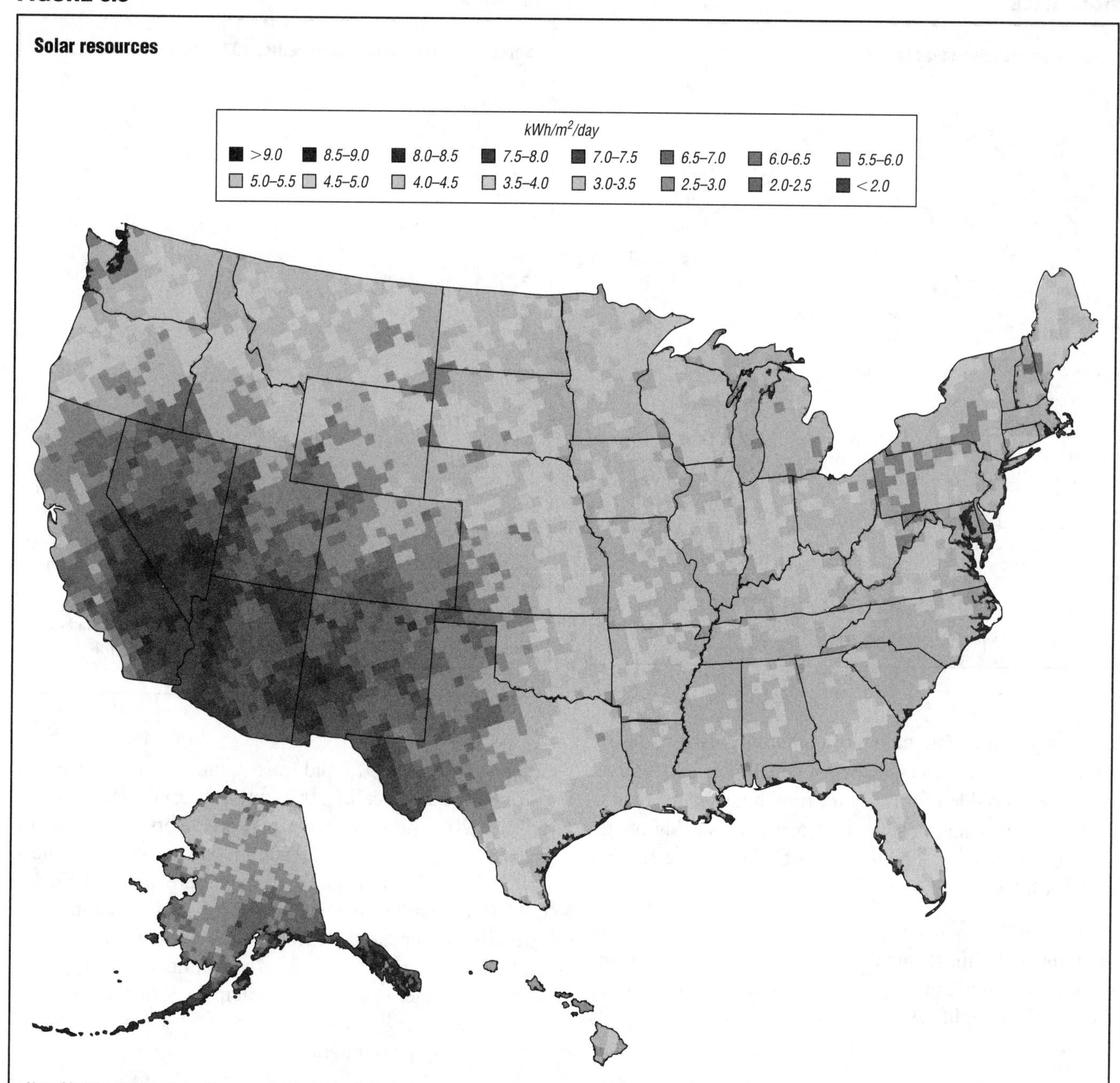

Note: Model estimates of monthly average daily total radiation using inputs derived from satellite and/or surface observations of cloud cover, aerosol optical depth, precipitable water vapor, albedo, atmospheric pressure and ozone resampled to 40km resolution.

SOURCE: "Direct Normal Solar Radiation (Two-Axis Tracking Concentrator)," in *Dynamic Maps, GIS Data, & Analysis Tools: Solar Maps*, National Renewable Energy Laboratory, May 2004, http://www.nrel.gov/gis/images/us_csp_annual_may2004.jpg (accessed May 3, 2006)

fluid is heated to more than 750° Fahrenheit; that heat is used to power an electric generator. In a distributed solar thermal system, the collected energy powers irrigation pumps, provides electricity for small communities, or captures normally wasted heat from the sun in industrial areas. In a central solar thermal system, the energy is collected at a central location and used by utility networks for a large number of customers.

Other systems include solar ponds and trough systems. Solar ponds are pools filled with water and salt. Because saltwater is denser than freshwater, the saltwater on the bottom absorbs the heat, which is trapped by the freshwater on top. Trough systems use U-shaped mirrors to concentrate the sunshine on water or on oil-filled tubes.

Photovoltaic Conversion Systems

The photovoltaic cell system converts sunlight directly into electricity without the use of mechanical generators. Photovoltaic cells have no moving parts, are easy to install, require little maintenance, and can last up

FIGURE 6.6

Active vs. passive solar houses

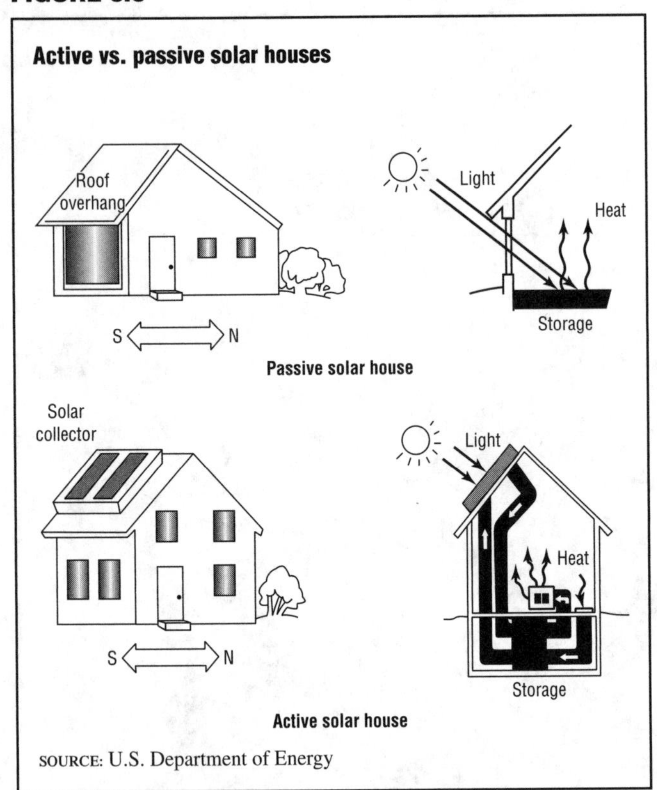

SOURCE: U.S. Department of Energy

FIGURE 6.7

Solar thermal collector shipments, 1974–84 and 1986–2003

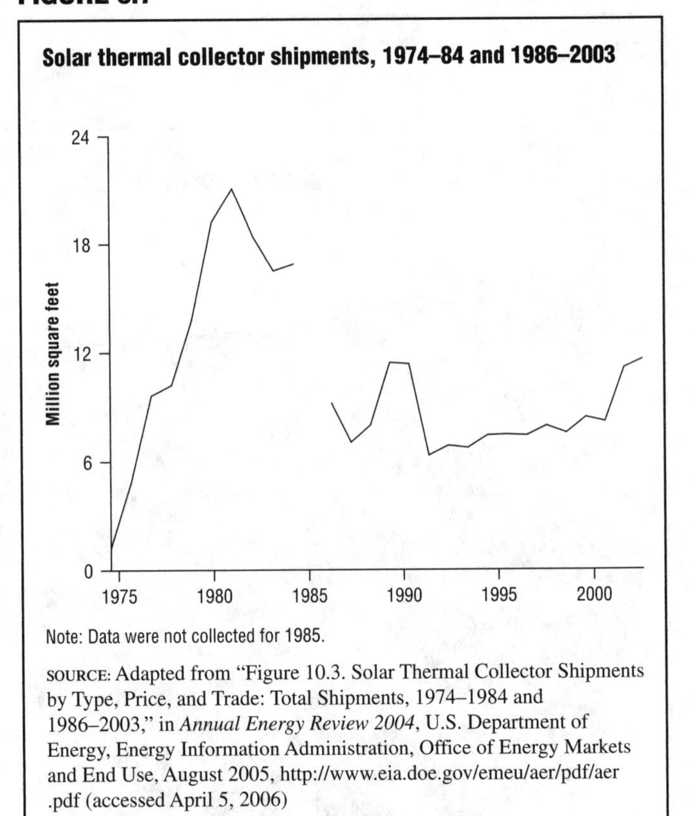

Note: Data were not collected for 1985.

SOURCE: Adapted from "Figure 10.3. Solar Thermal Collector Shipments by Type, Price, and Trade: Total Shipments, 1974–1984 and 1986–2003," in *Annual Energy Review 2004*, U.S. Department of Energy, Energy Information Administration, Office of Energy Markets and End Use, August 2005, http://www.eia.doe.gov/emeu/aer/pdf/aer.pdf (accessed April 5, 2006)

to twenty years. The cells are commonly used to power small devices, such as watches or calculators. On a larger scale they provide electricity for rural households, recreational vehicles, and businesses. Solar panels using photovoltaic cells have generated electricity for space stations and satellites for many years.

The cells produce the most power around noon, when sunlight is the most intense. They are usually connected to storage batteries that provide electricity during cloudy days and at night. A backup energy supply is usually required.

The use of photovoltaic cells is expanding around the world. Because they contain no turbines or other moving parts, operating costs are low, and maintenance is minimal. Above all, the fuel source (sunshine) is free and plentiful. The main disadvantage of photovoltaic cell systems is the high initial cost, although prices have fallen considerably. While toxic materials are often used in the construction of the cells, researchers are investigating new materials, recycling, and disposal.

Solar Energy Usage

Use of solar energy is difficult to measure because it is rarely connected to any kind of metered grid. However, shipments of solar equipment can be used as an indicator of use. According to the *Annual Energy Review 2004*, total shipments of solar thermal collectors peaked in 1981 at more than 21 million square feet, fell to 6.6 million square feet in 1991, and rose again to more than 11 million square feet in 2001 through 2003. (See Figure 6.7.) In 2003 most solar thermal collectors were sold for residential purposes in Sunbelt states, usually to heat water for swimming pools. (See Figure 6.8 and Figure 6.9.) The market for solar energy space heating has virtually disappeared. Only a small proportion of solar thermal collectors are used for commercial purposes, although some state and municipal power companies have solar energy systems they can use for additional power during peak hours.

Solar Power as an International Rural Solution

Getting electricity to rural areas has long been more expensive than serving cities. In the United States most farmers did not receive electrical power until 1935, when the Rural Electrification Administration provided low-cost financing to rural electric cooperatives. In places such as western China, the Himalayan foothills, and the Amazon basin today, the cost of connecting new rural customers to electricity grids remains very high.

While rural families may not have access to electrical grid systems, they do have sunlight: In most tropical countries, considerable sunlight falls on rooftops. Photovoltaic cells can be installed on them to run water pumps, lights, refrigerators, and communications equipment. Electricity produced by photovoltaic cells was initially too expensive—as much as a thousand times more than

FIGURE 6.8

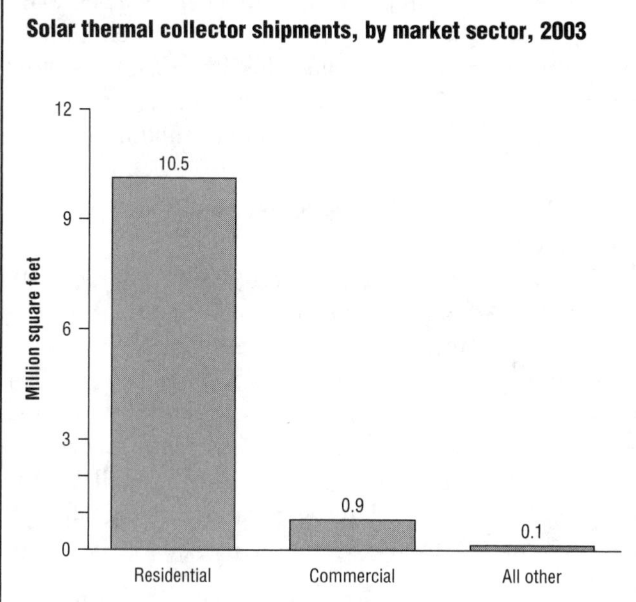

SOURCE: Adapted from "Figure 10.4. Solar Thermal Collector Shipments by End Use, Market Sector, and Type, 2003: Market Sector," in *Annual Energy Review 2004*, U.S. Department of Energy, Energy Information Administration, Office of Energy Markets and End Use, August 2005, http://www.eia.doe.gov/emeu/aer/pdf/aer.pdf (accessed April 5, 2006)

FIGURE 6.9

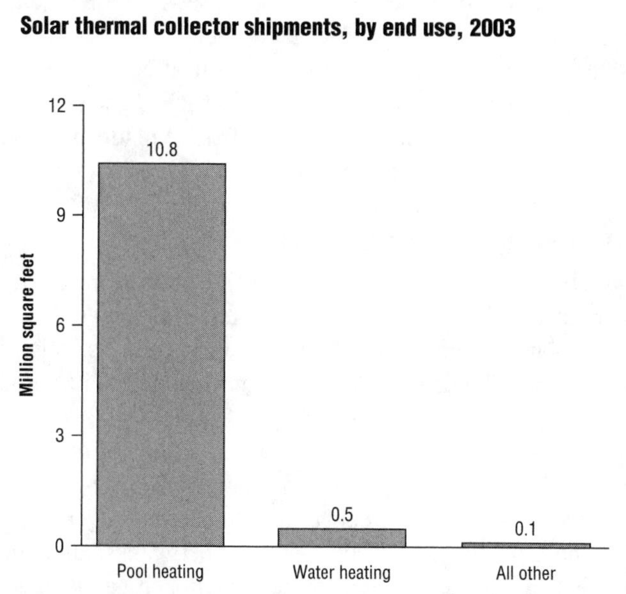

SOURCE: Adapted from "Figure 10.4. Solar Thermal Collector Shipments by End Use, Market Sector, and Type, 2003: End Use," in *Annual Energy Review 2004*, U.S. Department of Energy, Energy Information Administration, Office of Energy Markets and End Use, August 2005, http://www.eia.doe.gov/emeu/aer/pdf/aer.pdf (accessed April 5, 2006)

electricity produced by conventional power plants—but prices have continually fallen, making solar energy a competitive choice in some areas.

Advantages and Disadvantages of Solar Energy

The primary advantage of solar energy is its inexhaustible supply. It is especially useful in rural or remote areas that cannot be easily connected to an electrical power grid. Although solar power still costs more than three times as much as fossil fuel energy, utilities often turn to solar energy to provide "peaking power" on extremely hot or cold days. Building solar energy systems to provide peak power capacity is often cheaper than building the backup diesel generators that are often used.

Solar power's primary disadvantage is its reliance on a consistently sunny climate, which is possible in limited areas. It also requires a large amount of land for the most efficient collection of solar energy by electricity plants.

POWER FROM THE OCEAN

Oceans are not as easily controlled as rivers or water directed through canals into turbines, so unlocking their potential power is far more challenging. Three ideas undergoing experimentation are tidal plants, wave power, and ocean thermal energy conversion.

Tidal Power

Tidal plants use the movement of water as it ebbs and flows to generate power. A minimum tidal range of three to five yards is generally considered necessary for an economically feasible plant. (The tidal range is the difference in height between consecutive high and low tides.) The largest existing tidal facility is the 240-megawatt plant at the La Rance estuary in northern France, built in 1965. Canada built a smaller 40-megawatt unit at the Bay of Fundy, which has a fifteen-yard tidal range, the largest in the world. A larger plant for that site is under consideration. Russia has a small 400-kilowatt plant near Murmansk, close to the Barents Sea. The world's first offshore tidal turbine, in the ocean about a mile from Devon, England, began producing energy in 2003.

Wave Energy

One type of wave power plant is the oscillating water column; the first significant examples were built at Toftestallen, on Norway's Atlantic coast. In this system, the arrival of a wave forces water up a hollow 65-foot tower, displacing the air in the tower. The air rushes out the top through a turbine, whose rotors spin, generating electricity. When the wave falls back and the water level falls, air is sucked back in through the turbine, again generating electricity. Similar systems have been tested in China, India, Japan, Portugal, and Scotland.

A second type of wave power plant uses the overflow of high ocean waves. As the waves splash against the top of a dam, some of the water goes over and is trapped in a reservoir on the other side. The water is then directed through a turbine as it flows back to the sea.

Ocean Thermal Energy Conversion

Ocean thermal energy conversion uses the temperature difference between warm surface water and the cooler water in the ocean's depths to produce heat energy, which can power a heat engine to produce electricity. The systems can be installed on ships, barges, or offshore platforms with underwater cables that transmit electricity to shore.

HYDROGEN: A FUEL OF THE FUTURE?

Hydrogen, the lightest and most abundant chemical element, is the ideal fuel from the environmental point of view. Its combustion produces only water vapor, and it is entirely carbon free. Three-quarters of the mass of the universe is hydrogen, so in theory the supply is ample. However, the combustible form of hydrogen is a gas and is not found in nature. The many compounds containing hydrogen—water, for example—cannot be converted into pure hydrogen without the expenditure of considerable energy. Usually, the amount of energy required to make hydrogen gas is equal to the amount of energy obtained by the combustion of that gas. Therefore, with today's technology, little or nothing could be gained from an energy point of view.

Hydrogen fuel cells are similar to batteries, but they use hydrogen as a fuel. They can be used to produce electricity and heat and to power cars. Their development is also in experimental stages.

Research into the use of hydrogen as a fuel got a boost when President George W. Bush announced a hydrogen fuel initiative in his 2003 State of the Union Address. Congress appropriated $159 million for hydrogen and fuel cell research and development in fiscal year 2004; $223 million in fiscal year 2005; and $243 million in fiscal year 2006. Goals of the initiative include reducing the cost of hydrogen fuel to make it comparable to gasoline, resolving issues for safe in-vehicle storage of reserve fuel, and lowering the cost of hydrogen power systems to compete with internal combustion engines. Researchers project that hydrogen fuel cell vehicles could reduce U.S. demand for oil by about eleven million barrels per day—the amount the United States now imports"—by 2040 (http://www1.eere.energy.gov/hydrogenandfuelcells/presidents_initiative.html).

FUTURE TRENDS IN U.S. RENEWABLE ENERGY USE

In its *Annual Energy Outlook 2006*, the Energy Information Administration forecasted that total renewable fuel consumption, including ethanol for transportation, would increase from 6 quadrillion Btu in 2004 to 9.6 quadrillion Btu in 2030. About 60% of the projected demand would be for electricity generation. Renewable fuel is expected to remain a small contributor to overall electricity generation, rising slightly from 9% of the total generation in 2004 to 9.4% in 2030.

Hydropower is expected to remain the largest source of renewable electricity generation through 2030, but its share of total generation will fall from 6.8% in 2004 to 5.1% in 2030. The production of other renewables should increase steadily. (See Figure 6.10.) The largest source of renewable generation after hydropower is biomass, which is projected to more than double from 2004 to 2030. (See Figure 6.11.) The agency projected that wind power would increase from 0.4% of total generation in 2004 to 1.1% in 2030 and that high-output geothermal capacity would increase from 0.4% of total generation in 2004 to 0.9% in 2030.

Energy production from municipal solid waste and landfill gas is expected to stay static from 2004 to 2030, remaining at 0.5% of total generation. Solar energy is not expected to contribute much to the total of centrally generated electricity.

FIGURE 6.10

Grid-connected electricity generation from renewable energy sources, 1980–2030

[Billion kilowatt-hours]

— Conventional hydropower — Other renewables

History

Projections

400
300
200
100
0

1980 1995 2004 2015 2030

SOURCE: "Figure 63. Grid-Connected Electricity Generation from Renewable Energy Sources, 1980–2030 (Billion Kilowatthours)," in *Annual Energy Outlook 2006*, U.S. Department of Energy, Energy Information Administration, Office of Integrated Analysis and Forecasting, February 2006, http://www.eia.doe.gov/oiaf/aeo/pdf/0383(2006).pdf (accessed April 5, 2006)

FIGURE 6.11

Nonhydroelectric renewable electricity generation by energy source, selected years 2004–30

[Billion kilowatt-hours]

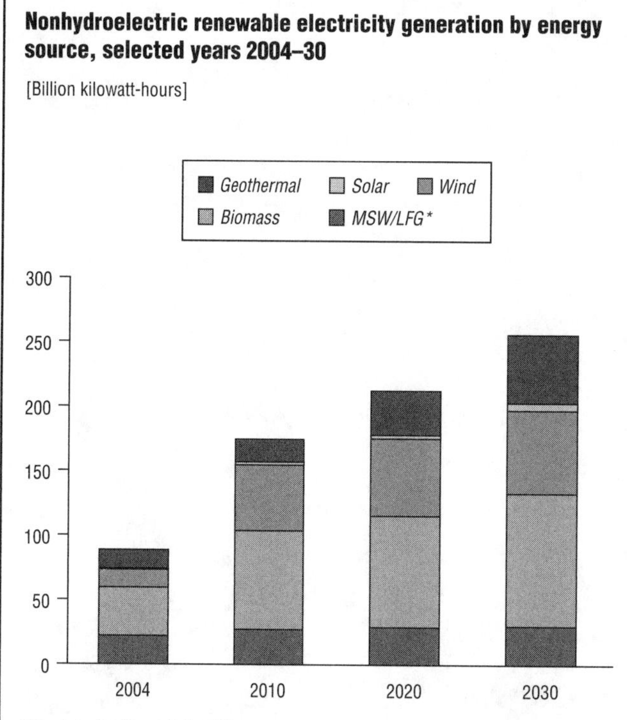

*Municipal solid waste/landfill gas.

SOURCE: "Figure 64. Nonhydroelectric Renewable Electricity Generation by Energy Source, 2004–2030 (Billion Kilowatthours)," in *Annual Energy Outlook 2006*, U.S. Department of Energy, Energy Information Administration, Office of Integrated Analysis and Forecasting, February 2006, http://www.eia.doe.gov/oiaf/aeo/pdf/0383(2006).pdf (accessed April 5, 2006)

CHAPTER 7

ENERGY RESERVES—OIL, GAS, COAL, AND URANIUM

Congress requires the Department of Energy to prepare estimates of the quantities of crude oil, natural gas, coal, and uranium that exist in the earth and can be used as fuel. These estimates, which include deposits in the United States and other parts of the world, are considered essential to the development, implementation, and evaluation of national energy policies. The estimates are also important because these resources are nonrenewable, which means they can be used up: They are formed much more slowly than it takes to consume them.

The focus of the estimates is recoverable reserves. "Proved reserves" are deposits of fuel in known locations that, based on the geological and engineering data, can be recovered using existing technology. Drilling or mining for these fuels also makes sense, given current economic conditions. Undiscovered recoverable resources, by contrast, are quantities of fuel that are thought to exist in favorable geologic settings. It would be feasible to retrieve these resources using existing technology, although it might not be feasible under current economic conditions.

CRUDE OIL

U.S. proved reserves of crude oil declined from 1994 through 1996, rose and then fell in 1997 and 1998, climbed through 2002, then fell again in 2003 and 2004. On December 31, 2004, crude oil reserves totaled 21.4 billion barrels—near the 1998 reserves level (*U.S. Crude Oil, Natural Gas, and Natural Gas Liquids Reserves 2004 Annual Report*, Energy Information Administration, 2005). Together, Texas, Alaska, California, and offshore areas in the Gulf of Mexico accounted for 77% of U.S. proved reserves in December 2004. (See Figure 7.1.) Of the four regions, only Texas reported an increase in proved reserves in 2004.

Proved reserves of crude oil rose in 1970 with the inclusion of Alaska's North Slope oil fields. As the oil has been extracted, reserves have steadily declined. In 1987 Alaska's proved reserves totaled 13.2 billion barrels of crude oil; by 2004 the state had only 4.3 billion barrels. (See Figure 7.1.) Proved reserves of crude oil fell by 119 million barrels in Alaska from 2003 through 2004.

The Gulf of Mexico federal offshore areas had about 4.1 billion barrels of proved reserves in 2004, down 410 million barrels from 2003. (See Figure 7.1.) Hurricane Ivan in September 2004 damaged some of the oil drilling platforms in the ocean, so a portion of the reserves became at least temporarily unrecoverable.

In 1996 scientists turned their attention to the dry lands of the Permian basin in western Texas and southeastern New Mexico. They found plenty of potential for crude oil; in fact, it became one of the most active onshore areas for new exploration. In 2004 Texas had about 4.6 billion barrels of crude oil proved reserves, up 30 million barrels from 2003. Proved reserves in New Mexico, however, dropped by 8 million barrels from 2003 through 2004.

NATURAL GAS

Figure 7.2 shows the distribution of U.S. proved reserves of dry natural gas in 2004, which totaled 192.5 trillion cubic feet, an increase of 3.5 billion cubic feet from 2003. The United States also had 201.2 trillion cubic feet of natural gas liquid proved reserves, a 2% increase from the volume reported in 2003 in the lower 48 states and a 1% increase in Alaska.

UNDISCOVERED OIL AND GAS RESOURCES

Other resources are believed to exist based on past geological experience, although they are not yet proved. The Energy Information Administration estimates that 105 billion barrels of crude oil, 682 trillion cubic feet of dry natural gas, and 8 billion barrels of natural gas liquids remain undiscovered in the United States.

FIGURE 7.1

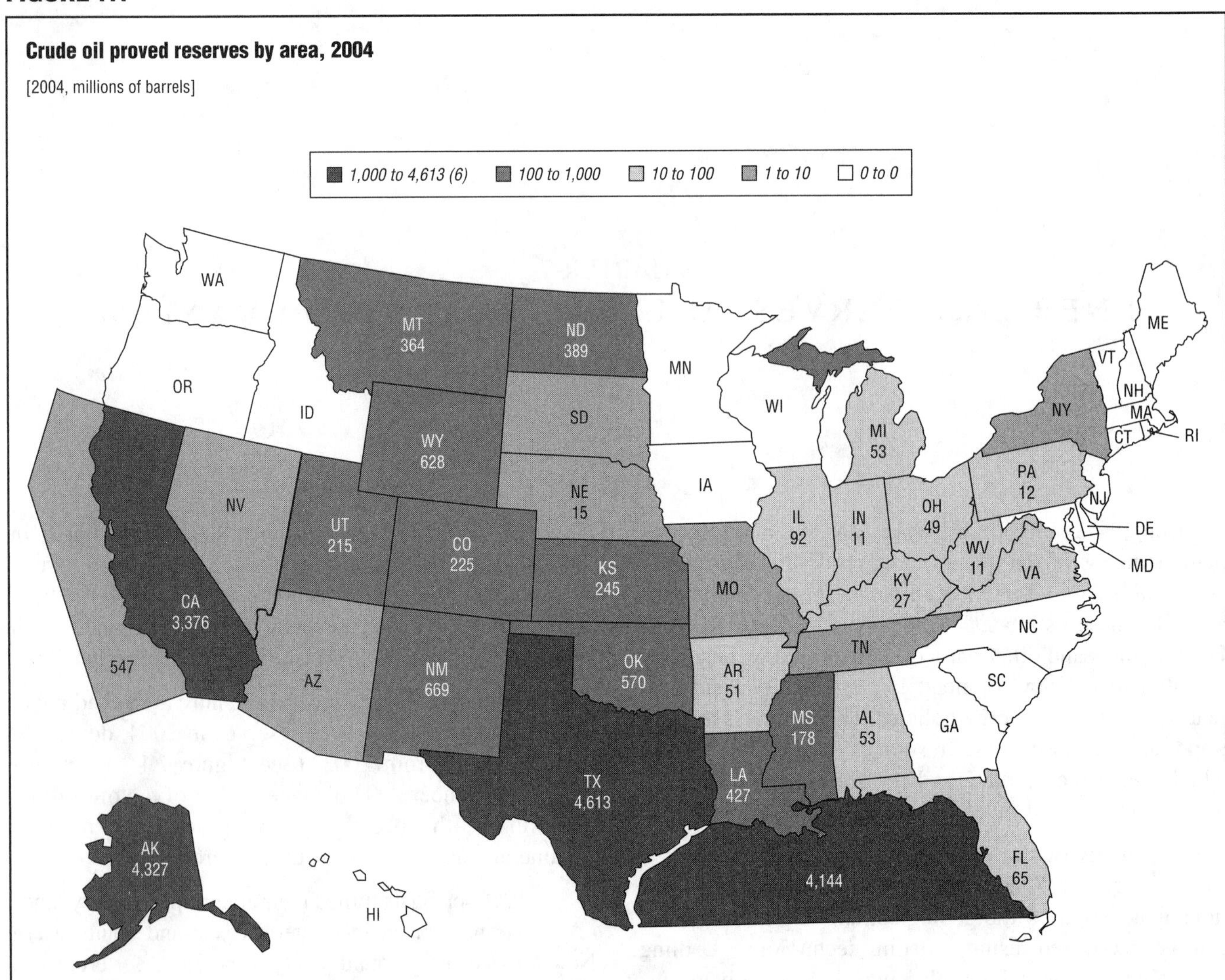

Notes: Four areas contain 77% of U.S. crude oil proved reserves. U.S. total is 21.371 billion barrels.

SOURCE: "Figure 16. Crude Oil Proved Reserves by Area, 2004," in *U.S. Crude Oil, Natural Gas, and Natural Gas Liquids Reserves 2004 Annual Report*, U.S. Department of Energy, Energy Information Administration, Office of Oil and Gas, November 2005, http://www.eia.doe.gov/pub/oil_gas/natural_gas/data_publications/crude_oil_natural_gas_reserves/current/pdf/arr.pdf (accessed April 13, 2006)

Looking for Oil and Gas

Finding oil and gas usually takes two steps. First, geological and geophysical exploration identifies areas where oil and gas are most likely to be found. Much of this exploration is seismic, using shock waves to determine the formations below the surface of the earth. Different rock formations transmit shock waves at different velocities, so they help determine if the geological features most often associated with oil and gas accumulations are present. After the seismic testing has been completed—and if it has been successful—exploratory wells are drilled to determine if oil or gas is present.

Drilling activity has declined dramatically since 1981, when 91,553 exploratory wells were drilled. (See Table 7.1.) Nearly 70% of them found oil and gas deposits. In 2004 only 34,925 were attempted, but 87.5% were successful. In 1981 oil companies had 3,970 rotary rigs in operation; by 2004 only 1,192 were operating. (See Figure 7.3.) Of this number, according to the Energy Information Administration, 165 rigs drilled for oil and 1,025 drilled for natural gas. There were 1,095 onshore rigs and 97 offshore rigs. The average depth of exploratory and development wells has steadily increased, from 3,635 feet in 1949 to 5,770 feet in 2004. Gas wells (averaging 6,045 feet in 2004) are typically deeper than oil wells (averaging 5,249 feet in 2004).

The Cost to Drill

In 2003 the average cost of drilling oil and gas wells was about $1,131,700 in real dollars—that is, adjusted for inflation—or $204 per foot. (See Table 7.2.) Historically, it has cost more to drill a gas well than an oil well because gas wells are deeper. In 2003, however, the cost of drilling an average gas well ($1,106,000) was not

FIGURE 7.2

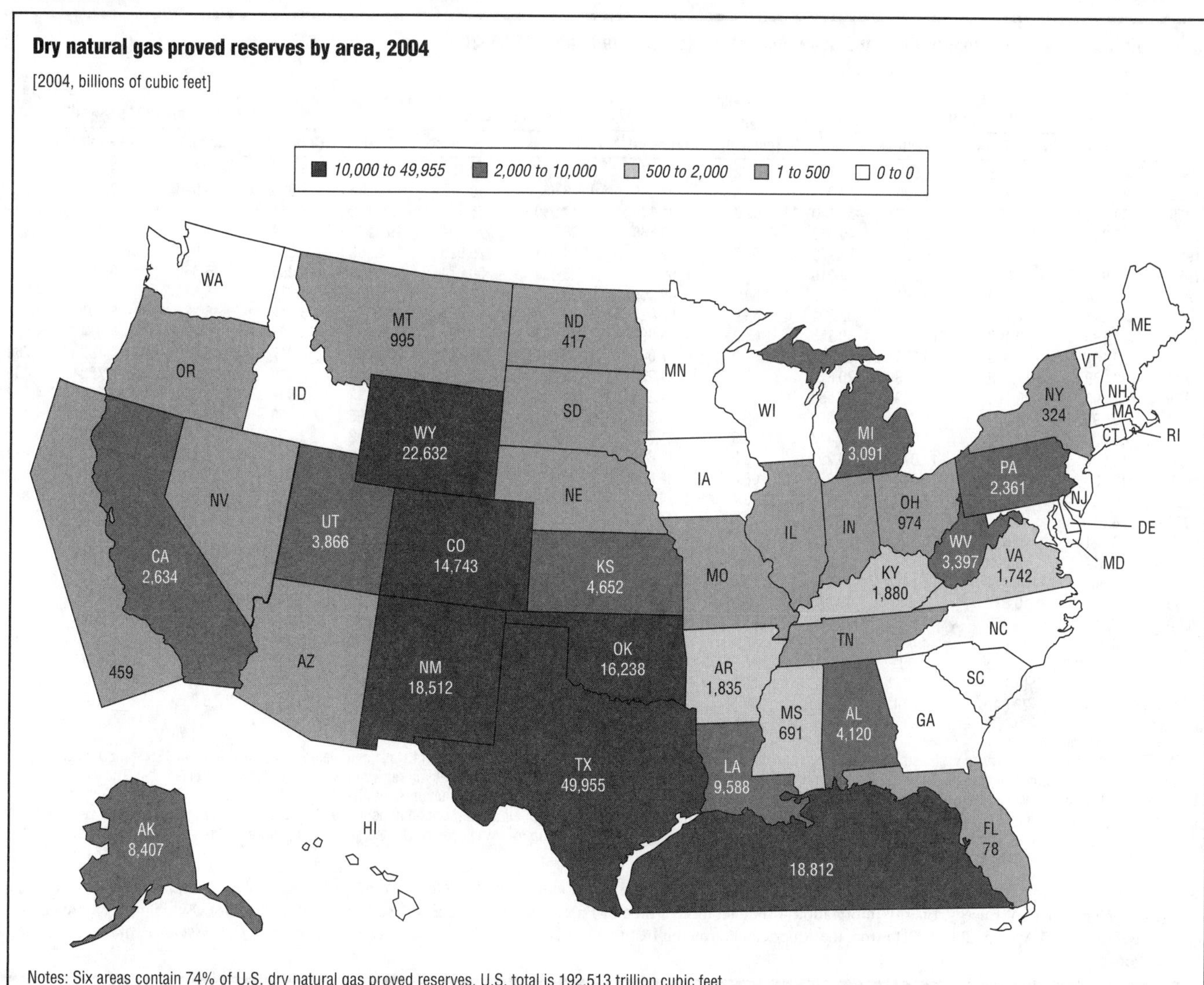

Notes: Six areas contain 74% of U.S. dry natural gas proved reserves. U.S. total is 192.513 trillion cubic feet.

SOURCE: "Figure 19. Dry Natural Gas Proved Reserves by Area, 2004," in *U.S. Crude Oil, Natural Gas, and Natural Gas Liquids Reserves 2004 Annual Report*, U.S. Department of Energy, Energy Information Administration, Office of Oil and Gas, November 2005, http://www.eia.doe.gov/pub/oil_gas/natural_gas/data_publications/crude_oil_natural_gas_reserves/current/pdf/arr.pdf (accessed April 13, 2006)

much higher than the cost of drilling an oil well ($1,037,300) because the average cost per foot of drilling an oil well (about $221) was higher than that of drilling a foot of gas well (about $190). Although drilling costs have fluctuated in recent years, it costs considerably more to drill a well today than it did in the 1960s and 1970s, not only because of inflation but also because all wells must now be drilled deeper.

The estimated expenditures on exploration for, and development of, oil and gas fields around the world by major U.S. companies peaked at $76.8 billion in 2000, a huge increase from each of the fifteen previous years. In 1984 the companies spent $65.3 billion, another high point. (See Table 7.3.) In 2003 U.S. energy companies spent $55.4 billion on oil and gas exploration around the world, with about half of that, $27.2 billion, spent for exploration in the United States.

Drilling in the Arctic National Wildlife Refuge

Controversy has developed over opening part of the Arctic National Wildlife Refuge in Alaska to oil exploration. Oil fields in Prudhoe Bay, directly west of the refuge, supply about 60% of Alaska's oil and 20% of the country's domestic oil, although production is dropping steadily as the oil is extracted. In 1980 Congress passed the Alaska National Interest Lands Conservation Act (PL 96-487), which set aside more than 104 million acres for parks and wilderness areas, including 19 million acres for the wildlife refuge. The conservation act did not, however, include the coastal plain.

TABLE 7.1

Crude oil and natural gas exploratory and development wells, selected years 1949–2004

Year	Wells drilled				Successful wells	Footage drilled (thousand feet)				Average depth (feet per well)			
	Crude oil	Natural gas	Dry holes	Total	(percent)	Crude oil	Natural gas	Dry holes	Total	Crude oil	Natural gas	Dry holes	Total
1949	21,352	3,363	12,597	37,312	66.2	79,428	12,437	43,754	135,619	3,720	3,698	3,473	3,635
1950	23,812	3,439	14,799	42,050	64.8	92,695	13,685	50,977	157,358	3,893	3,979	3,445	3,742
1955	30,432	4,266	20,452	55,150	62.9	121,148	19,930	85,103	226,182	3,981	4,672	4,161	4,101
1960	22,258	5,149	18,212	45,619	60.1	86,568	28,246	77,361	192,176	3,889	5,486	4,248	4,213
1965	18,065	4,482	16,226	38,773	58.2	73,322	24,931	76,629	174,882	4,059	5,562	4,723	4,510
1970	12,968	4,011	11,031	28,010	60.6	56,859	23,623	58,074	138,556	4,385	5,860	5,265	4,943
1972	11,378	5,440	10,891	27,709	60.7	49,269	30,006	58,556	137,831	4,330	5,516	5,377	4,974
1974	13,647	7,138	12,116	32,901	63.2	52,025	38,449	62,899	153,374	3,812	5,387	5,191	4,662
1976	17,688	9,409	13,758	40,855	66.3	68,892	49,113	68,977	186,982	3,895	5,220	5,014	4,577
1978	19,181	14,413	16,551	50,145	67.0	77,041	75,841	85,788	238,669	4,017	5,262	5,183	4,760
1980	32,639	17,333	20,638	70,610	70.8	124,350	91,484	98,820	314,654	3,810	5,278	4,788	4,456
1981	43,598	20,166	27,789	91,553	69.6	171,241	107,758	134,113	413,112	3,928	5,344	4,826	4,512
1982	39,199	18,979	26,219	84,397	68.9	148,881	106,627	122,787	378,295	3,798	5,618	4,683	4,482
1984	42,605	17,127	25,681	85,413	69.9	161,770	90,578	119,044	371,392	3,797	5,289	4,635	4,348
1986	19,097	8,516	12,678	40,291	68.5	76,622	44,727	60,507	181,856	4,012	5,252	4,773	4,514
1988	13,636	8,555	10,041	32,232	68.8	58,660	45,320	52,375	156,354	4,302	5,297	5,216	4,851
1990	12,198	11,044	8,313	31,555	73.7	54,480	55,869	43,352	153,701	4,466	5,059	5,215	4,871
1992	8,757	8,209	6,118	23,084	73.5	44,183	45,728	31,213	121,124	5,045	5,571	5,102	5,247
1994	6,721	9,538	5,307	21,566	75.4	36,090	59,412	29,306	124,809	5,370	6,229	5,522	5,787
1996	8,314	9,302	5,282	22,898	76.9	40,810[R]	57,872[R]	30,363[R]	129,045	4,909[R]	6,221[R]	5,748[R]	5,636
1998	7,064	11,144[R]	4,840	23,048[R]	79.0[R]	37,562[R]	74,129[R]	31,763[R]	143,454	5,317[R]	6,652[R]	6,562[R]	6,224
1999	4,176	10,877	3,412[R]	18,465[R]	81.5[R]	19,830[R]	58,064[R]	21,516[R]	99,410	4,749[R]	5,338[R]	6,306[R]	5,384[R]
2000[E]	7,358	16,455	4,025	27,838	85.5	34,844[R]	82,551[R]	23,997[R]	141,392	4,736[R]	5,017[R]	5,962[R]	5,079
2001[E]	8,060	22,083	4,084	34,227	88.1	41,861	118,518[R]	27,238[R]	187,616[R]	5,194[R]	5,367[R]	6,669[R]	5,482[R]
2002[E]	6,058	16,155[R]	3,581[R]	25,794[R]	86.1[R]	27,513[R]	90,469[R]	20,327[R]	138,310	4,542[R]	5,600[R]	5,677[R]	5,362[R]
2003[E]	7,284[R]	19,722[R]	3,687[R]	30,693[R]	88.0[R]	36,743[R]	118,216[R]	22,114[R]	177,074[R]	5,044[R]	5,994[R]	5,998[R]	5,769[R]
2004[E]	6,904	23,647	4,374	34,925	87.5	36,240	142,947	22,334	201,521	5,249	6,045	5,106	5,770

R=Revised. E=Estimate.

Notes: Data are for all wells. Service wells, stratigraphic tests, and core tests are excluded. For 1949–1959, data represent wells completed in a given year. For 1960–1969, data are for well completion reports received by the American Petroleum Institute during the reporting year. For 1970 forward, the data represent wells completed in a given year. The as-received well completion data for recent years are incomplete due to delays in the reporting of wells drilled. The Energy Information Administration (EIA) therefore statistically imputes the missing data to provide estimates of total well completions and footage where necessary. Totals may not equal sum of components due to independent rounding. Average depth may not equal average of components due to independent rounding For data not shown for 1951–1969, see http://www.eia.doe.gov/emeu/aer/resource.html. For related information, see http://www.eia.doe.gov/oil_gas/petroleum/info_glance/petroleum.html.

SOURCE: Adapted from "Table 4.5. Crude Oil and Natural Gas Exploratory and Development Wells, Selected Years, 1949–2004," in *Annual Energy Review 2004*, U.S. Department of Energy, Energy Information Administration, Office of Energy Markets and End Use, August 2005, http://www.eia.doe.gov/emeu/aer/pdf/aer.pdf (accessed April 5, 2006). Data from the American Petroleum Institute (API), *Quarterly Review of Drilling Statistics for the United States* for the years 1966–69.

The U.S. Department of the Interior, in a 1987 report to Congress, recommended that the 1.5-million-acre coastal plain be opened for exploration and extraction. It estimated that 3.2 billion barrels of recoverable oil exist in the area. It also predicted a 46% chance of recovering the oil, a high figure by industry standards. Vast quantities of natural gas are also likely to be found in the area.

Alaskan corporations supported the proposal because they wanted to share in the proceeds. Environmentalists strongly opposed the plan because of potential damage to the habitats and migration patterns of wildlife such as caribou, polar and grizzly bears, musk ox, wolves, Arctic foxes, and millions of nesting birds.

The government report did state that oil exploration could have a major effect on the migratory caribou herds, which number about 180,000 animals. While environmentalists estimated that 20% to 40% of the animals would be threatened, Department of Interior officials predicted that the caribou would change their migratory habits.

If the coastal plain were opened to drilling and major oil reserves were found, oil companies could operate there for several decades. Debate over opening the region to oil drilling has continued for decades, but as of mid-2006 it remained closed to development.

COAL

The Energy Information Administration estimated U.S. coal reserves at 496.1 billion short tons on January 1, 2004. (See Table 7.4.) About 42% of this coal, 210.8 billion short tons, is underground bituminous coal. Montana, Illinois, and Wyoming have the largest reserves of all types of coal.

In addition to untapped coal reserves, large stockpiles of coal are maintained by coal producers, distributors, and major consumers (such as electric utility companies and industrial plants) to compensate for possible

FIGURE 7.3

Crude oil and natural gas rotary rigs in operation by type, 1949–2004

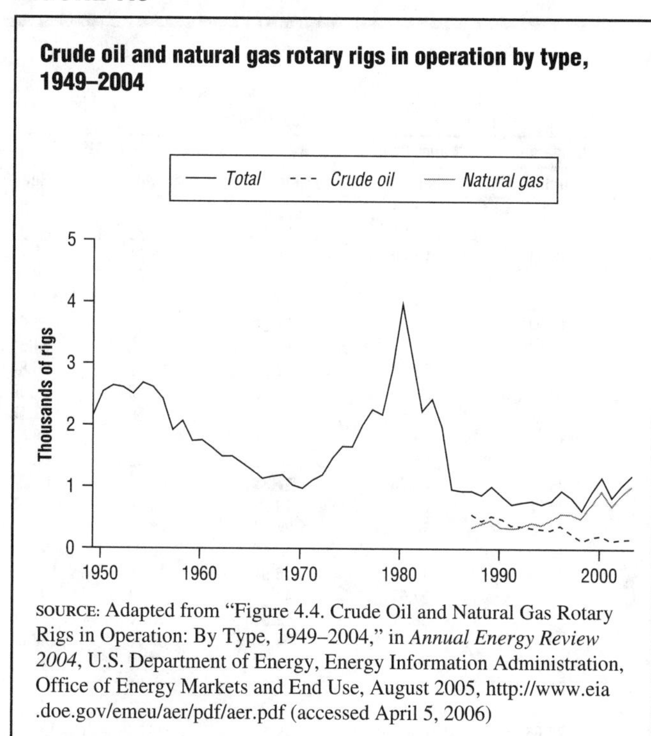

SOURCE: Adapted from "Figure 4.4. Crude Oil and Natural Gas Rotary Rigs in Operation: By Type, 1949–2004," in *Annual Energy Review 2004*, U.S. Department of Energy, Energy Information Administration, Office of Energy Markets and End Use, August 2005, http://www.eia.doe.gov/emeu/aer/pdf/aer.pdf (accessed April 5, 2006)

interruptions in supply. Although there is little seasonal change in demand for coal, supply can be affected by such factors as miners' strikes and bad weather. According to the *Annual Energy Review 2004* (Energy Information Administration, 2005), coal stockpiles totaled 147.2 million short tons in 2004. Electric utilities held slightly more than 72% of this coal, while coal producers and distributors stocked 24%.

URANIUM

The United States has enough uranium to fuel existing nuclear reactors for more than forty years, so exploration for new reserves has been reduced. In 2003 uranium reserves totaled 1.4 billion pounds of uranium oxide, mostly in Wyoming and New Mexico. The number of uranium mills producing uranium concentrate dropped from eleven at the end of 1997 to two at the end of 2003, but then rose to five at the end of December 2005, according to the Energy Information Administration in its quarterly *Domestic Uranium Production Report* (http://eia.doe.gov/cneaf/nuclear/dupr/qupd.html).

INTERNATIONAL RESERVES

Crude Oil

Estimated world crude oil reserves total approximately one trillion barrels (as of January 1, 2004), with most of it located in the Middle East. (See Table 7.5, which provides two different sets of estimates—one from PennWell's *Oil & Gas Journal* and one from Gulf Publishing Company's *World Oil*.) Saudi Arabia, Iraq, the United Arab Emirates, Kuwait, Iran, Canada, Venezuela, and Russia have the largest reserves. The United States has about twenty-two billion barrels of reserves, or 2% of the world's total oil reserves.

While countries with large proved reserves have fewer incentives to find new fields, companies from many nations have turned their attention to areas where production has been limited or nonexistent, such as Azerbaijan, Kazakhstan, and other areas around the Caspian Sea; the northeast Greenland Shelf; the Niger and Congo delta areas in Africa; and off Suriname in South America. The Energy Information Administration predicts increased offshore exploration, including in deep water (*International Energy Outlook 2004*, 2005). Offshore production is expected to increase near Algeria, Brazil, Canada, Colombia, Mexico, Nigeria, Venezuela, and in the North Sea. Deep-sea exploration is expensive, but could be profitable if oil prices rise.

Natural Gas

Depending on the estimate, Russia and the Middle East have 4,198.2 trillion or 4,880.2 trillion cubic feet of natural gas reserves (as of 2004), which amount to 69% or 71% of the world's estimated reserves. (See Table 7.5.) Russia has almost twice as much natural gas in reserve as any other country, while Iran and Qatar possess the largest natural gas reserves in the Middle East. Large reserves (more than 100 trillion cubic feet) are also located in the United Arab Emirates, Saudi Arabia, the United States, Algeria, Venezuela, Nigeria, and Iraq.

Coal

In 2002 worldwide recoverable reserves of coal were estimated at one trillion short tons. The three countries with the most plentiful coal reserves are the United States (271 billion short tons), Russia (173 billion short tons), and China (126 billion short tons). (See Figure 7.4.) (Note that the data for the United States are from December 31, 2003, while those for other countries are from December 31, 2002, the latest figures available.)

Uranium

The countries with the largest known uranium reserves as of June 2006, according to the World Nuclear Association (http://www.world-nuclear.org/info/printable_information_papers/inf75print.htm), are Australia, Kazakhstan, Canada, the United States, South Africa, Namibia, Brazil, Niger, the Russian Federation, and Uzbekistan. Australia is estimated to hold 1.1 million tons, or 24% of the world total, and Kazakhstan has 816,000 tons (17%). The United States possesses 7% of the total world reserves of uranium, which are estimated at 4.7 million tons.

TABLE 7.2

Costs of crude oil and natural gas wells drilled, selected years 1960–2003

	Costs per well (thousand dollars)					Costs per foot (dollars)				
	Crude oil	Natural gas	Dry holes	All		Crude oil	Natural gas	Dry holes	All	
Year	Nominal	Nominal	Nominal	Nominal	Real*	Nominal	Nominal	Nominal	Nominal	Real*
1960	52.2	102.7	44.0	54.9	261.1	13.22	18.57	10.56	13.01	61.83
1961	51.3	94.7	45.2	54.5	256.2	13.11	17.65	10.56	12.85	60.39
1962	54.2	97.1	50.8	58.6	271.8	13.41	18.10	11.20	13.31	61.71
1963	51.8	92.4	48.2	55.0	252.4	13.20	17.19	10.58	12.69	58.22
1964	50.6	104.8	48.5	55.8	252.2	13.12	18.57	10.64	12.86	58.11
1965	56.6	101.9	53.1	60.6	269.1	13.94	18.35	11.21	13.44	59.64
1966	62.2	133.8	56.9	68.4	295.1	15.04	21.75	12.34	14.95	64.51
1967	66.6	141.0	61.5	72.9	305.1	16.61	23.05	12.87	15.97	66.84
1968	79.1	148.5	66.2	81.5	327.0	18.63	24.05	12.88	16.83	67.56
1969	86.5	154.3	70.2	88.6	338.7	19.28	25.58	13.23	17.56	67.15
1970	86.7	160.7	80.9	94.9	344.6	19.29	26.75	15.21	18.84	68.42
1971	78.4	166.6	86.8	94.7	327.6	18.41	27.70	16.02	19.03	65.82
1972	93.5	157.8	94.9	106.4	352.8	20.77	27.78	17.28	20.76	68.82
1973	103.8	155.3	105.8	117.2	367.8	22.54	27.46	19.22	22.50	70.65
1974	110.2	189.2	141.7	138.7	399.5	27.82	34.11	26.76	28.93	83.31
1975	138.6	262.0	177.2	177.8	467.9	34.17	46.23	33.86	36.99	97.34
1976	151.1	270.4	190.3	191.6	476.7	37.35	49.78	36.94	40.46	100.66
1977	170.0	313.5	230.2	227.2	531.4	41.16	57.57	43.49	46.81	109.49
1978	208.0	374.2	281.7	280.0	611.8	49.72	68.37	52.55	56.63	123.76
1979	243.1	443.1	339.6	331.4	668.8	58.29	80.66	64.60	67.70	136.64
1980	272.1	536.4	376.5	367.7	680.4	66.36	95.16	73.70	77.02	142.52
1981	336.3	698.6	464.0	453.7	767.4	80.40	122.17	90.03	94.30	159.51
1982	347.4	864.3	515.4	514.4	820.0	86.34	146.20	104.09	108.73	173.34
1983	283.8	608.1	366.5	371.7	570.1	72.65	108.37	79.10	83.34	127.81
1984	262.1	489.8	329.2	326.5	482.5	66.32	88.80	67.18	71.90	106.27
1985	270.4	508.7	372.3	349.4	501.2	66.78	93.09	73.69	75.35	108.09
1986	284.9	522.9	389.2	364.6	511.7	68.35	93.02	76.53	76.88	107.90
1987	246.0	380.4	259.1	279.6	382.0	58.35	69.55	51.05	58.71	80.21
1988	279.4	460.3	366.4	354.7	468.6	62.28	84.65	66.96	70.23	92.78
1989	282.3	457.8	355.4	362.2	461.1	64.92	86.86	67.61	73.55	93.63
1990	321.8	471.3	367.5	383.6	470.2	69.17	90.73	67.49	76.07	93.23
1991	346.9	506.6	441.2	421.5	499.1	73.75	93.10	83.05	82.64	97.86
1992	362.3	426.1	357.6	382.6	442.9	69.50	72.83	67.82	70.27	81.35
1993	356.6	521.2	387.7	426.8	482.9	67.52	83.15	72.56	75.30	85.20
1994	409.5	535.1	491.5	483.2	535.4	70.57	81.90	86.60	79.49	88.07
1995	415.8	629.7	481.2	513.4	557.4	78.09	95.97	84.60	87.22	94.70
1996	341.0	616.0	541.0	496.1	528.6	70.60	98.67	95.74	88.92	94.74
1997	445.6	728.6	655.6	603.9	632.9	90.48	117.55	115.09	107.83	113.01
1998	566.0	815.6	973.2	769.1	797.2	108.88	127.94	157.79	128.97	133.69
1999	783.0	798.4	1,115.5	856.1	874.8	156.45	138.42	182.99	152.02	155.33
2000	593.4	756.9	1,075.4	754.6	754.6	125.96	138.39	181.83	142.16	142.16
2001	729.1	896.5	1,620.4	943.2	921.1[R]	153.72	172.05	271.63	181.94	177.68[R]
2002	882.8	991.9	1,673.4	1,054.2	1,012.8[R]	194.55	175.78	284.17	195.31	187.63[R]
2003	1,037.3	1,106.0	2,065.1	1,199.5	1,131.7	221.13	189.95	345.94	216.27	204.03

*In chained (2000) dollars, calculated by using gross domestic product implicit price deflators.
R=Revised.
Notes: The information reported for 1964 and prior years is not strictly comparable to that in more recent surveys. Average cost is the arithmetic mean and includes all costs for drilling and equipping wells and for surface-producing facilities. Wells drilled include exploratory and development wells; excludes service wells, stratigraphic tests, and core tests. For related information, see http://api-ec.api.org/newsplashpage/index.cfm.

SOURCE: Adapted from "Table 4.8. Costs of Crude Oil and Natural Gas Wells Drilled, 1960–2003," in *Annual Energy Review 2004*, U.S. Department of Energy, Energy Information Administration, Office of Energy Markets and End Use, August 2005, http://www.eia.doe.gov/emeu/aer/pdf/aer.pdf (accessed April 5, 2006). Data from American Petroleum Institute, *2003 Joint Association Survey on Drilling Costs.*

TABLE 7.3

Major U.S. energy companies' expenditures for crude oil and natural gas exploration and development by region, 1974–2003

[Billion dollars[a]]

	United States			Foreign								
Year	Onshore	Offshore	Total	Canada	OECD Europe[b]	Eastern Europe and former U.S.S.R.	Africa	Middle East	Other Eastern Hemisphere[c]	Other Western Hemisphere[d]	Total	Total
1974	NA	NA	8.7	NA	NA	—	NA	NA	NA	NA	3.8	12.5
1975	NA	NA	7.8	NA	NA	—	NA	NA	NA	NA	5.3	13.1
1976	NA	NA	9.5	NA	NA	—	NA	NA	NA	NA	5.2	14.7
1977	6.7	4.0	10.7	1.5	2.5	—	0.7	0.2	0.3	0.4	5.6	16.3
1978	7.5	4.3	11.8	1.6	2.6	—	0.8	0.3	0.4	0.6	6.4	18.2
1979	13.0	8.3	21.3	2.3	3.0	—	0.8	0.2	0.5	0.8	7.8	29.1
1980	16.8	9.4	26.2	3.1	4.3	—	1.4	0.2	0.8	1.0	11.0	37.2
1981	19.9	13.0	33.0	1.8	5.0	—	2.1	0.3	1.9	1.3	12.4	45.4
1982	27.2	11.9	39.1	1.9	6.3	—	2.1	0.4	2.4	1.1	14.2	53.3
1983	16.0	11.1	27.1	1.6	4.3	—	1.7	0.5	2.0	0.6	10.7	37.7
1984	32.1	16.0	48.1	5.4	5.5	—	3.4	0.5	2.0	0.5	17.3	65.3
1985	20.0	8.5	28.5	1.9	3.7	—	1.6	0.9	1.3	0.7	10.1	38.6
1986	12.5	4.9	17.4	1.1	3.2	—	1.1	0.3	1.2	0.6	7.5	24.9
1987	9.7	4.5	14.3	1.9	3.0	—	0.8	0.4	2.8	0.5	9.2	23.5
1988	12.9	8.1	21.0	5.4	4.3	—	0.8	0.4	1.4	0.7	13.0	34.1
1989	9.0	6.0	15.0	6.3	3.5	—	1.0	0.4	2.3	0.6	14.1	29.1
1990	10.2	4.9	15.1	1.8	6.6	—	1.4	0.6	2.4	0.7	13.6	28.7
1991	9.6	4.6	14.2	1.7	6.8	—	1.5	0.5	2.4	0.7	13.7	27.9
1992	7.3	3.0	10.3	1.1	6.8	—	1.4	0.6	2.4	0.6	12.9	23.2
1993	7.2	3.7	10.9	1.6	5.5	0.3	1.5	0.7	2.5	0.6	12.5	23.5
1994	7.8	4.8	12.6	1.8	4.4	0.3	1.4	0.4	2.8	0.7	11.9	24.5
1995	7.7	4.7	12.4	1.9	5.2	0.4	2.0	0.4	2.4	0.9	13.2	25.6
1996	7.9	6.7	14.6	1.6	5.6	0.5	2.8	0.5	4.1	1.6	16.6	31.3
1997	13.0	8.8	21.8	2.0	7.1	0.6	3.0	0.6	3.0	1.6	17.9	39.8
1998	13.5	11.0	24.4	4.8	8.6	1.3	3.1	0.9	3.9	3.7	26.4	50.8
1999	6.6	6.9	13.5	2.1	4.1	0.6	3.1	0.4	3.4	3.8	17.5	31.0
2000	27.1	21.0	48.0	4.9	7.5	0.9	2.7	0.6	6.8	5.4	28.8	76.8
2001	24.2	9.6	33.9	15.3	5.4	0.9	5.5	0.7	5.0	3.1	35.9	69.8
2002	22.3	9.5	31.8	6.7	9.8	1.3	5.1	0.8	6.2	1.6	31.4	63.2
2003	14.7	12.5	27.2	4.9	5.7	2.1	9.2	1.0	4.2	1.1	28.2	55.4

[a]Nominal dollars.
[b]The European members of the Organization for Economic Cooperation and Development (OECD) are Austria, Belgium, Denmark, Finland, France, Germany, Greece, Iceland, Ireland, Italy, Luxembourg, the Netherlands, Norway, Portugal, Spain, Sweden, Switzerland, Turkey, and the United Kingdom, and, for 1997 forward, Czech Republic, Hungary, and Poland.
[c]This region includes areas that are eastward of the Greenwich prime meridian to 180° longitude and that are not included in other domestic or foreign classifications.
[d]This region includes areas that are westward of the Greenwich prime meridian to 180° longitude and that are not included in other domestic or foreign classifications.
NA=Not available. —=Not applicable.
Notes: "Major U.S. energy companies" are the top publicly-owned, U.S.-based crude oil and natural gas producers and petroleum refiners that form the financial reporting system (FRS). Totals may not equal sum of components due to independent rounding. For related information, see http://www.eia.doe.gov/emeu/finance.

SOURCE: "Table 4.10. Major U.S. Energy Companies' Expenditures for Crude Oil and Natural Gas Exploration and Development by Region, 1974–2003 (Billion Dollars)," in *Annual Energy Review 2004*, U.S. Department of Energy, Energy Information Administration, Office of Energy Markets and End Use, August 2005, http://www.eia.doe.gov/emeu/aer/pdf/aer.pdf (accessed April 5, 2006)

TABLE 7.4

Coal demonstrated reserve base, January 1, 2004

[Billion short tons]

		Bituminous coal		Subbituminous coal		Lignite	Total		
Region and state	**Anthracite**	**Underground**	**Surface**	**Underground**	**Surface**	**Surface[a]**	**Underground**	**Surface**	**Total**
Appalachian	**7.3**	**71.4**	**23.1**	**0.0**	**0.0**	**1.1**	**75.4**	**27.6**	**102.9**
Alabama	0.0	1.1	2.1	0.0	0.0	1.1	1.1	3.2	4.3
Kentucky, eastern	0.0	1.4	9.4	0.0	0.0	0.0	1.4	9.4	10.8
Ohio	0.0	17.6	5.8	0.0	0.0	0.0	17.6	5.8	23.4
Pennsylvania	7.2	19.6	0.9	0.0	0.0	0.0	23.4	4.3	27.7
Virginia	0.1	1.1	0.6	0.0	0.0	0.0	1.2	0.6	1.8
West Virginia	0.0	29.5	3.9	0.0	0.0	0.0	29.5	3.9	33.5
Other[b]	0.0	1.1	0.3	0.0	0.0	0.0	1.1	0.3	1.5
Interior	**0.1**	**117.5**	**27.4**	**0.0**	**0.0**	**13.0**	**117.6**	**40.4**	**158.0**
Illinois	0.0	88.0	16.6	0.0	0.0	0.0	88.0	16.6	104.6
Indiana	0.0	8.8	0.8	0.0	0.0	0.0	8.8	0.8	9.6
Iowa	0.0	1.7	0.5	0.0	0.0	0.0	1.7	0.5	2.2
Kentucky, western	0.0	16.0	3.6	0.0	0.0	0.0	16.0	3.6	19.6
Missouri	0.0	1.5	4.5	0.0	0.0	0.0	1.5	4.5	6.0
Oklahoma	0.0	1.2	0.3	0.0	0.0	0.0	1.2	0.3	1.6
Texas	0.0	0.0	0.0	0.0	0.0	12.5	0.0	12.5	12.5
Other[c]	0.1	0.3	1.1	0.0	0.0	0.5	0.4	1.6	1.9
Western	**s**	**21.9**	**2.3**	**121.3**	**60.2**	**29.5**	**143.2**	**92.0**	**235.2**
Alaska	0.0	0.6	0.1	4.8	0.6	s	5.4	0.7	6.1
Colorado	s	7.8	0.6	3.8	0.0	4.2	11.6	4.8	16.4
Montana	0.0	1.4	0.0	69.6	32.6	15.8	71.0	48.4	119.3
New Mexico	s	2.7	0.9	3.5	5.1	0.0	6.2	6.0	12.2
North Dakota	0.0	0.0	0.0	0.0	0.0	9.1	0.0	9.1	9.1
Utah	0.0	5.2	0.3	0.0	0.0	0.0	5.2	0.3	5.5
Washington	0.0	0.3	0.0	1.0	s	s	1.3	0.0	1.3
Wyoming	0.0	3.8	0.5	38.7	21.8	0.0	42.5	22.3	64.8
Other[d]	0.0	0.0	0.0	s	s	0.4	0.0	0.4	0.4
U.S. total	**7.5**	**210.8**	**52.8**	**121.3**	**60.2**	**43.5**	**336.2**	**159.9**	**496.1**
States east of the Mississippi River	7.3	184.3	44.1	0.0	0.0	1.1	188.3	48.6	236.8
States west of the Mississippi River	0.1	26.5	8.7	121.3	60.2	42.4	148.0	111.3	259.3

[a]Lignite resources are not mined underground in the United States.
[b]Georgia, Maryland, North Carolina, and Tennessee.
[c]Arkansas, Kansas, Louisiana, and Michigan.
[d]Arizona, Idaho, Oregon, and South Dakota.
s=Less than 0.05 billion short tons.
Notes: Data represent known measured and indicated coal resources meeting minimum seam and depth criteria, in the ground as of January 1, 2004. These coal resources are not totally recoverable. Net recoverability with current mining technologies ranges from 0 percent (in far northern Alaska) to more than 90 percent. Fifty-four percent of the demonstrated reserve base of coal in the United States is estimated to be recoverable. Totals may not equal sum of components due to independent rounding. For related information, see http://www.eia.doe.gov/fuelcoal.html.

SOURCE: "Table 4.11. Coal Demonstrated Reserve Base, January 1, 2004 (Billion Short Tons)," in *Annual Energy Review 2004*, U.S. Department of Energy, Energy Information Administration, Office of Energy Markets and End Use, August 2005, http://www.eia.doe.gov/emeu/aer/pdf/aer.pdf (accessed April 5, 2006)

TABLE 7.5

World crude oil and natural gas reserves, January 1, 2004

Region and country	Crude oil (billion barrels)		Natural gas (trillion cubic feet)	
	Oil & gas journal	World oil	Oil & gas journal	World oil
North America	**216.5**	**41.4**	**263.1**	**268.9**
Canada	178.9[a]	5.0	59.1	59.1
Mexico	15.7	14.6	15.0	20.7
United States	21.9	21.9	189.0	189.0
Central and South America	**98.8**	**75.2**	**249.4**	**240.9**
Argentina	2.8	2.7	23.4	21.6
Bolivia	0.4	0.5	24.0	27.6
Brazil	8.5	10.6	8.5	8.7
Colombia	1.8	1.5	4.5	4.0
Ecuador	4.6	5.0	0.3	0.4
Peru	0.3	0.9	8.7	8.8
Trinidad and Tobago	1.0	0.8	25.9	19.1
Venezuela	77.8	52.5	148.0	149.2
Other	1.5	0.8	6.1	1.5
Western Europe	**18.4**	**16.4**	**185.1**	**170.1**
Denmark	1.3	1.3	2.6	2.8
Germany	0.4	0.3	10.8	7.7
Italy	0.6	0.5	8.0	4.8
Netherlands	0.1	0.1	62.0	55.1
Norway	10.4	9.4	74.8	74.7
United Kingdom	4.7	4.3	22.2	21.8
Other	0.8	0.6	4.7	3.2
Eastern Europe and former U.S.S.R.	**79.2**	**89.0**	**1,964.2**	**2,693.2**
Hungary	0.1	0.1	1.2	2.4
Kazakhstan	9.0	NA	65.0	NA
Romania	1.0	0.5	3.6	5.0
Russia	60.0	65.4	1,680.0	2,340.5
Other[c]	9.1	23.0	214.4	345.3
Middle East	**726.8**	**686.3**	**2,518.2**	**2.539.7**
Bahrain	0.1	NA	3.3	NA
Iran	125.8	105.0	940.0	935.0
Iraq	115.0	115.0	110.0	112.6
Kuwait[b]	99.0	99.4	55.5	56.6
Oman	5.5	5.7	29.3	31.0
Qatar	15.2	27.4	910.0	913.4
Saudi Arabia[b]	261.9	261.8	231.1	238.5
Syria	2.5	2.4	8.5	18.0
United Arab Emirates	97.8	66.2	212.1	204.1
Yemen	4.0	2.9	16.9	17.0
Other	s	0.7	1.6	13.5
Africa	**87.0**	**104.6**	**453.5**	**443.2**
Algeria	11.3	14.0	160.0	171.5
Angola	5.4	8.8	1.6	4.0
Cameroon	0.4	NA	3.9	NA
Congo (Brazzaville)	1.5	1.4	3.2	4.2
Egypt	3.7	3.6	58.5	7.1
Libya	36.0	30.5	46.4	46.0
Nigeria	25.0	33.0	159.0	180.0
Tunisia	0.3	0.5	2.8	2.6
Other	3.4	12.8	18.1	27.8

TABLE 7.5

World crude oil and natural gas reserves, January 1, 2004 [CONTINUED]

Region and country	Crude oil (billion barrels)		Natural gas (trillion cubic feet)	
	Oil & gas journal	World oil	Oil & gas journal	World oil
Asia and Oceania	**38.3**	**37.7**	**445.1**	**449.9**
Australia	3.5	4.0	90.0	142.9
Brunei	1.4	1.1	13.8	8.3
China	18.3	15.5	53.3	47.9
India	5.4	4.0	30.1	14.6
Indonesia	4.7	5.5	90.3	67.7
Japan	0.1	NA	1.4	NA
Malaysia	3.0	3.1	75.0	57.6
New Zealand	0.1	0.1	1.3	1.5
Pakistan	0.3	0.3	26.8	28.2
Papua New Guinea	0.2	0.3	12.2	13.3
Thailand	0.6	0.5	13.3	12.8
Other	0.9	3.5	37.4	55.3
World	**1,265.0**	**1,050.7**	**6,078.6**	**6,805.8**

[a]Comprises 4.5 billion barrels of conventional crude oil and condensate and 174.4 billion barrels of bitumen in Alberta's oil sands.
[b]Data for Kuwait and Saudi Arabia include one-half of the reserves in the neutral zone between Kuwait and Saudi Arabia.
[c]Albania, Azerbaijan, Belarus, Bulgaria, Czech Republic, Georgia, Kyrgyzstan, Lithuania, Poland, Slovakia, Tajikistan, Turkmenistan, Ukraine, Uzbekistan.
NA=Not available. s=Less than 0.05 billion barrels.
Notes: All reserve figures except those for the former U.S.S.R. and natural gas reserves in Canada are proved reserves recoverable with present technology and prices at the time of estimation. Former U.S.S.R. and Canadian natural gas figures include proved, and some probable reserves. Totals may not equal sum of components due to independent rounding. For related information, see http://www.eia.doe.gov/international.

SOURCE: "Table 11.4. World Crude Oil and Natural Gas Reserves, January 1, 2004," in *Annual Energy Review 2004*, U.S. Department of Energy, Energy Information Administration Office of Energy Markets and End Use, August 2005, http://www.eia.doe.gov/emeu/aer/pdf/aer.pdf (accessed April 5, 2006)

FIGURE 7.4

World recoverable reserves of coal in top reserves countries, 2002

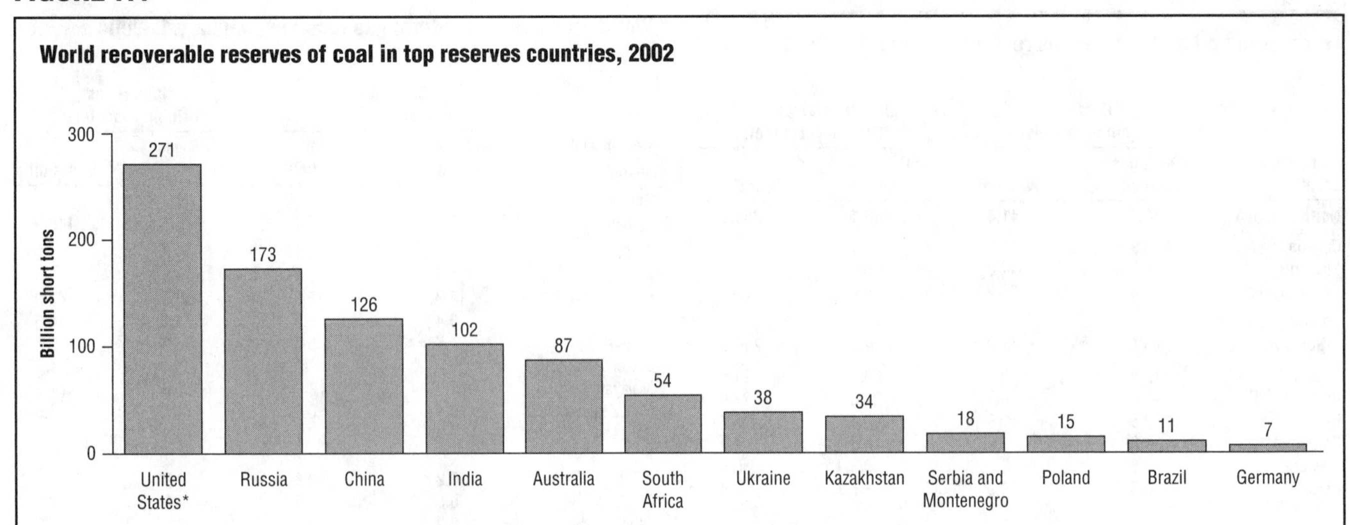

*U.S. reserves are at end of 2003, one year later than other data in this figure.
Note: Data are at end of year.

SOURCE: Adapted from "Figure 11.13. World Recoverable Reserves of Coal, 2002: Top Reserves Countries," in *Annual Energy Review 2004*, U.S. Department of Energy, Energy Information Administration, Office of Energy Markets and End Use, August 2005, http://www.eia.doe.gov/emeu/aer/pdf/aer.pdf (accessed April 5, 2006)

CHAPTER 8
ELECTRICITY

Since 1879, when Thomas Edison flipped the first switch to light Menlo Park, New Jersey, the use of electrical power has become nearly universal in the United States.

WHAT IS ELECTRICITY?

Electricity is a form of energy resulting from the movement of charged particles, such as electrons (negatively charged subatomic particles) and protons (positively charged subatomic particles). Static electricity, for example, is caused by friction: When one material rubs against another it transfers charged particles. The zap you might feel and the spark you might see when you drag your feet along the carpet and then touch a metal doorknob demonstrate static electricity—electrons being transferred between you and the doorknob.

Electric current is the flow of electric charge; it is measured in amperes (amps). Electrical power is the rate at which energy is transferred by electric current. A watt is the standard measure of electrical power, named after the Scottish engineer James Watt (1736–1819). The term *wattage* refers to the amount of electrical power required to operate a particular appliance or device. A kilowatt is a unit of electrical power equal to 1,000 watts, while a kilowatt-hour is a unit of electrical work equal to that done by one kilowatt acting for one hour.

Electrical Capacity

The generating capacity of an electrical power plant, measured in watts, indicates its ability to produce electrical power. A 1,000-kilowatt generator running at full capacity for one hour supplies 1,000 kilowatt-hours of power. That generator operating continuously for an entire year could produce 8.76 million kilowatt-hours of electricity (1,000 kilowatts × 24 hours per day × 365 days per year). However, no generator can operate at 100% capacity during an entire year because of legal restrictions and downtime for routine maintenance and outages. On average about one-fourth of the generating capacity of an electrical plant is not available at any given time.

Electricity demands vary daily and seasonally, so the continuous operation of electrical generators is usually not necessary. Utilities depend on steam, nuclear energy, and large hydroelectric plants to meet routine demand. Auxiliary gas, turbine, internal combustion, and smaller hydroelectric plants are used during short periods of high demand.

An Electric Power System

An electric power system has several components. Figure 8.1 illustrates a simple electric system. Generating units (power plants) produce electricity, transmission lines carry electricity over long distances, and distribution lines deliver the electricity to customers. Substations connect the pieces of the system together, while energy control centers coordinate the operation of all the components.

U.S. ELECTRICITY USAGE

In 2004 net generation of electricity totaled nearly 4 trillion kilowatt-hours. Table 8.1 shows that electricity use in the United States—measured by retail sales of utility companies—has increased continually since 1949. In the thirty years between 1974 and 2004 retail sales of electricity increased 108%, from 1.7 trillion kilowatt-hours to 3.6 trillion kilowatt-hours.

According to the Energy Information Administration (*Annual Energy Review 2004*, 2005, http://tonto.eia.doe.gov/FTPROOT/multifuel/038404.pdf), coal has been and continues to be the most-used raw source for electricity production in the United States. It accounted for approximately 2 trillion kilowatt-hours of electricity in 2004, about half of the electricity generated. (See Figure 8.2

FIGURE 8.1

A simple electric system

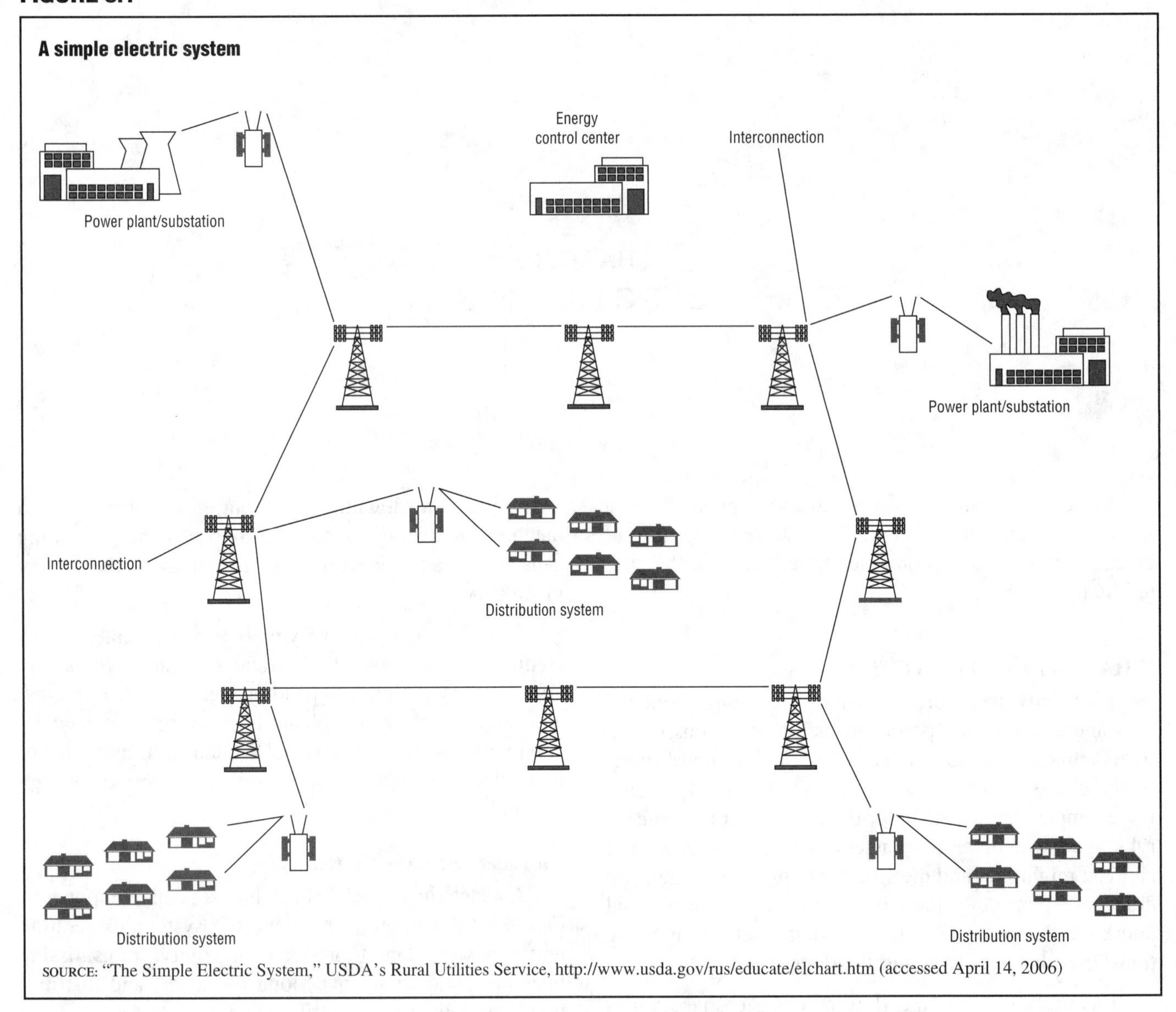

SOURCE: "The Simple Electric System," USDA's Rural Utilities Service, http://www.usda.gov/rus/educate/elchart.htm (accessed April 14, 2006)

and Figure 8.3.) Nuclear power was the second-largest source of electricity (789 billion kilowatt-hours, or 20% of the total), followed by natural gas (619 billion kilowatt-hours, or 18%) and hydroelectric power, a renewable energy source (265 billion kilowatt-hours, or 7%). Very little electricity (59 billion kilowatt-hours, nearly 4%) was generated by all other renewable sources combined, such as geothermal, solar, and wind power.

From 1949 through the early 1990s the industrial sector was the largest consumer of electricity in the United States. Since then sales to the residential sector have been higher. After 1998 sales in the commercial sector became higher than sales in the industrial sector as well. (See Table 8.2.) In 2004 about 1.3 trillion kilowatt-hours went to residential users, 1.2 trillion kilowatt-hours to commercial customers, and 1 trillion kilowatt-hours to industrial users.

Consumption of electricity in general is growing because electricity is being used increasingly to perform tasks that were once done with coal, natural gas, or human muscle: manufacturing steel, assembling cars, and milking cows. Electricity is used extensively in technology fields, such as the computer industry, and residential and commercial customers need electricity to run appliances and machinery, such as air conditioners.

THE ELECTRIC BILL

The cost of electricity is affected by the amount of energy used to create the electricity and move it to the consumer. In 2004, for example, about 40.8 quadrillion Btu of energy were consumed by U.S. utilities to generate 14.2 quadrillion Btu of electricity. After accounting for energy used by the power plants themselves, only 13.5 quadrillion Btu were net generation—the amount available for transmission to customers. (See Figure 8.2.)

TABLE 8.1

Electricity overview, selected years 1949–2004

[Billion kilowatthours]

Year	Net generation				Imports[a]		Exports[a]		T & D losses[e] and unaccounted for[f]	End use		
	Electric power sector[b]	Commercial sector[c]	Industrial sector[d]	Total	From Canada	Total	To Canada	Total		Retail sales[g]	Direct use[h]	Total
1949	291	NA	5	296	NA	2	NA	s	43	255	NA	255
1950	329	NA	5	334	NA	2	NA	s	44	291	NA	291
1955	547	NA	3	550	NA	5	NA	s	58	497	NA	497
1960	756	NA	4	759	NA	5	NA	1	76	688	NA	688
1965	1,055	NA	3	1,058	NA	4	NA	4	104	954	NA	954
1970	1,532	NA	3	1,535	NA	6	NA	4	145	1,392	NA	1,392
1972	1,750	NA	3	1,753	NA	10	NA	3	166	1,595	NA	1,595
1974	1,867	NA	3	1,870	NA	15	NA	3	177	1,706	NA	1,706
1976	2,038	NA	3	2,041	NA	11	NA	2	194	1,855	NA	1,855
1978	2,206	NA	3	2,209	NA	21	NA	1	211	2,018	NA	2,018
1980	2,286	NA	3	2,290	NA	25	NA	4	216	2,094	NA	2,094
1982	2,241	NA	3	2,244	NA	33	NA	4	187	2,086	NA	2,086
1984	2,416	NA	3	2,419	NA	42	NA	3	173	2,286	NA	2,286
1986	2,487	NA	3	2,490	NA	41	NA	5	158	2,369	NA	2,369
1988	2,704	NA	3	2,707	NA	39	NA	7	161	2,578	NA	2,578
1990	2,901	6	131	3,038	16	18	16	16	203[R]	2,713	125[R]	2,837[R]
1992	2,934	6	143	3,084	26	28	2	3	212[R]	2,763	134[R]	2,897[R]
1994	3,089	8	151	3,248	45	47	1	2	211[R]	2,935	146[R]	3,081[R]
1996	3,284	9	151	3,444	42	43	2	3	231[R]	3,101	153[R]	3,254[R]
1998	3,457	9	154	3,620	40	40	12	14	221	3,264	161	3,425
1999	3,530	9	156	3,695	43	43	13	14	240[R]	3,312	172[R]	3,484[R]
2000	3,638	8	157	3,802	49	49	13	15	244[R]	3,421	171[R]	3,592[R]
2001	3,580	7	149	3,737	38	39	16	16	226[R]	3,370	163[R]	3,532[R]
2002	3,698	7	153	3,858	36	36	13	14	253[R]	3,463	166[R]	3,629[R]
2003	3,721[R]	7[R]	155[R]	3,883[R]	29	30	24	24	233[R]	3,488[R]	168[R]	3,656[R]
2004[P]	3,794	7	152	3,953	33	34	22	23	248	3,551	166	3,717

[a]Electricity transmitted across U.S. borders with Canada and Mexico.
[b]Electricity-only and combined-heat-and-power (CHP) plants within the NAICS (North American Industry Classification System) 22 category whose primary business is to sell electricity, or electricity and heat, to the public. Through 1988, data are for electric utilities only; beginning in 1989, data are for electric utilities and independent power producers.
[c]Commercial combined-heat-and-power (CHP) and commercial electricity-only plants.
[d]Industrial combined-heat-and-power (CHP) and industrial electricity-only plants. Through 1988, data are for industrial hydroelectric power only.
[e]Transmission and distribution losses (electricity losses that occur between the point of generation and delivery to the customer).
[f]Data collection frame differences.
[g]Electricity retail sales to ultimate customers by electric utilities and, beginning in 1996, other energy service providers.
[h]Use of electricity that is 1) self-generated, 2) produced by either the same entity that consumes the power or an affiliate, and 3) used in direct support of a service or industrial process located within the same facility or group of facilities that house the generating equipment. Direct use is exclusive of station use.
R=Revised. P=Preliminary. NA=Not available. s=Less than 0.5 billion kilowatthours.
Notes: Totals may not equal sum of components due to independent rounding. For data not shown for 1951–1969, see http://www.eia.doe.gov/emeu/aer/elect.html. For related information, see http://www.eia.doe.gov/fuelelectric.html.

SOURCE: Adapted from "Table 8.1. Electricity Overview, Selected Years, 1949–2004 (Billion Kilowatthours)," in *Annual Energy Review 2004*, U.S. Department of Energy, Energy Information Administration, Office of Energy Markets and End Use, August 2005, http://www.eia.doe.gov/emeu/aer/pdf/aer.pdf (accessed April 5, 2006)

Almost 27.3 quadrillion Btu were lost when fuel was converted, and about 1.3 quadrillion Btu were lost during transmission and distribution (labeled *T & D losses* in Figure 8.2). In the end, for every three units of energy that were converted to create electricity in 2004, slightly less than one unit actually reached the end user.

Between 1960 and 1970, the price of electricity declined, but it began to increase during the 1970s because of an oil embargo by the Organization of Petroleum Exporting Countries. (See Figure 8.4.) From the mid-1980s to 2004 the price of electricity dropped because prices of energy resources declined. Prices often varied by location. As Figure 8.5 shows, in 2004 electricity was most expensive in Hawaii, New York, California, New Hampshire, Vermont, Alaska, Rhode Island, Massachusetts, New Jersey, and Connecticut. According to the *Electric Power Annual 2004* (Energy Information Administration, 2005, http://www.eia.doe.gov/cneaf/electricity/epa/epa_sum.html), the average price of electricity sold to the residential sector was $0.09 per kilowatt-hour in 2004, while the commercial sector paid $0.082 per kilowatt-hour. Industrial users paid only $0.053 per kilowatt-hour in 2004 because the huge amounts of electricity they use allowed them to receive volume discounts. The average price for all sectors across the United States in 2004 was $0.076 per kilowatt-hour.

DEREGULATION OF ELECTRIC UTILITIES

Regulated by the government for decades, electric utilities are in the midst of a highly controversial shift toward unregulated markets and increased competition. In 1978 Congress passed the Public Utilities Regulatory

FIGURE 8.2

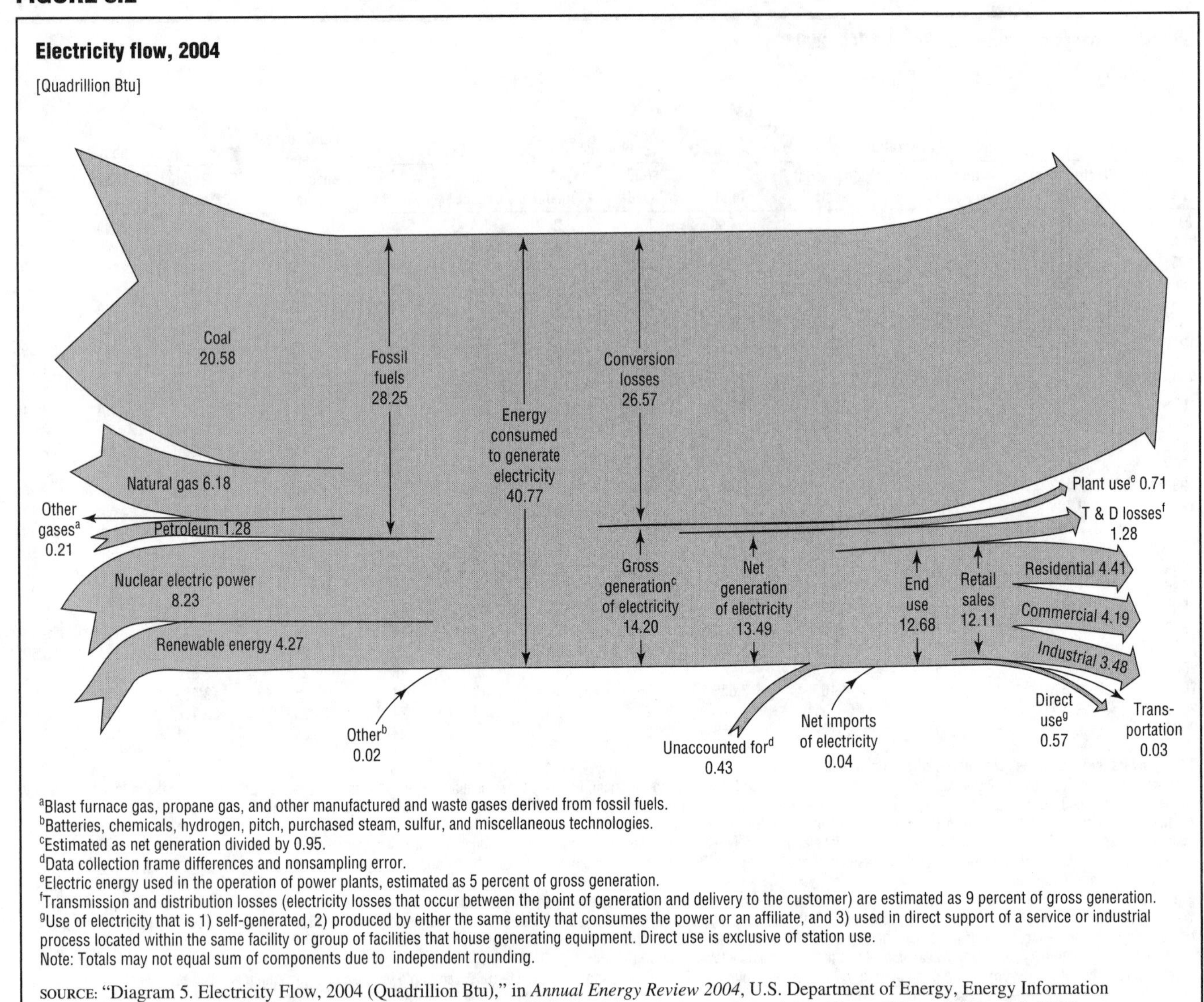

Electricity flow, 2004

[Quadrillion Btu]

[a]Blast furnace gas, propane gas, and other manufactured and waste gases derived from fossil fuels.
[b]Batteries, chemicals, hydrogen, pitch, purchased steam, sulfur, and miscellaneous technologies.
[c]Estimated as net generation divided by 0.95.
[d]Data collection frame differences and nonsampling error.
[e]Electric energy used in the operation of power plants, estimated as 5 percent of gross generation.
[f]Transmission and distribution losses (electricity losses that occur between the point of generation and delivery to the customer) are estimated as 9 percent of gross generation.
[g]Use of electricity that is 1) self-generated, 2) produced by either the same entity that consumes the power or an affiliate, and 3) used in direct support of a service or industrial process located within the same facility or group of facilities that house generating equipment. Direct use is exclusive of station use.
Note: Totals may not equal sum of components due to independent rounding.

SOURCE: "Diagram 5. Electricity Flow, 2004 (Quadrillion Btu)," in *Annual Energy Review 2004*, U.S. Department of Energy, Energy Information Administration, Office of Energy Markets and End Use, August 2005, http://www.eia.doe.gov/emeu/aer/pdf/aer.pdf (accessed April 5, 2006)

Policies Act (PL 95-617), which required utilities to buy electricity from private companies when that would be cheaper than building their own power plants. The Energy Policy Act of 1992 (PL 102-486) gave other electricity generators greater access to the market, resulting in widespread debates about regulatory, economic, energy, and environmental policies. State public utility commissions conducted proceedings and crafted rules related to competition.

California was a leader in deregulation. In the summer of 2000, however, the state experienced rolling electrical blackouts, and electricity bills doubled for many customers. Fearful of similar blackouts and price spikes, most other states had slowed or stopped their efforts to deregulate their electricity markets by the spring of 2001. At that time, twenty-four states and the District of Columbia had begun deregulation. Then, during an investigation of Enron Corporation, documents were found that showed how Enron's electricity traders had boosted profits by using strategies that added to electricity costs and congestion on transmission lines. As a result, public confidence in power companies, in general, and deregulation, in particular, eroded. As of February 2003 only seventeen states plus the District of Columbia were actively engaged in restructuring their utilities. (See Figure 8.6, which shows the most recent data available.) In addition, five states had delayed deregulation, and California had suspended its restructuring activities. Restructuring was not active in twenty-seven states.

INTERNATIONAL ELECTRICITY USAGE

World Production

In 2003 approximately 15.8 trillion kilowatt-hours of electricity were generated around the world, including 10.4 trillion kilowatt-hours generated by fossil fuels,

FIGURE 8.3

Electricity net generation by major sources, 1949–2004

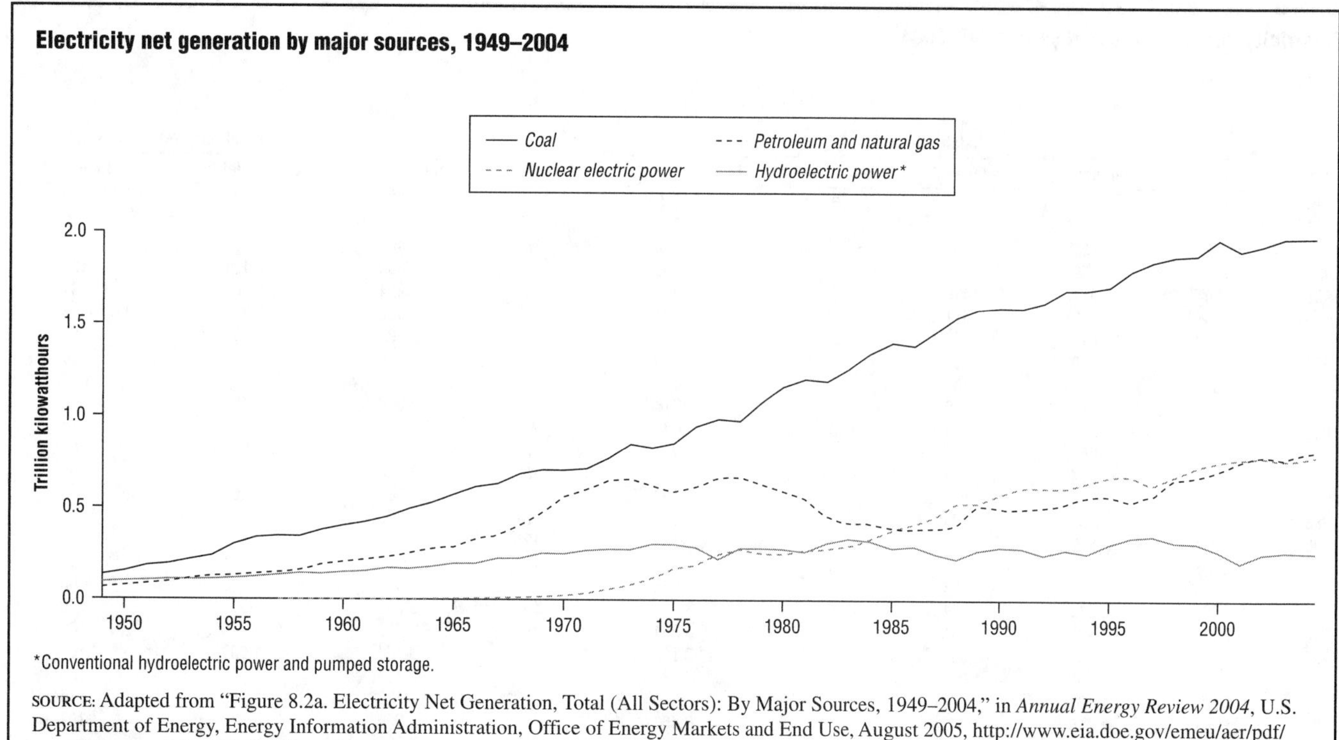

*Conventional hydroelectric power and pumped storage.

SOURCE: Adapted from "Figure 8.2a. Electricity Net Generation, Total (All Sectors): By Major Sources, 1949–2004," in *Annual Energy Review 2004*, U.S. Department of Energy, Energy Information Administration, Office of Energy Markets and End Use, August 2005, http://www.eia.doe.gov/emeu/aer/pdf/aer.pdf (accessed April 5, 2006)

2.6 trillion kilowatt-hours generated by hydroelectric power, 2.5 trillion kilowatt-hours from nuclear power sources, and 0.3 trillion kilowatt-hours from wood, waste, wind, and other sources. (See Figure 8.7.) According to the *Annual Energy Review 2004*, the United States accounted for 24.7% of this production; China, 11.4%; Japan, 6.3%; and Russia, 5.7%. Figure 8.8 shows net generation of electricity by type and by region of the world.

World Consumption

According to the *International Energy Annual 2004* (Energy Information Administration, 2005, http://www.eia.doe.gov/iea/elec.html), total world electricity consumption increased from 9 trillion kilowatt-hours in 1986 to 14.8 trillion kilowatt-hours in 2003. North America, Central America, and South America consumed 5.2 trillion kilowatt-hours, or 35% of the total; Asia and Oceania used 4.4 trillion kilowatt-hours (30%); Europe, 3.1 trillion (about 21%); and Eurasia, 1.2 trillion (7.9%). The Middle East and Africa each consumed about 3% of the world's electricity.

TRENDS IN THE U.S. ELECTRICAL POWER INDUSTRY

In *Annual Energy Outlook 2006* (2006, http://www.eia.doe.gov/oiaf/aeo/index.html), the Energy Information Administration predicted that from 2004 to 2025, total electricity consumption would grow at a rate of 1.6% annually. This compares with 7% growth per year during the 1960s. Several factors led to this diminished growth in consumption. For example, in the 1960s consumers were just beginning to switch to electrical appliances, which increased the amount of electricity each household used; by 2004 appliances were commonplace and new purchases often replaced existing appliances and did not drive up total consumption. The efficiency of appliances had also improved by 2004, so each one used less electricity. Commercial demand was expected to grow by 2.2% per year between 2004 and 2025 because of growth in commercial floor space, while industrial demand was expected to increase by 1.6% per year as industrial output rises.

Demand for electricity in the United States has always been related to economic growth. However, electricity use is expected to grow more slowly than the gross domestic product, a measure of economic growth. Figure 8.9 shows how electricity sales are related more to economic growth than to population growth. Note that the phrase *five-year moving average* means that each point on the graph is an average for that year's data plus the previous four years' data. This type of averaging is used to determine long-term averages without the weight of such cyclical influences as the weather.

The rate of growth of consumption carries financial risks for electric companies. If the industry underestimates future needs for electricity, consumers could experience power shortages or losses. Excessive projections of the

TABLE 8.2

Electricity end use, selected years 1949–2004

[Billion kilowatt-hours]

Year	Retail sales[a]					Direct use[f]	Total end use[g]	Discontinued retail sales series	
	Residential	Commercial[b]	Industrial[c]	Transportation[d]	Total retail sales[e]			Commercial (old)[h]	Other (old)[i]
1949	67	59[E]	123	6[E]	255	NA	255	45	20
1950	72	66[E]	146	7[E]	291	NA	291	51	22
1955	128	103[E]	260	6[E]	497	NA	497	79	29
1960	201	159[E]	324	3[E]	688	NA	688	131	32
1965	291	231[E]	429	3[E]	954	NA	954	200	34
1970	466	352[E]	571	3[E]	1,392	NA	1,392	307	48
1972	539	413[E]	641	3[E]	1,595	NA	1,595	359	56
1974	578	440[E]	685	3[E]	1,706	NA	1,706	385	58
1976	606	492[E]	754	3[E]	1,855	NA	1,855	425	70
1978	674	531[E]	809	3[E]	2,018	NA	2,018	461	73
1980	717	559	815	3	2,094	NA	2,094	488	74
1982	730	609	745	3	2,086	NA	2,086	526	86
1984	780	664	838	4	2,286	NA	2,286	583	85
1986	819	715	831	4	2,369	NA	2,369	631	89
1988	893	784	896	5	2,578	NA	2,578	699	90
1990	924	838	946	5	2,713	125[R]	2,837[R]	751	92
1992	936	850	973	5	2,763	134[R]	2,897[R]	761	93
1994	1,008	913	1,008	5	2,935	146[R]	3,081[R]	820	98
1996	1,083	980	1,034	5	3,101	153[R]	3,254[R]	887	98
1998	1,130	1,078	1,051	5	3,264	161	3,425	979	104
1999	1,145	1,104	1,058	5	3,312	172[R]	3,484[R]	1,002	107
2000	1,192	1,159	1,064	5	3,421	171[R]	3,592[R]	1,055	109
2001	1,203	1,197	964	5	3,370	163[R]	3,532[R]	1,089	114
2002	1,267	1,218	972	6[R]	3,463	166[R]	3,629[R]	1,116	107
2003	1,273[R]	1,200[R]	1,008[R]	7[R]	3,488[R]	168[R]	3,656[R]	—	—
2004[P]	1,293	1,229	1,021	8	3,551	166[E]	3,717	—	—

[a]Electricity retail sales to ultimate customers reported by electric utilities and, beginning in 1996, other energy service providers.
[b]Commercial sector, including public street and highway lighting, interdepartmental sales, and other sales to public authorities.
[c]Industrial sector. Through 2002, excludes agriculture and irrigation; beginning in 2003, includes agriculture and irrigation.
[d]Transportation sector, including sales to railroads and railways.
[e]The sum of "residential," "commercial," "industrial," and "transportation."
[f]Use of electricity that is 1) self-generated, 2) produced by either the same entity that consumes the power or an affiliate, and 3) used in direct support of a service or industrial process located within the same facility or group of facilities that house the generating equipment. Direct use is exclusive of station use.
[g]The sum of "total retail sales" and "direct use."
[h]"Commercial (old)" is a discontinued series—data are for the commercial sector, excluding public street and highway lighting, interdepartmental sales, and other sales to public authorities.
[i]"Other (old)" is a discontinued series—data are for public street and highway lighting, interdepartmental sales, other sales to public authorities, agriculture and irrigation, and transportation including railroads and railways.
R=Revised. P=Preliminary. E=Estimate. NA=Not available. —=Not applicable.
Note: Totals may not equal sum of components due to independent rounding. For data not shown for 1951–1969, see http://www.eia.doe.gov/emeu/aer/elect.html. For related information, see http://www.eia.doe.gov/fuelelectric.html.

SOURCE: Adapted from "Table 8.9. Electricity End Use, Selected Years, 1949–2004 (Billion Kilowatt-hours)," *Annual Energy Review 2004*, U.S. Department of Energy, Energy Information Administration, Office of Energy Markets and End Use, August 2005, http://www.eia.doe.gov/emeu/aer/pdf/aer.pdf (accessed April 5, 2006)

nation's needs, by contrast, could mean billions of dollars spent on unneeded equipment.

The Energy Information Administration estimated that the United States would need 347 gigawatts (a gigawatt, pronounced "jigawatt," equals one billion watts) of new generating capacity from 2005 to 2030 to meet growing demand for electricity and to replace aging power plants, most of it after 2015. From 2005 to 2030, sixty-five gigawatts of capacity are expected to be taken out of production, mainly old fossil-fired plants that are not competitive with newer types of fossil-fired plants.

The agency noted that the high cost of natural gas, petroleum, and coal in 2004–05 led to a jump in electricity prices. By 2015, however, as new sources of natural gas and coal are brought on line, electricity prices should fall to $0.071 per kilowatt-hour. After 2015, natural gas and petroleum prices would rise but electricity producers would rely more on coal for power generation. The result would be a slow rise in electricity prices to $0.075 cents per kilowatt-hour in 2030.

Continued concerns about pollution and global warming could result in tightened environmental emission standards, which could, in turn, have an impact on electrical utility expansion, supply, and prices. Advances in solar and wind turbine technology could make renewable sources of electrical power more economical. Some energy experts and environmentalists claim that increased efficiency and conservation efforts are the most sensible alternatives to new construction or to the burning of more fossil fuels in existing plants.

FIGURE 8.4

Average real[a] retail prices of electricity sold by electric utilities, 1960–2004

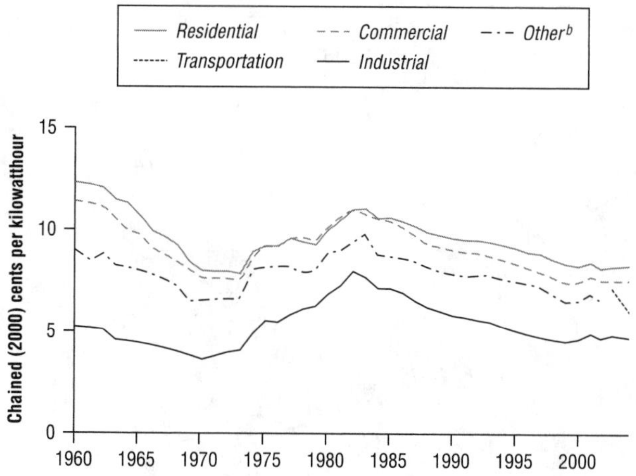

Residential | Commercial | Other[b] | Transportation | Industrial

Chained (2000) cents per kilowatthour

0, 5, 10, 15

1960 1965 1970 1975 1980 1985 1990 1995 2000

[a]In chained (2000) dollars, calculated by using gross domestic product implicit price deflators.

[b]Public streeet and highway lighting, other sales to public authorities, sales to railroads and railways, and interdepartmental sales.

SOURCE: Adapted from "Figure 8.10. Average Retail Prices of Electricity: By Sector[1], Real Prices, 1960–2004," in *Annual Energy Review 2004*, U.S. Department of Energy, Energy Information Administration, Office of Energy Markets and End Use, August 2005, http://www.eia.doe.gov/emeu/aer/pdf/aer.pdf (accessed April 5, 2006)

FIGURE 8.5

Average electricity rates by state, 2004

[Cents per kilowatt-hour]

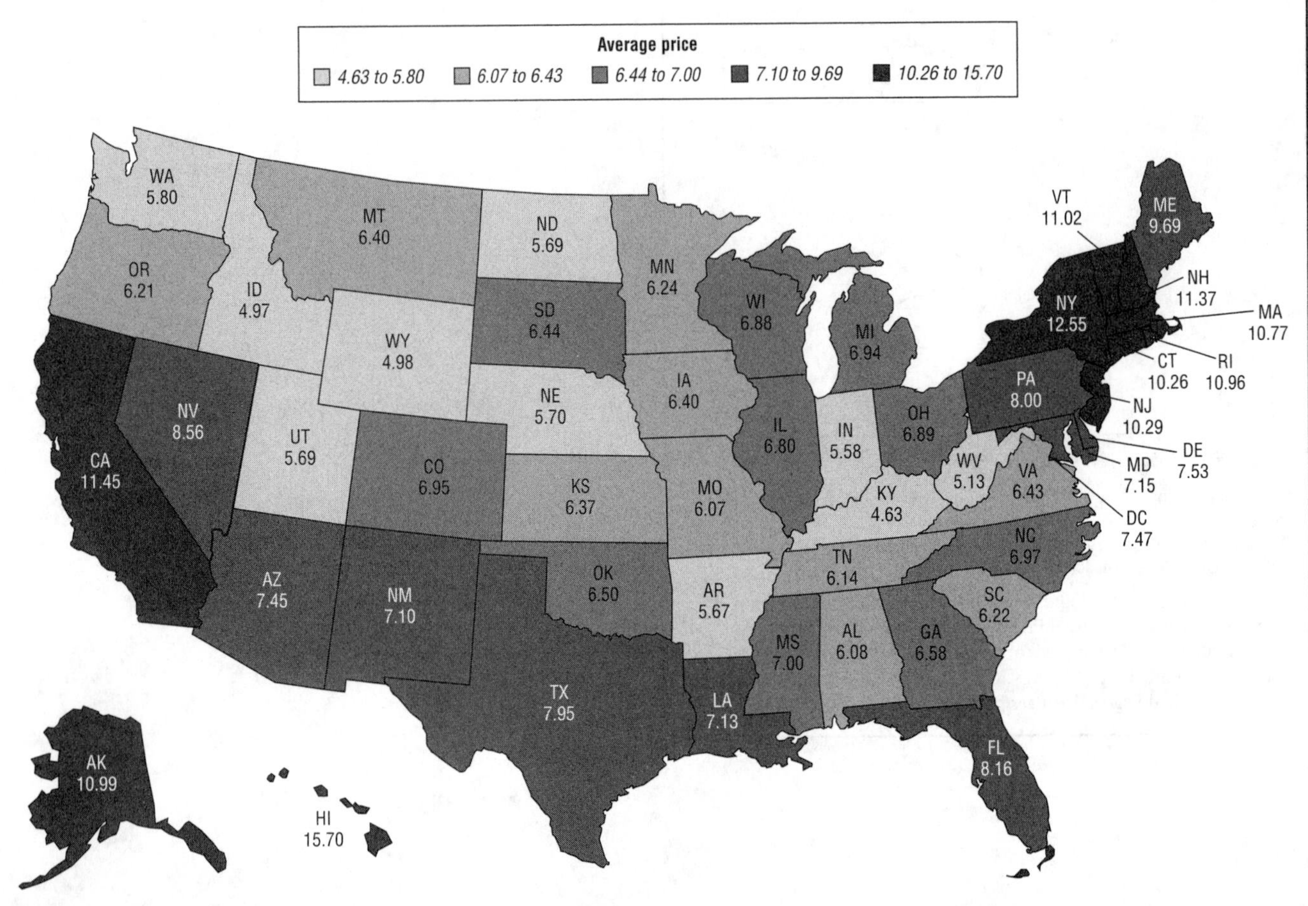

Note: U.S. total average price per kilowatt-hour is 7.62 cents.

SOURCE: "Figure 7.4. U.S. Electric Power Industry Average Retail Price of Electricity by State, 2004 (Cents per Kilowatt-hour)," in *Electric Power Annual 2004*, U.S. Department of Energy, Energy Information Administration, Office of Coal, Nuclear, Electric and Alternate Fuels, November 2005, http://www.eia.doe.gov/cneaf/electricity/epa/epa.pdf (accessed April 14, 2006)

FIGURE 8.6

Status of state electric industry restructuring activity, as of February 2003

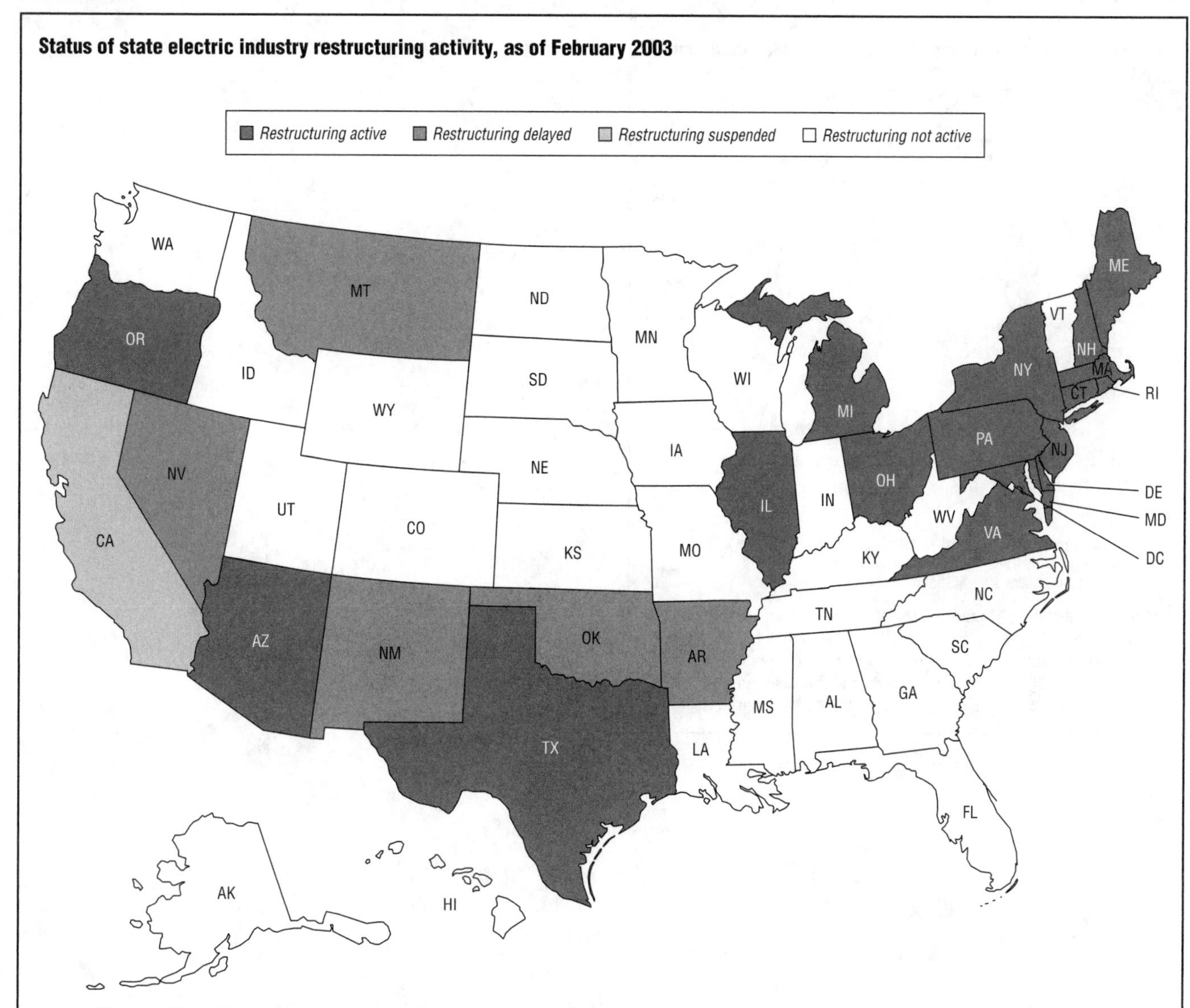

SOURCE: "Status of State Electric Industry Restructuring Activity as of February 2003," U.S. Department of Energy, Energy Information Administration, http://www.eia.doe.gov/cneaf/electricity/chg_str/restructure.pdf (accessed April 14, 2006)

FIGURE 8.7

World net generation of electricity by type, 1980, 1990, and 2003

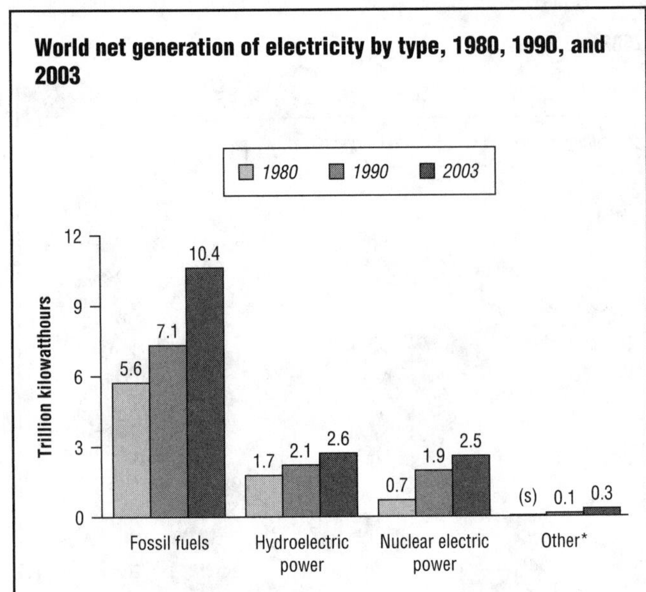

*Wood, waste, geothermal, solar, wind, batteries, chemicals, hydrogen, pitch, purchased steam, sulfur, and miscellaneous technologies.
(s)=Less than 0.05 trillion kilowatt-hours.

SOURCE: Adapted from "Figure 11.16. World Net Generation of Electricity: Net Generation by Type, 1980, 1990, and 2003," in *Annual Energy Review 2004*, U.S. Department of Energy, Energy Information Administration, Office of Energy Markets and End Use, August 2005, http://www.eia.doe.gov/emeu/aer/pdf/aer.pdf (accessed April 5, 2006)

FIGURE 8.8

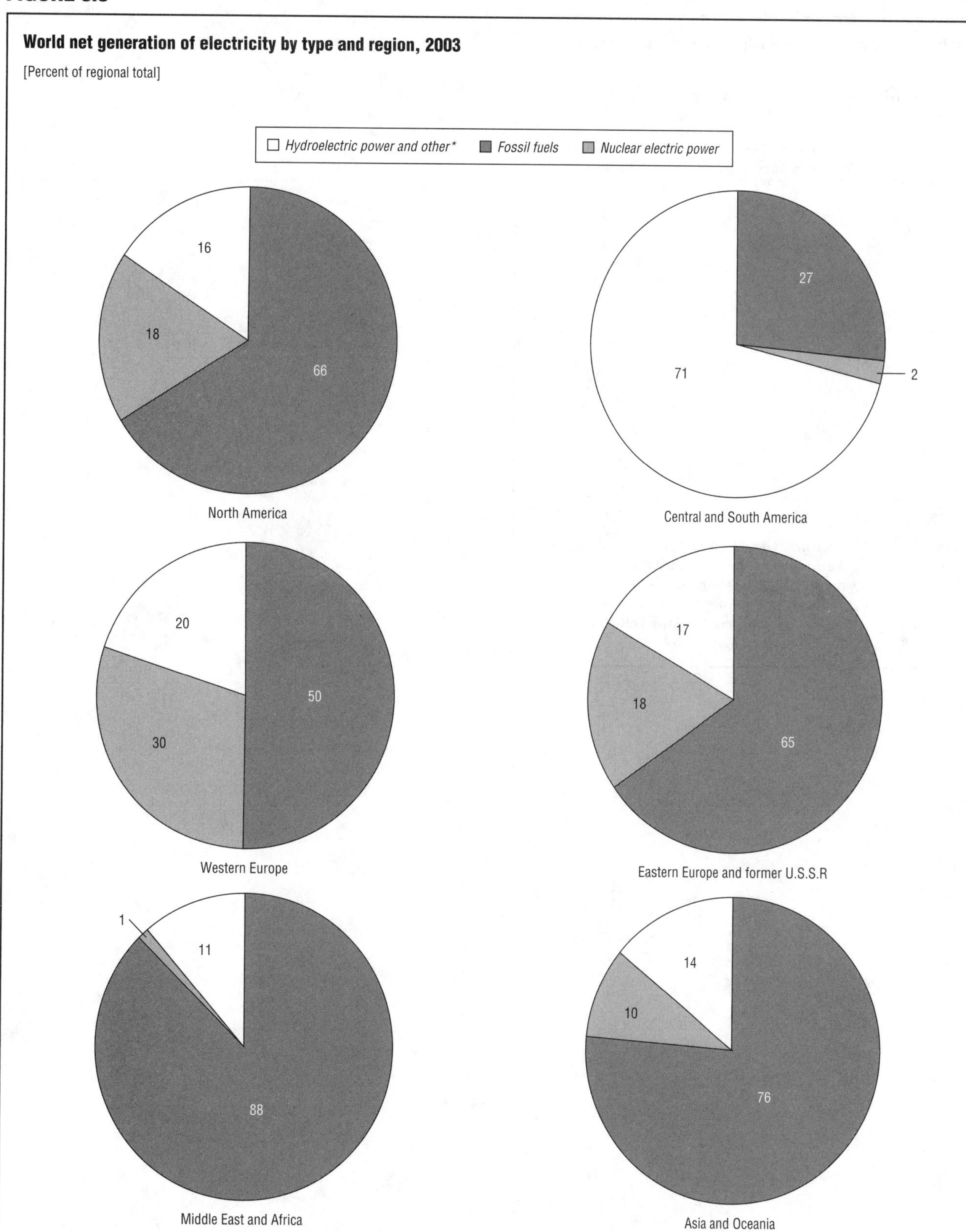

*Wood, waste, geothermal, solar, wind, batteries, chemicals, hydrogen, pitch, purchased steam, sulfur, and miscellaneous technologies.

SOURCE: Adapted from "Figure 11.16. World Net Generation of Electricity: Net Generation by Type by Region, 2003 (Percent of Regional Total)," in *Annual Energy Review 2004*, U.S. Department of Energy, Energy Information Administration, Office of Energy Markets and End Use, August 2005, http://www.eia.doe.gov/emeu/aer/pdf/aer.pdf (accessed April 5, 2006)

FIGURE 8.9

Population, gross domestic product, and electricity sales, 1965–2025

[Five-year moving average annual percent growth]

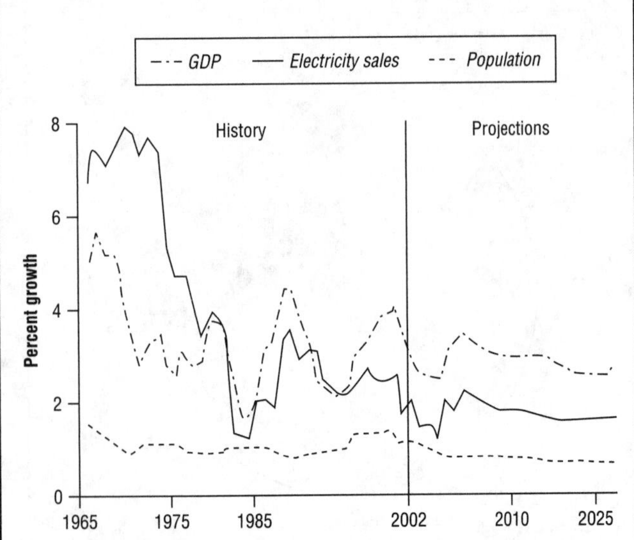

SOURCE: "Figure 67. Population, Gross Domestic Product, and Electricity Sales, 1965–2025 (5-Year Moving Average Annual Percent Growth)," in *Annual Energy Outlook 2004*, U.S. Department of Energy, Energy Information Administration, Office of Integrated Analysis and Forecasting, January 2004, http://tonto.eia.doe.gov/FTPROOT/forecasting/0383(2004).pdf (accessed May 7, 2006)

CHAPTER 9
ENERGY CONSERVATION

ENERGY CONSERVATION AND EFFICIENCY

Energy conservation is the efficient use of energy, without necessarily curtailing the services that energy provides. Conservation occurs when societies develop efficient technologies that reduce energy needs. Environmental concerns, such as acid rain and the potential for global warming, have increased public awareness about the importance of energy conservation.

Energy efficiency can be measured by two indicators. The first is energy consumption per person (per capita) per year. Annual per person energy consumption in the United States was 215 million Btu in 1949. It topped out at 360 million Btu in 1978 and 1979; dropped to 313 million Btu by 1983; and then slowly rose until it reached 352 million Btu in 2000. It leveled off until 2004, when the annual rate of consumption per capita was 340 million Btu. (See Figure 9.1.)

The second indicator of efficiency is energy consumption per dollar of gross domestic product, the total value of goods and services produced by a nation. When a country grows in its energy efficiency, it uses less energy to produce the same amount of goods and services. In 1949 nearly 19,600 Btu of energy were consumed for each dollar of gross domestic product. (See Figure 9.2.) In 1970 about 17,990 Btu of energy were consumed per dollar, and by 2004 only 9,200 Btu were used per dollar of gross domestic product.

ENERGY CONSERVATION, PUBLIC HEALTH, AND THE ENVIRONMENT

People living in cities with high levels of pollution have higher risks of mortality from certain diseases than those living in less polluted cities. Energy-related emissions generate a vast majority of these polluting chemicals. (Table 4.3 in Chapter 4 shows some air pollutants and their sources.) According to the American Lung Association, air pollution has been related to such diseases as asthma, bronchitis, emphysema, and lung cancer. The group estimated the annual health costs of exposure to the most serious air pollutants to be in the billions. Clean and efficient energy technologies, they say, represent a cost-effective investment in public health.

Global warming is long-term climate change—a worldwide temperature increase—caused by the "greenhouse effect." This occurs when heat is trapped within the atmosphere by high levels of carbon dioxide, methane, nitrogen oxide, hydrofluorocarbons, sulfur dioxides, and perfluorocarbons. Just as the glass of a greenhouse or the windows of a car trap heat, the "greenhouse gases" keep Earth warmer than it would be if the atmosphere contained only oxygen and nitrogen.

The Intergovernmental Panel on Climate Change

In 1988 the United Nations established the Intergovernmental Panel on Climate Change, a group of 2,000 of the world's leading scientists. It reported in 1995 that global warming is real, serious, and accelerating. The most likely cause, the group said, is primarily the burning of coal, oil, and gasoline, which has increased the amount of carbon dioxide and other greenhouse gases in the atmosphere. Deforestation is another factor, because it reduces the amount of carbon dioxide absorbed and stored in plants.

In its *Third Assessment Report* in 2001 (another is scheduled for publication in 2007), the group said it had a clearer understanding of the causes and consequences of global warming, largely because so much climate research and environmental monitoring had been undertaken. The group described the effect that global warming would have on weather patterns, water resources, the seasons, ecosystems, and extreme climate events, and it urged governments to move quickly with policies to protect the planet.

FIGURE 9.1

Energy consumption per person, 1949–2004

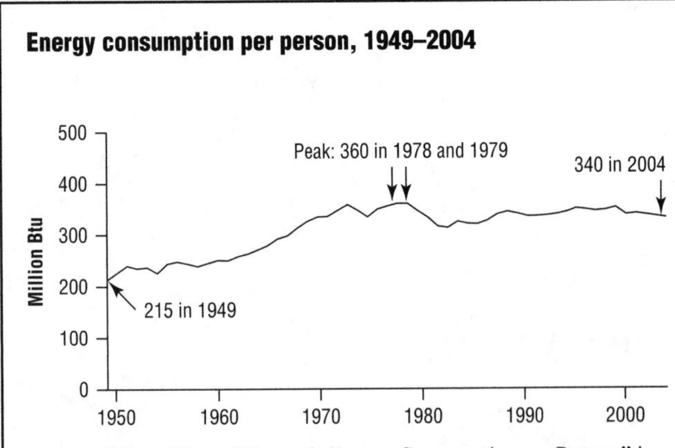

SOURCE: Adapted from "Figure 2. Energy Consumption per Person," in *Annual Energy Review 2004*, U.S. Department of Energy, Energy Information Administration, Office of Energy Markets and End Use, August 2005, http://www.eia.doe.gov/emeu/aer/pdf/aer.pdf (accessed April 5, 2006)

FIGURE 9.2

Energy consumption per dollar of gross domestic product, 1949–2004

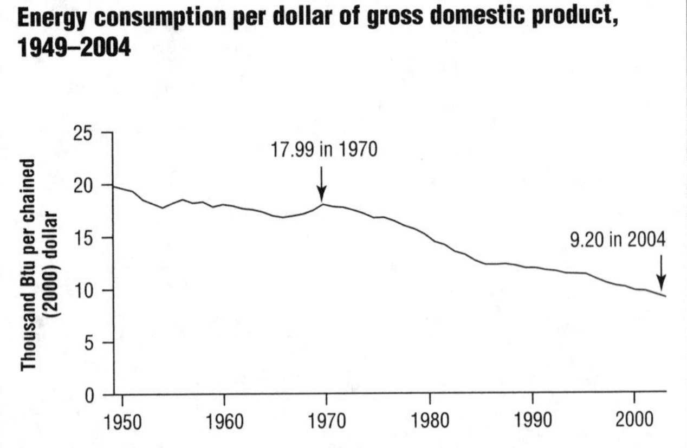

SOURCE: Adapted from "Figure 3. Energy Use per Dollar of Gross Domestic Product," in *Annual Energy Review 2004*, U.S. Department of Energy, Energy Information Administration, Office of Energy Markets and End Use, August 2005, http://www.eia.doe.gov/emeu/aer/pdf/aer.pdf (accessed April 5, 2006)

The Kyoto Protocol

In December 1997 the United Nations convened a 160-nation conference on global warming in Kyoto, Japan, to develop a treaty on climate change that would place binding caps on industrial emissions. The initial draft of the treaty, called the Kyoto Protocol to the United Nations Framework Convention on Climate Change (or simply the Kyoto Protocol), bound industrialized nations to reducing their emissions of six greenhouse gases below 1990 levels. It marked the first time nations made such sweeping pledges to cut emissions.

Each country had a different target to reach by 2012: the United States was to cut emissions by 7%, most European nations by 8%, and Japan by 6%. Reductions were to begin by 2008. Developing nations were not required to make such pledges. The United States had proposed a program of voluntary pledges by developing nations, but that section was deleted, as was a tough system of enforcement. Instead, each country was to decide for itself how to achieve its goal. The draft treaty provided market-driven tools for reducing emissions. For example, nations would be allowed to sell emissions "credits" to other nations. The draft treaty also set up a Clean Development Fund to help poorer nations with technology to reduce their emissions.

Getting the treaty ratified proved difficult, however. President Bill Clinton signed the protocol but the U.S. Senate did not ratify it. To save the treaty, diplomats from 178 nations met in Bonn, Germany, in July 2001 and drafted a compromise treaty. In October 2001, 2,000 delegates from 160 countries worked for 12 days in Marrakech, Morocco, to complete a final draft of the Kyoto Protocol.

President George W. Bush rejected that final draft of the Kyoto Protocol, characterizing it as "fatally flawed." He said implementation of the treaty would harm the U.S. economy and unfairly require only the industrial nations to cut emissions. In particular, he noted that neither China, the world's second-largest emitter of greenhouse gases, nor India, another high emitter, were bound by the protocol. He said at the time:

> Our country, the United States, is the world's largest emitter of man-made greenhouse gases. We account for almost 20% of the world's man-made greenhouse emissions. We also account for about one-quarter of the world's economic output. We recognize the responsibility to reduce our emissions. We also recognize the other part of the story—that the rest of the world emits 80% of all greenhouse gases. And many of those emissions come from developing countries. . . .
>
> We recognize our responsibility and will meet it—at home, in our hemisphere, and in the world. My Cabinet-level working group on climate change is recommending a number of initial steps, and will continue to work on additional ideas. The working group proposes the United States help lead the way by advancing the science on climate change, advancing the technology to monitor and reduce greenhouse gases, and creating partnerships within our hemisphere and beyond to monitor and measure and mitigate emissions.

When Russia ratified the treaty in November 2004 the Kyoto Protocol had support from countries whose emissions totaled 55% of the world's greenhouse gases, which was the minimum needed for the treaty to go into effect. At the end of that year 130 countries had signed on, including all the European Union members, Japan, and Norway. The Kyoto Protocol took effect on February 16, 2005, without the support of the United States and Australia.

U.S. attitudes about the Kyoto Protocol varied widely. Most business leaders said the treaty went too far and was too costly for the U.S. economy, while environmentalists said the treaty's standards did not go far enough. Some experts doubted that any action emerging from Kyoto would be sufficient to prevent the doubling of greenhouse gases. In May 2005—in response to the Bush administration's policies—mayors from 132 cities in the United States joined a bipartisan coalition to fight global warming on the local level. The coalition pledged to have their cities meet Kyoto Protocol requirements for the United States: a reduction of heat-trapping gas emissions to 7% below 1990 levels by 2012.

Levels of greenhouse gas emissions for the United States for 1990 and from 1998 to 2004 are shown in Table 9.1 in teragrams of carbon dioxide equivalents (Tg $C0_2$ Eq.). A teragram is a trillion grams. Each gas in the table is reported by its global-warming potential in Tg, which allows these numbers to be compared. Higher numbers indicate greater harm done to the environment by the gas. Conversely, numbers in parentheses are for sources, such as land-use change and forestry, that reduce greenhouse gas emissions by the carbon equivalents shown. In 2004 emissions of greenhouse gases in the United States reached 6,294.3 Tg, which was 1,085.7 Tg, or 21%, higher than emissions in 1990.

ENERGY EFFICIENCY IN TRANSPORTATION

The U.S. transportation system plays a central role in the economy. Highway transportation is dependent on vehicles with internal combustion engines, which are fueled almost exclusively by petroleum. According to the Energy Information Administration of the U.S. Department of Energy (*Annual Energy Review 2004*, 2005), the transportation sector accounted for 28% of all energy consumed in the United States in 2004. Americans used 27.8 quadrillion Btu of energy for transportation that year (see Figure 1.9 in Chapter 1), of which petroleum made up 97%. Despite the improvements in transportation efficiency in recent decades, the agency expects the transportation sector to consume 37.3 quadrillion Btu in 2030 (*International Energy Outlook 2006*, 2006).

Automotive Efficiency

Policy makers interested in the energy used for transportation have an array of conservation options. (See Table 9.2.) However, not all options are mutually supportive. For example, efforts to promote a freer flow of automobile traffic, such as high-occupancy vehicle (HOV) lanes or free parking for car pools, may sabotage efforts to shift travelers to mass transit or to reduce trip lengths and frequency.

In the United States the automobile dominates the transportation sector; cars and light-duty vehicles used 83% of all transportation energy in 2006, as reported by the Bureau of Transportation Statistics (*National Transportation Statistics, 2006*, 2006). The *Annual Energy Review 2004* also noted that motor gasoline, which is divided among passenger cars, light- and heavy-duty trucks, aircraft, and miscellaneous other modes of transportation, consumed about 68% of the oil used in the United States in 2004.

The major growth in fuel use since the 1970s has been that consumed by pickup trucks and, in more recent years, that consumed by vans and sport-utility vehicles. In 2004 the amount used in pickup trucks, vans, and sport-utility vehicles was approximately 75% of the amount used in automobiles and motorcycles. Meanwhile, the use of automobile fuel has remained fairly constant because increases in fuel efficiency have offset the growth in car miles traveled. Boosting efficiency of trucks, vans, and sport-utility vehicles will become increasingly important in controlling the demand for oil.

THE CORPORATE AVERAGE FUEL ECONOMY STANDARDS. The 1973 oil embargo by the Organization of the Petroleum Exporting Countries painfully reminded the United States how dependent it had become on foreign sources of fuel. It prompted Congress to pass the 1975 Energy Policy and Conservation Act (PL 94-163), which set the initial Corporate Average Fuel Economy standards. The standards were modified in 1980 with the Automobile Fuel Efficiency Act (PL 96-425).

The standards required domestic automakers to increase the average mileage of new cars sold to 27.5 miles per gallon (mpg) by 1985. Manufacturers could still sell large, less efficient cars, but to meet the average fuel efficiency rates, they also had to sell smaller, more efficient cars. Automakers that failed to meet each year's standards were fined; those that managed to surpass the rates earned credits that they could use in years when they fell below the requirements. While keeping their cars relatively large and roomy, companies managed to improve mileage with such innovations as electronic fuel injection, which supplied fuel to an automotive engine more efficiently than its predecessor, the carburetor.

The standards have had a significant effect. Fuel economy of all motor vehicles (which includes passenger cars, vans, pickup trucks, sport-utility vehicles, and trucks) increased from 11.9 miles per gallon (mpg) in 1973 to 17 mpg in 2003. (See Table 9.3.) Greater gains have been made in the economy of passenger cars. In 1974, just after the oil embargo, cars averaged 13.6 mpg; in 2003 the average new-car fuel economy was 22.3 mpg.

TABLE 9.1

Trends in greenhouse gas emissions and sinks, 1990 and 1998–2004

[Teragrams of carbon dioxide equivalents]

Gas/source	1990	1998	1999	2000	2001	2002	2003	2004
CO_2	**5,005.3**	**5,620.2**	**5,695.0**	**5,864.5**	**5,795.2**	**5,815.9**	**5,877.7**	**5,988.0**
Fossil fuel combustion	4,696.6	5,271.8	5,342.4	5,533.7	5,486.9	5,501.8	5,571.1	5,656.6
Non-energy use of fuels	117.2	152.8	160.6	140.7	131.0	136.5	133.5	153.4
Iron and steel production	85.0	67.7	63.8	65.3	57.8	54.6	53.3	51.3
Cement manufacture	33.3	39.2	40.0	41.2	41.4	42.9	43.1	45.6
Waste combustion	10.9	17.1	17.6	17.9	18.6	18.9	19.4	19.4
Ammonia production and urea application	19.3	21.9	20.6	19.6	16.7	18.5	15.3	16.9
Lime manufacture	11.2	13.9	13.5	13.3	12.8	12.3	13.0	13.7
Limestone and dolomite use	5.5	7.4	8.1	6.0	5.7	5.9	4.7	6.7
Natural gas flaring	5.8	6.6	6.9	5.8	6.1	6.2	6.1	6.0
Aluminum production	7.0	6.4	6.5	6.2	4.5	4.6	4.6	4.3
Soda ash manufacture and consumption	4.1	4.3	4.2	4.2	4.1	4.1	4.1	4.2
Petrochemical production	2.2	3.0	3.1	3.0	2.8	2.9	2.8	2.9
Titanium dioxide production	1.3	1.8	1.9	1.9	1.9	2.0	2.0	2.3
Phosphoric acid production	1.5	1.6	1.5	1.4	1.3	1.3	1.4	1.4
Ferroalloy production	2.0	2.0	2.0	1.7	1.3	1.2	1.2	1.3
CO_2 consumption	0.9	0.9	0.8	1.0	0.8	1.0	1.3	1.2
Zinc production	0.9	1.1	1.1	1.1	1.0	0.9	0.5	0.5
Lead production	0.3	0.3	0.3	0.3	0.3	0.3	0.3	0.3
Silicon carbide consumption	0.1	0.2	0.1	0.1	0.1	0.1	0.1	0.1
Net CO_2 flux from land use, land-use change and forestry[a]	*(910.4)*	*(744.0)*	*(765.7)*	*(759.5)*	*(768.0)*	*(768.6)*	*(774.8)*	*(780.1)*
International bunker fuels[b]	*113.5*	*114.6*	*105.2*	*101.4*	*97.8*	*89.5*	*84.1*	*94.5*
Biomass combustion[b]	*216.7*	*217.2*	*222.3*	*226.8*	*200.5*	*194.4*	*202.1*	*211.2*
CH_4	**618.1**	**579.5**	**569.0**	**566.9**	**560.3**	**559.8**	**564.4**	**556.7**
Landfills	172.3	144.4	141.6	139.0	136.2	139.8	142.4	140.9
Natural gas systems	126.7	125.4	121.7	126.7	125.6	125.4	124.7	118.8
Enteric fermentation	117.9	116.7	116.8	115.6	114.6	114.7	115.1	112.6
Coal mining	81.9	62.8	58.9	56.3	55.5	52.5	54.8	56.3
Manure management	31.2	38.8	38.1	38.0	38.9	39.3	39.2	39.4
Wastewater treatment	24.8	32.6	33.6	34.3	34.7	35.8	36.6	36.9
Petroleum systems	34.4	29.7	28.5	27.8	27.4	26.8	25.9	25.7
Rice cultivation	7.1	7.9	8.3	7.5	7.6	6.8	6.9	7.6
Stationary sources	7.9	6.8	7.0	7.3	6.6	6.2	6.5	6.4
Abandoned coal mines	6.0	6.9	6.9	7.2	6.6	6.0	5.8	5.6
Mobile sources	4.7	3.8	3.6	3.5	3.3	3.2	3.0	2.9
Petrochemical production	1.2	1.7	1.7	1.7	1.4	1.5	1.5	1.6
Iron and steel production	1.3	1.2	1.2	1.2	1.1	1.0	1.0	1.0
Agricultural residue burning	0.7	0.8	0.8	0.8	0.8	0.7	0.8	0.9
Silicon carbide production	+	+	+	+	+	+	+	+
International bunker fuels[b]	*0.2*	*0.2*	*0.1*	*0.1*	*0.1*	*0.1*	*0.1*	*0.1*
N_2O	**394.9**	**440.6**	**419.4**	**416.2**	**412.8**	**407.4**	**386.1**	**386.7**
Agricultural soil management	266.1	301.1	281.2	278.2	282.9	277.8	259.2	261.5
Mobile sources	43.5	54.8	54.1	53.1	50.0	47.5	44.8	42.8
Manure management	16.3	17.4	17.4	17.8	18.1	18.0	17.5	17.7
Nitric acid production	17.8	20.9	20.1	19.6	15.9	17.2	16.7	16.6
Human sewage	12.9	14.9	15.4	15.5	15.6	15.6	15.8	16.0
Stationary sources	12.3	13.4	13.4	13.9	13.5	13.2	13.6	13.7
Settlements remaining settlements	5.6	6.2	6.2	6.0	5.8	6.0	6.2	6.4
Adipic acid production	15.2	6.0	5.5	6.0	4.9	5.9	6.2	5.7
N_2O product usage	4.3	4.8	4.8	4.8	4.8	4.8	4.8	4.8
Waste combustion	0.5	0.4	0.4	0.4	0.5	0.5	0.5	0.5
Agricultural residue burning	0.4	0.5	0.4	0.5	0.5	0.4	0.4	0.5
Forest land remaining forest land	0.1	0.4	0.5	0.4	0.4	0.4	0.4	0.4
International bunker fuels[b]	*1.0*	*1.0*	*0.9*	*0.9*	*0.9*	*0.8*	*0.8*	*0.9*

The Environmental Protection Agency computes the data in a different way, using a procedure called a three-year moving average. A graph of their results (see Figure 9.3) shows that the fuel economy of cars and trucks increased rapidly from 1975 to the mid-1980s. The increase slowed through the late 1980s, declined through the mid-1990s, and then remained relatively stable.

The total fuel economy of automobiles is expected to increase as more fuel-efficient cars enter the market and older, less fuel-efficient autos drop out of operation. However, new-car fuel economy has risen only slightly since 1986, and nearly all gains in automobile efficiency have been offset by increased weight and power in new vehicles since 1988.

TABLE 9.1

Trends in greenhouse gas emissions and sinks, 1990 and 1998–2004 [CONTINUED]

[Teragrams of carbon dioxide equivalents]

Gas/source	1990	1998	1999	2000	2001	2002	2003	2004
HFCs, PFCs, and SF_6	**90.8**	**133.4**	**131.5**	**134.7**	**124.9**	**132.7**	**131.0**	**143.0**
Substitution of ozone depleting substances	0.4	54.5	62.8	71.2	78.6	86.2	93.5	103.3
HCFC-22 production	35.0	40.1	30.4	29.8	19.8	19.8	12.3	15.6
Electrical transmission and distribution	28.6	16.7	16.1	15.3	15.3	14.5	14.0	13.8
Semiconductor manufacture	2.9	7.1	7.2	6.3	4.5	4.4	4.3	4.7
Aluminum production	18.4	9.1	9.0	9.0	4.0	5.3	3.8	2.8
Magnesium production and processing	5.4	5.8	6.0	3.2	2.6	2.6	3.0	2.7
Total	**6,109.0**	**6,773.7**	**6,814.9**	**6,982.3**	**6,893.1**	**6,915.8**	**6,959.1**	**7,074.4**
Net emissions (sources and sinks)	**5,198.6**	**6,029.6**	**6,049.2**	**6,222.8**	**6,125.1**	**6,147.2**	**6,184.3**	**6,294.3**

+ Does not exceed 0.05 Tg CO_2 Eq.
[a]Parentheses indicate negative values or sequestration. The net CO_2 flux total includes both emissions and sequestration, and constitutes a sink in the United States. Sinks are only included in net emissions total.
[b]Emissions from International Bunker Fuels and Biomass Combustion are not included in totals.
Note: Totals may not sum due to independent rounding.

SOURCE: "Table ES-2. Recent Trends in U.S. Greenhouse Gas Emissions and Sinks (TgCO_2 Eq.)," in *Inventory of U.S. Greenhouse Gas Emissions and Sinks: 1990–2004 Executive Summary*, U.S. Environmental Protection Agency, April 2006, http://yosemite.epa.gov/oar/globalwarming.nsf/UniqueKeyLookup/RAMR6MBLP4/$File/06ES.pdf (accessed April 14, 2006)

Fuel economy is expected to increase for the light-truck automotive group that includes sport-utility vehicles, vans, and pickup trucks. It is the fastest-growing segment of the auto industry, accounting for about half of the U.S. light-vehicle market in 2004 (see Figure 9.4) and producing most of the profits of the major auto companies. In 1996 Congress authorized the Department of Transportation to set a standard of 20.7 mpg for light trucks. That standard was increased to 21 mpg for model year 2005, to 21.6 mpg for model year 2006, and to 22.2 mpg for model year 2007.

In 1999 the Environmental Protection Agency imposed regulations tightening emissions standards on cars, minivans, small pickup trucks, and sport-utility vehicles weighing less than 8,500 pounds. This was the first time that sport-utility vehicles and other light-duty trucks became subject to the same national pollution standards as cars. The standard of an average of 0.07 grams per mile for nitrogen oxides went into effect in 2004. Standards for hydrocarbons, nitrogen oxides, carbon monoxide, and particulates were phased in, beginning through 2008.

The potential for increased fuel economy in large trucks is considered huge, because their current fuel economy is so much lower than that of automobiles. In 2004 the National Commission on Energy Policy noted that tractor-trailers consumed two-thirds of all truck fuel—1.5 million barrels per day in 2000. Research sponsored by the commission determined that the average fuel economy for new tractor-trailers could be raised by 29% in 2008 and by 58% in 2015. If the heavy-truck fleet were to reach a fuel efficiency of 10 mpg through technological improvements, projected oil demand would drop by 300,000 barrels per day.

Cheap gasoline prices throughout the 1990s took away the sense of urgency surrounding fuel efficiency, which was demonstrated by the high growth of large-vehicle sales. In addition, after repeal of the federal law that set the speed limit at 55 miles per hour, many states allowed higher speed limits, which lowered fuel efficiency.

Alternative-Fuel Vehicles

NUMBERS AND TYPES. In 1995 almost 250,000 alternative-fuel vehicles were on U.S. roads. By 2004 that total had increased to 547,904. (See Table 9.4.) These totals include vehicles originally manufactured to run on alternative fuels as well as gasoline or diesel vehicles that had been converted. The manufacture of new alternative-fuel vehicles has increased steadily.

A number of different types of fuels are used in these vehicles, as shown in Table 9.4:

- Liquefied petroleum gas is a mixture of propane and butane. Thirty-five percent of all alternative-fuel vehicles ran on liquefied petroleum gas in 2004.
- Ethanol is ethyl alcohol, a grain alcohol, mixed with gasoline and sold as gasohol. The 85% formulation of gasohol was the second most common fuel for these vehicles in 2004, powering 27% of them.
- Compressed natural gas is natural gas that is stored in pressurized tanks. When burned, it releases one-tenth the carbon monoxide, hydrocarbon, and nitrogen of

TABLE 9.2

Transportation conservation options

Transportation conservation options
Improve the technical efficiency of vehicles
1. Higher fuel economy requirements—CAFE standards (R)
2. Reducing congestion: smart highways (E,I), flextime (E,R), better signaling (I), improved maintenance of roadways (I), time of day charges (E), improved air traffic controls (I,R), plus options that reduce vehicular traffic
3. Higher fuel taxes (E)
4. Gas guzzler taxes, or feebate schemes (E)
5. Support for increased R&D (E,I)
6. Inspection and maintenance programs (R)
Increase load factor
1. HOV lanes (I)
2. Forgiven tolls (E), free parking for carpools (E)
3. Higher fuel taxes (E)
4. Higher charges on other vmt trip-dependent factors (E): parking (taxes, restrictions, end of tax treatment as business cost), tolls, etc.
Change to more efficient modes
1. Improvements in transit service
a. New technologies—maglev, high speed trains (E,I)
b. Rehabilitation of older systems (I)
c. Expansion of service—more routes, higher frequency (I)
d. Other service improvements (I)—dedicated busways, better security, more bus stop shelters, more comfortable vehicles
2. Higher fuel taxes (E)
3. Reduced transit fares through higher US. transit subsidies (E)*
4. Higher charges on other vmt/trip-dependent factors for less efficient modes (E)—tolls, parking
5. Shifting urban form to higher density, more mixed use, greater concentration through zoning changes (R), encouragement of "infill" development (E,R,I), public investment in infrastructure (I), etc.
Reduce number or length of trips
1. Shifting urban form to higher density, more mixed use, greater concentration (E,R,I)
2. Promoting working at home or at decentralized facilities (E,I)
3. Higher fuel taxes (E)
4. Higher charges on other vmt/trip-dependent factors (E)
Shift to alternative fuels
1. Fleet requirements for alternative fuel-capable vehicles and actual use of alternative fuels (R)
2. Low-emission/zero emission vehicle (LEV/ZEV) requirements (R)
3. Various promotions (E): CAFE credits, emission credits, tax credits, etc.
4. Higher fuel taxes that do not apply to alternative fuels (E), or subsidies for the alternatives (E)
5. Support for increased R&D (E,I)
6. Public investment—government fleet investments (I)
Freight options
1. RD&D of technology improvements (E, I)

*U.S. transit subsidies, already among the highest in the developed world, may merely promote inefficiencies.
Notes: CAFE=corporate average fuel economy; E=economic incentive; HOV=high-occupancy vehicle; I=public investment; maglev=trains supported by magnetic levitation; R=regulatory action; R&D=research and development; RD&D=research, development, and demonstration; vmt=vehicle-miles traveled.

SOURCE: "Table 5-1. Transportation Conservation Options," in *Saving Energy in U.S. Transportation*, U.S. Congress, Office of Technology Assessment, July 1994, http://www.wws.princeton.edu/ota/disk1/1994/9432/9432.PDF (accessed April 14, 2006)

gasoline. It powered 26% of alternative-fuel vehicles in 2004.

- Electricity, used by 10% of alternative-fuel vehicles in 2004, can be used for battery-powered, fuel-cell, or hybrid vehicles.
- Methanol is a liquid fuel that can be produced from natural gas, coal, or biomass (plant material, vegetation, or agricultural waste). The 85% formulation of methanol was used by only 0.8% of all alternative-fuel vehicles in 2004. Its use is declining.
- Liquefied natural gas is natural gas (mostly methane) that has been liquefied by reducing its temperature to −260° Fahrenheit. It was used by only 0.6% of all alternative-fuel vehicles in 2004.
- Biodiesels (not listed in Table 9.4) are liquid biofuels made from soybean, rapeseed, or sunflower oil. They can also be made from animal tallow and such agricultural by-products as rice hulls.

The largest numbers of alternative-fuel vehicles are located in California, Texas, New York, and Oklahoma. (See Table 9.5.) In 2004 U.S. cities had 9,116 alternative-fuel buses on their streets. Most were transit buses, rather than school buses, intercity buses, or trolleys ("Table 35: Number of Onroad Alternative Fuel Buses in Use by Transit Agencies, by Bus Type, Fuel Type and Configuration, 2004," Energy Information Administration, http://www.eia.doe.gov/cneaf/alternate/page/atftables/t3504.xls).

ALTERNATIVE-FUEL VEHICLES AND THE MARKETPLACE. A fuel supply must be readily available if alternative-fuel vehicles are to become a viable transportation option. Ideally, an infrastructure for supplying alternative fuels would be developed simultaneously with the vehicles. Table 9.6 shows the types and numbers of alternative-fuel stations available in each state. In April 2006 more than 5,000 alternative fueling sites were in operation in the United States. Privately owned vehicles used 319 million gallons, or 71% of the total 447.2 million gallons of alternative fuel consumed in the United States during 2004; state and local vehicles used 115.8 million gallons (26%), and federal vehicles used 12.4 million gallons (3%). (See Table 9.7.)

Chrysler Corporation stopped making natural-gas vehicles after model year 1997 because it had lost money, selling only 4,000 of them after production began in 1992. General Motors, which had suspended sales of natural-gas vehicles in 1994, resumed sales in 1997. Ford began selling some natural-gas versions of its cars and trucks in 1995. Commercial fleets (airport vehicles, school and transit buses, and taxicabs), not retail customers, are the main buyers of natural-gas vehicles.

In the early days of the automobile, electric cars outnumbered vehicles with internal-combustion engines. With the introduction of technology for producing low-cost gasoline, however, electric vehicles fell out of favor. However, as cities became choked with air pollution, the idea of an efficient electric car reemerged. To make it acceptable to the public, several considerations have had to be addressed: How many miles could an electric car be driven before it needed to be recharged? How lightweight would the vehicle need to be? And could the electric car

TABLE 9.3

Motor vehicle mileage, fuel consumption, and fuel rates, selected years 1949–2003

Year	Passenger cars[a]			Vans, pickup trucks, and sport utility vehicles[b]			Trucks[c]			All motor vehicles[d]		
	Mileage (miles per vehicle)	Fuel consumption (gallons per vehicle)	Fuel rate (miles per gallon)	Mileage (miles per vehicle)	Fuel consumption (gallons per vehicle)	Fuel rate (miles per gallon)	Mileage (miles per vehicle)	Fuel consumption (gallons per vehicle)	Fuel rate (miles per gallon)	Mileage (miles per vehicle)	Fuel consumption (gallons per vehicle)	Fuel rate (miles per gallon)
1949	9,388	627	15.0	[e]	[e]	[e]	9,712	1,080	9.0	9,498	726	13.1
1950	9,060	603	15.0	[e]	[e]	[e]	10,316	1,229	8.4	9,321	725	12.8
1955	9,447	645	14.6	[e]	[e]	[e]	10,576	1,293	8.2	9,661	761	12.7
1960	9,518	668	14.3	[e]	[e]	[e]	10,693	1,333	8.0	9,732	784	12.4
1965	9,603	661	14.5	[e]	[e]	[e]	10,851	1,387	7.8	9,826	787	12.5
1970	9,989	737	13.5	8,676	866	10.0	13,565	2,467	5.5	9,976	830	12.0
1972	10,171	754	13.5	9,534	922	10.3	14,780	2,657	5.6	10,279	857	12.0
1973	9,884	737	13.4	9,779	931	10.5	15,370	2,775	5.5	10,099	850	11.9
1974	9,221	677	13.6	9,452	862	11.0	14,995	2,708	5.5	9,493	788	12.0
1976	9,418	681	13.8	10,127	934	10.8	15,438	2,764	5.6	9,774	806	12.1
1978	9,500	665	14.3	10,968	948	11.6	18,045	3,263	5.5	10,077	816	12.4
1980	8,813	551	16.0	10,437	854	12.2	18,736	3,447	5.4	9,458	712	13.3
1982	9,050	535	16.9	10,276	762	13.5	19,931	3,647	5.5	9,644	686	14.1
1984	9,248	530	17.4	11,151	797	14.0	22,550	3,967	5.7	10,017	691	14.5
1986	9,464	543	17.4	10,764	738	14.6	22,143	3,821	5.8	10,143	692	14.7
1988	9,972	531	18.8	11,465	745	15.4	22,485	3,736	6.0	10,721	688	15.6
1990	10,504	520	20.2	11,902	738	16.1	23,603	3,953	6.0	11,107	677	16.4
1992	10,857	517	21.0	12,381	717	17.3	25,373	4,210	6.0	11,558	683	16.9
1994	10,992	531	20.7	12,156	701	17.3	25,838	4,202	6.1	11,683	698	16.7
1996	11,330	534	21.2	11,811	685	17.2	26,092	4,221	6.2	11,813	700	16.9
1998	11,754	544	21.6	12,173	707	17.2	25,397	4,135	6.1	12,211	721	16.9
1999	11,848	553	21.4	11,957	701	17.0	26,014	4,352	6.0	12,206	732	16.7
2000	11,976	547	21.9	11,672	669	17.4	25,617	4,391	5.8	12,164	720	16.9
2001	11,831	534	22.1	11,204	636	17.6	26,602	4,477	5.9	11,887	695	17.1
2002	12,202[R]	555[R]	22.0[R]	11,364[R]	650[R]	17.5[R]	27,071[R]	4,642[R]	5.8	12,171[R]	719[R]	16.9[R]
2003[P]	12,242	550	22.3	11,467	647	17.7	27,286	4,750	5.7	12,210	716	17.0

[a]Through 1989, includes motorcycles.
[b]Includes a small number of trucks with 2 axles and 4 tires, such as step vans.
[c]Single-unit trucks with 2 axles and 6 or more tires, and combination trucks.
[d]Includes buses and motorcycles, which are not separately displayed.
[e]Included in "trucks."
R=Revised. P=Preliminary.
Notes: For data not shown for 1951–1969, see http://www.eia.doe.gov/aer/consump.html. For related information, see http://www.fhwa.dot.gov/policy/ohpi/hss/index.htm.

SOURCE: Adapted from "Table 2.8. Motor Vehicle Mileage, Fuel Consumption, and Fuel Rates, Selected Years, 1949–2003," in *Annual Energy Review 2004*, U.S. Department of Energy, Energy Information Administration, Office of Energy Markets and End Use, August 2005, http://www.eia.doe.gov/emeu/aer/pdf/aer.pdf (accessed April 5, 2006)

keep up with the speed and driving conditions of busy freeways and highways?

Electric vehicles can be battery powered, run on fuel cells, or be hybrids, which are powered by both an electric motor and a small conventional engine. EV1, a two-seater by General Motors (GM), was the first commercially available electric car. In 1999 the company introduced its second-generation electric car, the Gen II, which used a lead-acid battery pack and had a driving range of approximately ninety-five miles. The Gen II was also offered with an optional nickel-metal hydride battery pack, which increased its range to 130 miles. However, after the California Air Resources Board relaxed automobile-emissions requirements, phasing them in through 2017, rather than by 2003, GM found that it could no longer market the electric cars effectively. When leases on the cars ran out in 2003, GM began reclaiming them.

Fuel-cell electric vehicles use an electrochemical process that converts a fuel's energy into usable electricity. Fuel cells produce very little sulfur and nitrogen dioxide and generate less than half the carbon dioxide of internal-combustion engines. Rather than needing to be recharged, they are simply refueled. Hydrogen, natural gas, methanol, and gasoline can all be used with a fuel cell.

DaimlerChrysler's Mercedes-Benz division unveiled the first drivable fuel-cell car in 1999. Called the New Electric Car, it produced zero emissions, ran on liquid hydrogen, and traveled 280 miles on a full 11-gallon tank. Several models have been road tested since. Ecostar, an alliance between Ford, DaimlerChrysler, and Ballard Power Systems, is also working on developing new fuel cells to power vehicles. In 2006 about 300 fuel-cell cars were being tested worldwide. Hydrogen fuel-cell cars are expected to be available to consumers in 2010.

Hybrid cars have both an electric motor and a small internal-combustion engine. A sophisticated computer system automatically shifts from the electric motor to

FIGURE 9.3

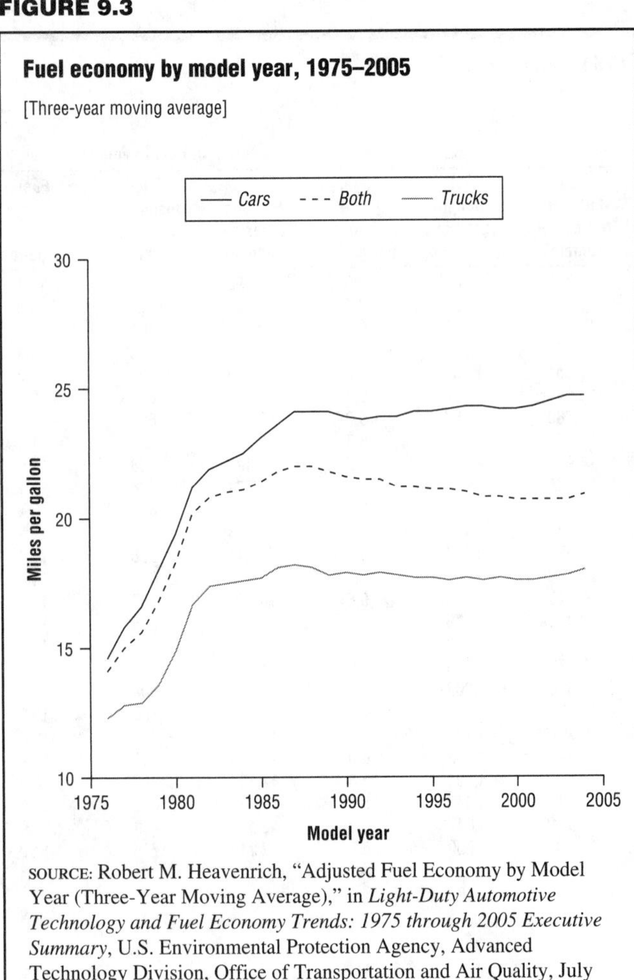

SOURCE: Robert M. Heavenrich, "Adjusted Fuel Economy by Model Year (Three-Year Moving Average)," in *Light-Duty Automotive Technology and Fuel Economy Trends: 1975 through 2005 Executive Summary*, U.S. Environmental Protection Agency, Advanced Technology Division, Office of Transportation and Air Quality, July 2005, http://www.epa.gov/otaq/cert/mpg/fetrends/420s05001.pdf (accessed April 14, 2006)

FIGURE 9.4

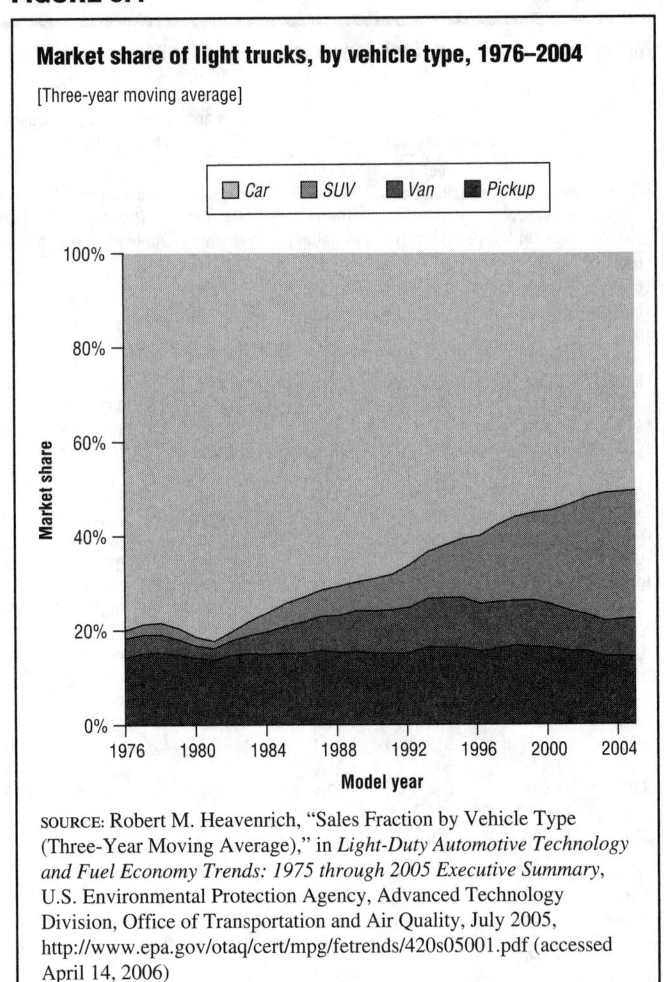

SOURCE: Robert M. Heavenrich, "Sales Fraction by Vehicle Type (Three-Year Moving Average)," in *Light-Duty Automotive Technology and Fuel Economy Trends: 1975 through 2005 Executive Summary*, U.S. Environmental Protection Agency, Advanced Technology Division, Office of Transportation and Air Quality, July 2005, http://www.epa.gov/otaq/cert/mpg/fetrends/420s05001.pdf (accessed April 14, 2006)

the gas engine, as needed, for optimum driving. The electric motor is recharged while the car is driving and braking. Because the gasoline engine does only part of the work, fuel economy is high. The engines are also designed to produce very low emissions.

In 2002 two hybrid passenger cars were introduced in the United States: the Toyota Prius, a sedan with front and back seating, and the two-passenger Honda Insight. Both cars were sold in Japan for several years before being introduced to the U.S. market. In model year 2005 Ford offered the first hybrid sport-utility vehicle, the Escape, which won the 2004 North American Truck of the Year award. The vehicle is reported to get 35 mpg with city driving, traveling about 400 miles on a 15-gallon tank. By model year 2006 many auto manufacturers—including Honda, Ford, Toyota, Lexus, and Mercury—offered gas-electric hybrids.

MANDATING ALTERNATIVE FUELS. Several laws have been passed to encourage or mandate the use of vehicles powered by fuels other than gasoline. The Clean Air Act Amendments of 1990 (PL 101-549) required certain businesses and local governments with fleets of ten or more vehicles in twenty-one metropolitan areas to phase in alternative-fuel vehicles over time—20% of those fleets had to be alternative-fuel vehicles by 1998. While great strides have been made, compliance with the mandates cannot be determined because reporting and enforcement methods are inadequate.

The Energy Policy Act of 1992 (PL 102-486) was passed in the wake of the 1991 Persian Gulf War to conserve energy and increase the proportion of energy supplied domestically. It required that 75% of all vehicles purchased by the federal government in 1999 and thereafter be alternative-fuel vehicles. Agency budget cuts and inadequate enforcement have slowed compliance with these regulations. Still, many municipal governments and the U.S. Postal Service have put into operation fleets of natural-gas vehicles, such as garbage trucks, transit buses, and postal vans.

Air Travel Efficiency

Flying carries an environmental price, as it is an energy-intensive form of transportation. In much of the

TABLE 9.4

Estimated number of alternative-fueled vehicles in use, by fuel, 1995–2004

Fuel	1995	1996	1997	1998	1999	2000	2001	2002	2003 (Projected)	2004 (Projected)	Average annual growth rate (percent)
Liquefied petroleum gases (LPG)	172,806	175,585	175,679	177,183	178,610	181,994	185,053	187,680	190,438	194,389	1.3
Compressed natural gas (CNG)	50,218	60,144	68,571	78,782	91,267	100,750	111,851	120,839	132,988	143,742	12.4
Liquefied natural gas (LNG)	603	663	813	1,172	1,681	2,090	2,576	2,708	3,030	3,134	20.1
Methanol, 85 percent (M85)[a]	18,319	20,265	21,040	19,648	18,964	10,426	7,827	5,873	4,917	4,592	−14.3
Methanol, neat (M100)	386	172	172	200	198	0	0	0	0	0	0.0
Ethanol, 85 percent (E85)[a, b]	1,527	4,536	9,130	12,788	24,604	87,570	100,303	120,951	133,776	146,195	78.8
Ethanol, 95 percent (E95)[a]	136	361	347	14	14	4	0	0	0	0	0.0
Electricity[c]	2,860	3,280	4,453	5,243	6,964	11,830	17,847	33,047	45,656	55,852	39.1
Non-LPG subtotal	**74,049**	**89,421**	**104,526**	**117,847**	**143,692**	**212,670**	**240,404**	**283,418**	**320,367**	**353,515**	**19.0**
Total	**246,855**	**265,006**	**280,205**	**295,030**	**322,302**	**394,664**	**425,457**	**471,098**	**510,805**	**547,904**	**9.3**

[a]The remaining portion of 85-percent methanol and both ethanol fuels is gasoline.
[b]In 1997, some vehicle manufacturers began including E85-fueling capability in certain model lines of vehicles. For 2002, the EIA estimated that the number of E-85 vehicles that are capable of operating on E85, gasoline, or both, is about 4.1 million. Many of these alternative-fueled vehicles (AFVs) are sold and used as traditional gasoline-powered vehicles. In this table, AFVs in use include only those E85 vehicles believed to be intended for use as AFVs. These are primarily fleet-operated vehicles.
[c]Excludes gasoline-electric hybrids.
Notes: Estimates for 2003, are based on plans or projections. Estimates for historical years may be revised in future reports if new information becomes available.

SOURCE: "Table 1. Estimated Number of Alternative-Fueled Vehicles in Use in the United States, by Fuel, 1995–2004," in *Alternatives to Traditional Transportation Fuels 2003 Estimated Data*, U.S. Department of Energy, Energy Information Administration, February 2004, http://www.eia.doe.gov/cneaf/alternate/page/datatables/aft1-13_03.html (accessed May 7, 2006)

industrialized world, air travel is replacing more energy-efficient rail or bus travel. According to the Bureau of Transportation Statistics (part of the Department of Transportation), jet fuel consumption rose from 12.7 billion gallons in 1995 to 13 billion gallons in 2003.

Jet fuel consumption can affect global warming. Airplanes spew nearly four million tons of nitrogen oxide into the air each year, much of it while cruising in the tropospheric zone five to seven miles above the planet, where ozone is formed. (See Figure 9.5.) The Environmental Protection Agency has estimated that air traffic accounts for about 3% of all global greenhouse warming. The Intergovernmental Panel on Climate Change, which was established by the United Nations, noted that emissions deposited directly into the atmosphere do greater harm than those released at ground level.

Although each generation of airplane engines gets cleaner and more fuel efficient, there are also other engines in the airline industry—those in the trucks, cars, and carts that service airplane fleets. Electric utility companies, including the Edison Electric Institute and the Electric Power Research Institute, launched a program in 1993 to electrify airports. By converting terminal transport buses, food trucks, and baggage-handling carts to electricity, airports could reduce air pollution considerably. As of December 2005, only a few U.S. airports and airlines were operating significant numbers of electric ground-support equipment. However, airport electrification implementation and research were ongoing.

RESIDENTIAL AND COMMERCIAL CONSERVATION

Total energy use in buildings in the United States has increased because the numbers of people, households, and offices has increased. However, energy use per unit area (commercial) or per person (residential) has roughly stabilized through a variety of efficiency measures. The sources of energy in buildings have changed dramatically. Use of fuel oil has dropped, with natural gas making up most of the difference. At the same time, other energy demands have risen. Electronic office equipment, such as computers, fax machines, printers, and copiers, have sharply increased electricity loads in commercial buildings. According to the *Annual Energy Review 2004*, energy use in the residential and commercial sectors accounted for an increasing share of total U.S. energy consumption: 29% in 1950, 33% in 1970, and 39% in 2004.

Building Efficiency

Energy conservation in buildings in both the residential and commercial sectors has improved considerably since the early 1980s. Among the techniques for reducing energy use are advanced window designs, "daylighting" (letting light in from the outside by adding a skylight or building a large building around an atrium), solar water heating, landscaping, and planting trees.

Residential energy consumption has been reduced by building more efficient new housing and appliances, improving energy efficiency in existing housing, and building more multiple-family units. Also, many people have migrated to the South and West, where their combined use of heating and cooling has generally been lower than usage in other parts of the country.

TABLE 9.5

Estimated number of alternative-fueled vehicles in use, by state and fuel type, 2002

State	Liquefied petroleum gases	Natural gas	Methanol	Ethanol	Electricity	Total
Alabama	4,289	1,341	0	2,713	636	8,979
Alaska	145	401	0	720	11	1,277
Arizona	1,082	7,243	201	1,583	1,662	11,771
Arkansas	2,199	340	0	300	0	2,839
California	21,537	24,990	4,787	9,517	10,670	71,501
Colorado	5,611	2,694	3	3,491	126	11,925
Connecticut	379	2,762	1	1,849	156	5,147
Delaware	85	489	10	783	11	1,378
District of Columbia	7	1,462	50	1,408	316	3,243
Florida	4,171	4,152	6	7,856	357	16,542
Georgia	4,418	4,484	39	2,076	4,550	15,567
Hawaii	842	0	0	1,467	204	2,513
Idaho	1,581	3,412	0	240	0	5,233
Illinois	5,259	3,120	17	6,916	89	15,401
Indiana	1,426	3,397	0	1,670	91	6,584
Iowa	2,179	18	27	1,903	12	4,139
Kansas	3,565	748	1	1,649	22	5,985
Kentucky	2,214	1,191	0	2,313	0	5,718
Louisiana	1,117	896	3	1,309	0	3,325
Maine	158	77	0	134	21	390
Maryland	2,570	3,634	7	2,901	45	9,157
Massachusetts	249	1,006	36	1,331	78	2,700
Michigan	4,822	991	48	4,840	1,606	12,307
Minnesota	2,162	509	0	3,361	0	6,032
Mississippi	1,193	140	0	543	0	1,876
Missouri	2,642	476	95	3,878	11	7,102
Montana	2,980	268	0	309	0	3,557
Nebraska	4,338	370	0	1,095	11	5,814
Nevada	1,487	3,111	0	973	0	5,571
New Hampshire	718	42	0	169	167	1,096
New Jersey	358	2,723	4	2,681	190	5,956
New Mexico	6,069	1,969	11	2,140	435	10,624
New York	6,213	13,100	88	3,723	9,299	32,423
North Carolina	4,560	559	0	4,539	112	9,770
North Dakota	1,310	155	0	354	0	1,819
Ohio	2,487	2,647	26	4,537	242	9,939
Oklahoma	17,839	3,322	0	1,122	0	22,283
Oregon	3,084	1,034	20	1,528	212	5,878
Pennsylvania	1,107	2,299	108	4,008	89	7,611
Rhode Island	122	331	0	391	0	844
South Carolina	3,047	362	0	4,051	0	7,460
South Dakota	1,374	44	0	384	0	1,802
Tennessee	2,623	763	0	3,068	200	6,654
Texas	39,279	9,961	162	6,706	82	56,190
Utah	3,227	1,961	8	1,966	0	7,162
Vermont	366	5	0	199	178	748
Virginia	927	4,735	7	3,740	1,086	10,495
Washington	4,397	1,925	73	2,760	11	9,166
West Virginia	39	378	0	595	0	1,012
Wisconsin	1,459	1,207	35	3,075	37	5,813
Wyoming	2,368	303	0	87	22	2,780
U.S. total	**187,680**	**123,547**	**5,873**	**120,951**	**33,047**	**471,098**

Notes: Natural gas includes compressed natural gas (CNG) and liquefied natural gas (LNG). Methanol includes M85 and M100. Ethanol includes E85 and E95. Excludes gasoline-electric hybrids. Totals may not equal sum of components due to independent rounding.

SOURCE: "Table 4. Estimated Number of Alternative-Fueled Vehicles in Use, by State and Fuel Type, 2002," in *Alternatives to Traditional Transportation Fuels 2003 Estimated Data*, U.S. Department of Energy, Energy Information Administration, February 2004, http://www.eia.doe.gov/cneaf/alternate/page/datatables/aft1-13_03.html (accessed May 7, 2006)

In the residential sector the largest share of energy savings has been the result of better construction, higher quality insulation, and more energy-efficient windows and doors. According to the Office of Technology Assessment (*Building Energy Efficiency*, 1992), roughly one-fourth the energy used to heat and cool buildings is lost through poor insulation and poorly insulated windows. Before the 1973 energy crisis, 70% of new windows sold were single-glazed (only a single pane of glass). By 1990, because of changes in building codes and public interest, 80% of windows sold were double-glazed, cutting energy loss in half. Double-glazed windows have two panes of glass sandwiched together with a small space in between. The glass may be specially treated or the space between the panes may be filled with a gas, either of which increases the insulating effectiveness of the window.

TABLE 9.6

Alternative fuel station counts, by state and fuel type, as of April 16, 2006

State	CNG	E85	LPG	ELEC	BD	HY	LNG	Totals by state
Alabama	1	0	74	0	0	0	0	75
Alaska	0	0	12	0	0	0	0	12
Arizona	30	5	74	18	4	1	4	136
Arkansas	4	0	57	0	0	0	0	61
California	179	3	257	406	18	9	30	902
Colorado	21	11	72	4	22	0	0	130
Connecticut	11	0	19	4	1	0	0	35
Delaware	1	0	3	0	3	0	0	7
District of Columbia	1	0	0	0	0	1	0	2
Florida	22	2	70	7	4	0	0	105
Georgia	16	6	51	0	17	0	0	90
Hawaii	0	0	6	11	3	0	0	20
Idaho	8	1	28	0	2	0	1	40
Illinois	11	96	73	0	9	0	0	189
Indiana	11	18	42	0	10	0	0	81
Iowa	0	41	29	0	8	0	0	78
Kansas	3	8	49	0	4	0	0	64
Kentucky	0	5	36	0	6	0	0	47
Louisiana	8	0	14	0	0	0	0	22
Maine	1	0	6	0	2	0	0	9
Maryland	13	4	19	0	3	0	0	39
Massachusetts	9	0	28	28	1	0	0	66
Michigan	15	6	88	0	13	2	0	124
Minnesota	3	203	34	0	2	0	0	242
Mississippi	0	0	40	0	6	0	0	46
Missouri	6	25	88	0	2	0	0	121
Montana	2	5	31	0	6	0	0	44
Nebraska	1	27	23	0	1	0	0	52
Nevada	16	1	25	0	10	1	0	53
New Hampshire	3	0	14	10	10	0	0	37
New Jersey	15	0	11	0	1	0	0	27
New Mexico	8	3	60	0	2	0	0	73
New York	37	6	28	0	4	0	0	75
North Carolina	11	9	67	0	36	0	0	123
North Dakota	4	22	16	0	0	0	0	42
Ohio	12	7	75	0	14	0	0	108
Oklahoma	53	4	72	1	5	0	0	135
Oregon	14	1	34	4	14	0	0	67
Pennsylvania	31	1	63	0	11	0	0	106
Rhode Island	6	0	4	2	0	0	0	12
South Carolina	5	31	34	2	24	0	0	96
South Dakota	0	33	22	0	0	0	0	55
Tennessee	6	5	59	0	9	0	0	79
Texas	29	4	628	2	10	0	2	675
Utah	63	3	27	0	3	0	0	96
Vermont	1	0	6	1	5	0	0	13
Virginia	12	2	25	0	10	0	0	49
Washington	14	2	60	0	18	0	0	94
West Virginia	2	2	8	0	0	0	0	12
Wisconsin	18	13	56	0	2	0	0	89
Wyoming	11	4	33	0	13	0	0	61
Totals by fuel	**748**	**619**	**2,750**	**500**	**348**	**14**	**37**	**5,016**

Notes: CNG=Compressed natural gas. E85=85% Ethanol. LPG=Propane. ELEC=Electric. BD=Biodiesel. HY=Hydrogen. LNG=Liquefied natural gas.

SOURCE: "Alternative Fueling Station Counts by State and Fuel Type," U.S. Department of Energy, Alternative Fuels Data Center, April 16, 2006, http://www.eere.energy.gov/afdc/infrastructure/station_counts.html (accessed April 16, 2006)

Overall, however, energy consumption per household has remained fairly steady since 1982. Technology gains have been offset by an increase in the size of new homes and more demand for energy services. (See Figure 9.6.)

As in the residential sector, improved technology, materials, and construction methods have helped to slow the growth of energy use in commercial buildings. Glass-walled buildings, especially, have undergone transformation: Certain types of glass are now chosen for their ability to divert the heat of the sun and reduce the amount of energy needed for cooling. Like homes, many commercial structures are now designed to take advantage of breezes and to deflect the coldest winter winds.

Home Appliance Efficiency

The number of households in the United States is increasing, which is increasing the demand for energy-intensive products and services such as air-conditioning. According to

TABLE 9.7

Estimated consumption of alternative transportation fuels, by vehicle ownership, 2000, 2002, and 2004

[Thousand gasoline-equivalent gallons]

Fuel	2000 Federal	2000 State/local	2000 private	2000 Total	2002 Federal	2002 State/local	2002 private	2002 Total	2004 Federal	2004 State/local	2004 private	2004 Total
Liquefied petroleum gases (LPG)	458	19,730	192,388	212,576	213	20,097	202,833	223,143	164	19,277	222,927	242,36
Compressed natural gas (CNG)	6,294	34,061	46,390	86,745	6,142	52,482	62,046	120,670	7,449	73,217	78,798	159,464
Liquefied natural gas (LNG)	101	6,031	1,127	7,259	124	7,797	1,461	9,382	146	9,035	1,687	10,868
Methanol, 85 percent (M85)*	2	193	390	585	2	103	232	337	1	82	174	257
Methanol, neat (M100)	0	0	0	0	0	0	0	0	0	0	0	0
Ethanol, 85 percent (E85)*	2,661	5,482	3,928	12,071	3,495	9,215	5,073	17,783	4,331	12,713	5,361	22,40
Ethanol, 95 percent (E95)*	0	13	0	13	0	0	0	0	0	0	0	0
Electricity	639	595	1,824	3,058	191	1,165	5,918	7,274	291	1,461	10,084	11,836
Total	**10,155**	**66,105**	**246,047**	**322,307**	**10,167**	**90,859**	**277,563**	**378,589**	**12,382**	**115,785**	**319,031**	**447,198**

*The remaining portion of 85-percent methanol and both ethanol fuels is gasoline. Consumption data include the gasoline portion of the fuel.

Notes: Fuel quantities are expressed in a common base unit of gasoline-equivalent gallons to allow comparisons of different fuel types. Gasoline-equivalent gallons do not represent gasoline displacement. Gasoline equivalent is computed by dividing the lower heating value of the alternative fuel by the lower heating value of gasoline and multiplying this result by the alternative fuel consumption value. Lower heating value refers to the Btu content per unit of fuel excluding the heat produced by condensation of water vapor in the fuel. Totals may not equal sum of components due to independent rounding. Estimates for 2004, are based on plans or projections. Estimates for historical years maybe revised in future reports if new information becomes available.

SOURCE: "Table 13. Estimated Consumption of Alternative Transportation Fuels in the United States, by Vehicle Ownership, 2000, 2002, and 2004," in *Alternatives to Traditional Transportation Fuels 2003 Estimated Data*, U.S. Department of Energy, Energy Information Administration, February 2004, http://www.eia.doe.gov/cneaf/alternate/page/datatables/aft1-13_03.html (accessed May 7, 2006)

FIGURE 9.5

Layers of the atmosphere

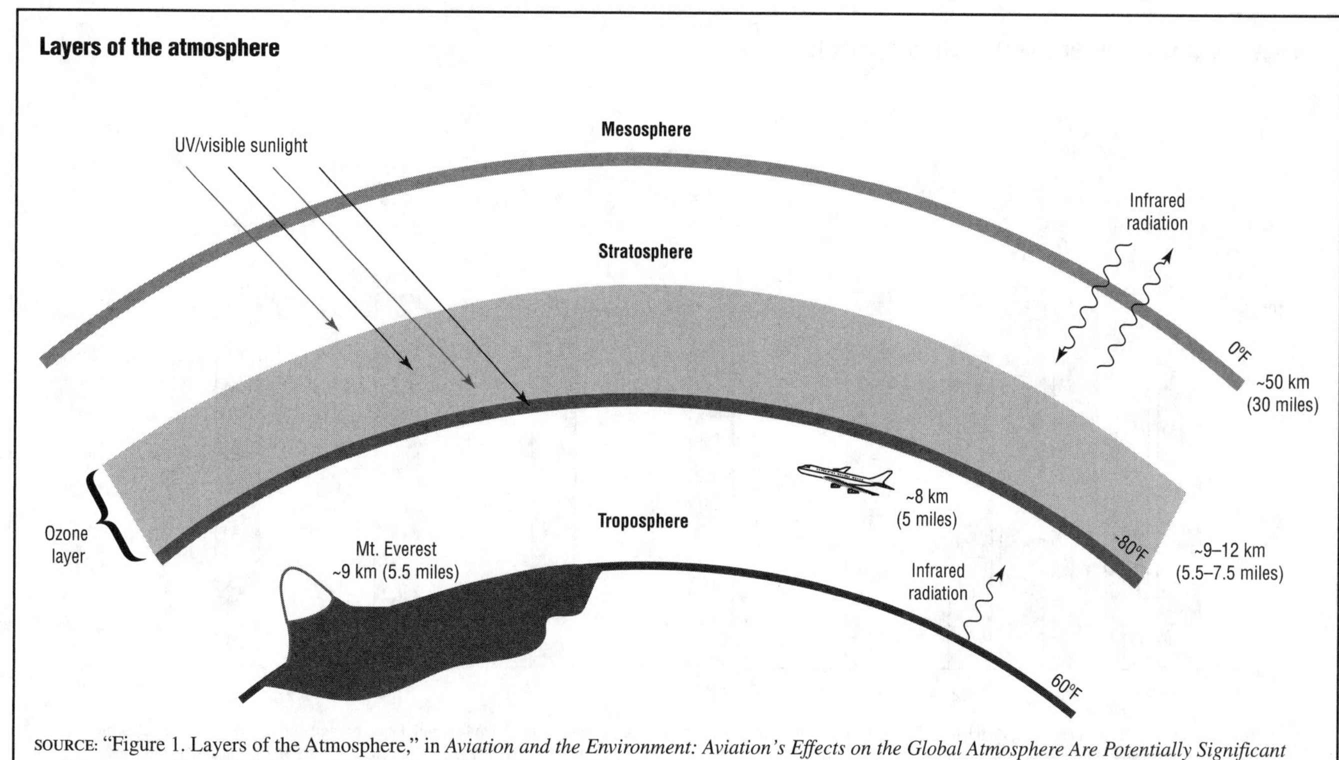

SOURCE: "Figure 1. Layers of the Atmosphere," in *Aviation and the Environment: Aviation's Effects on the Global Atmosphere Are Potentially Significant and Expected to Grow*, U.S. General Accounting Office, February 2000, http://www.gao.gov/new.items/rc00057.pdf (accessed April 16, 2006)

the Energy Information Administration, residential energy use accounted for slightly more than 21% of total national energy consumption in 2004. In 2001 (the most recent year for which the agency has compiled data) space heating used 47% of the total residential energy consumed, down from 51% in 1997; appliances 30%, up from 27%; water heating 17%, down from 19%; and air conditioners 6%, up from 4%.

The number of electrical appliances in U.S. households has increased steadily over the past few of decades. (See Figure 9.7, which is based on the most recent data compiled by the Energy Information Administration.) By 2001 about 99% of American homes had color televisions, 86% had microwave ovens, 79% had clothes washers, and 56% had personal computers.

In 1987 Congress passed the National Appliance Energy Conservation Act (PL 100-12), which gave the Department of Energy the authority to formulate minimum efficiency requirements for thirteen classes of consumer products. It could also revise and update those standards as technologies and economic conditions changed. Table 9.8 shows the products affected and the years in which appliance efficiency standards were established or revised for each, as well as the future effective dates of standards.

Energy efficiency has increased for all major household appliances but most dramatically for refrigerators and freezers. Since 1972 the energy efficiency of new refrigerators and freezers has more than tripled because of better insulation, motors, compressors, and accessories such as automatic defrost (Brian Halweill, "Good Stuff? A Behind-the-Scenes Guide to the Things We Buy: Appliances," World Watch Institute, 2006, http://www.worldwatch.org/pubs/goodstuff/appliances/this.href). These improvements have been accomplished at relatively low cost to manufacturers. In addition, efficiency labels are now required on appliances, which makes purchasing efficient models easier.

According the Department of Energy, by the early 2000s, air conditioners and heat pumps, another major group of appliances, had shown a 30% to 50% improvement in energy efficiency since the mid-1970s. Although this improvement in energy efficiency was less than the improvements in refrigerators and freezers, it is significant because these appliances are large energy users.

Efficiency of water heaters and furnaces improved between 5% and 20% from the mid-1970s to the early 2000s. However, the technological improvements in these appliances are relatively costly compared with the overall price of the products, so energy-conserving models have higher retail prices. This discourages many consumers from purchasing them, even though the more efficient models may save money in the long run. In addition, many water heaters and furnaces are bought by builders, who have little incentive to pay more for the most efficient models, or by homeowners in emergency situations, when availability and installation seem much more important than energy efficiency.

FIGURE 9.6

Energy consumption per household, selected years 1978–2001

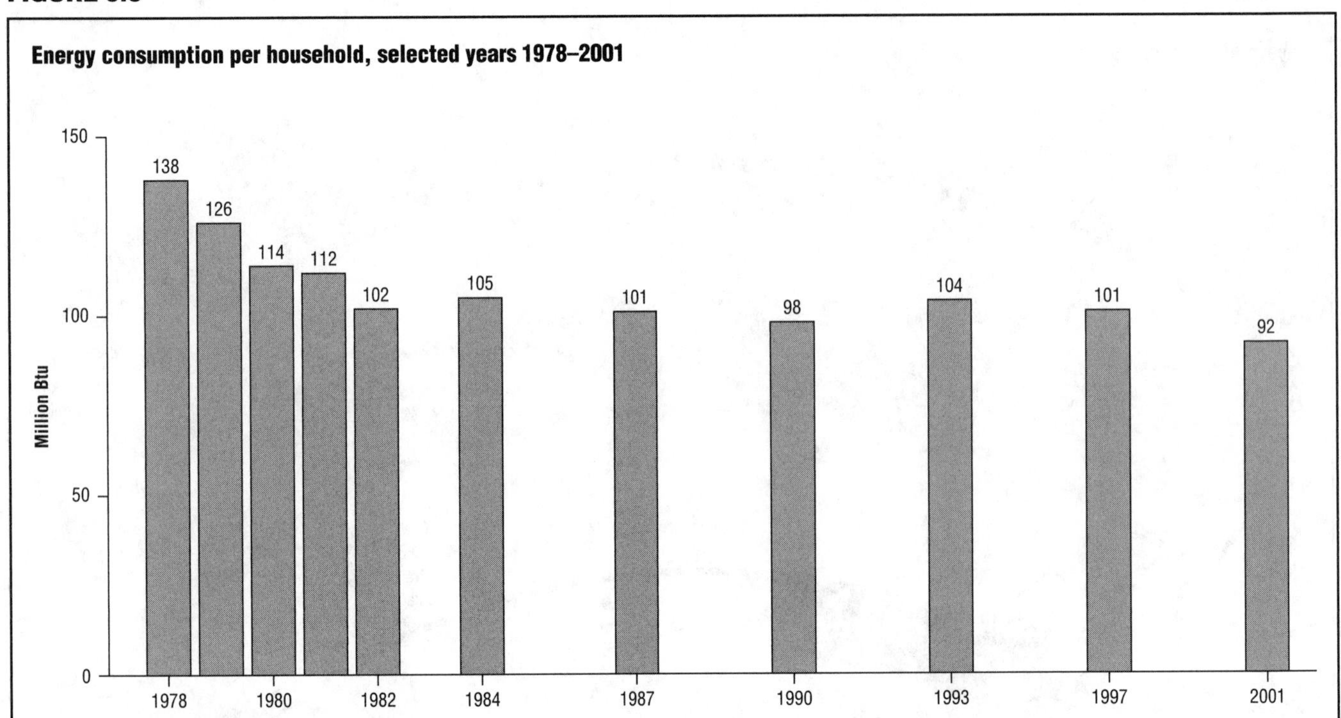

Notes: Data include natural gas, electricity, distillate fuel oil, kerosene, and liquefied petroleum gases; data do not include wood. For years not shown, there are no data available. Data for 1978 through 1984 are for April of the year shown through March of the following year; data for 1987, 1990, 1993, 1997, and 2001 are for the calendar year.

SOURCE: Adapted from "Figure 2.4. Household Energy Consumption: Consumption per Household, Selected Years, 1978–2001," in *Annual Energy Review 2004*, U.S. Department of Energy, Energy Information Administration, Office of Energy Markets and End Use, August 2005, http://www.eia.doe.gov/emeu/aer/pdf/aer.pdf (accessed April 5, 2006)

FIGURE 9.7

Households with selected electric appliances, 1980 and 2001

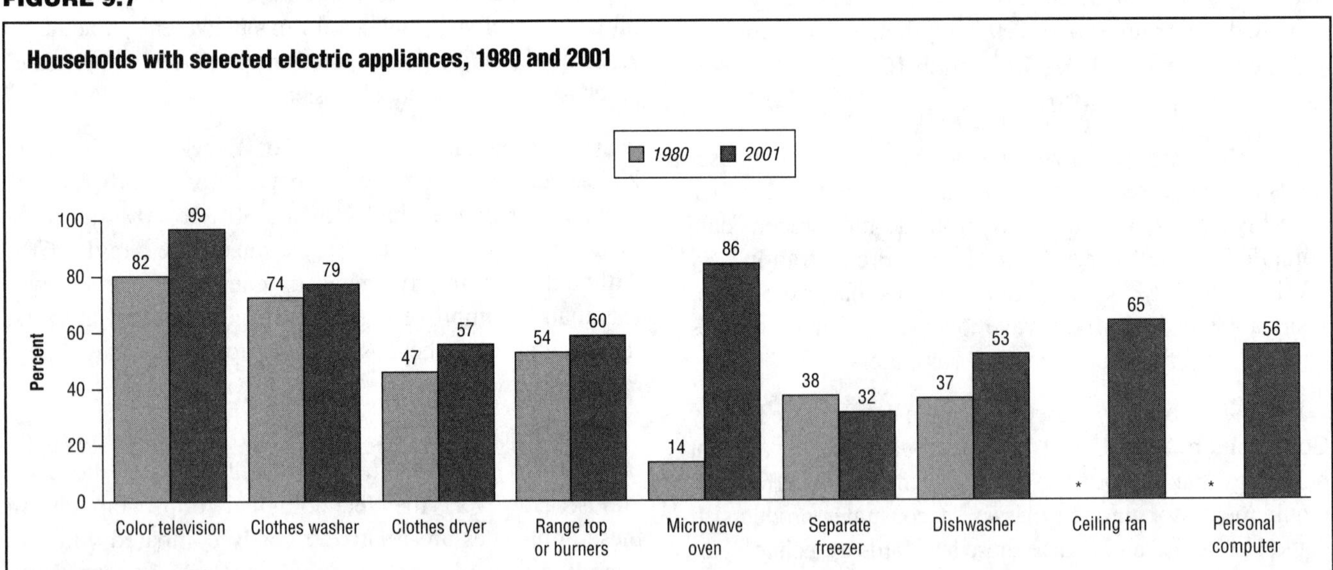

*Not collected in 1980.

SOURCE: Adapted from "Figure 2.6. Households with Selected Appliances and Types of Main Heating Fuel: Households with Selected Electric Appliances, 1980 and 2001," in *Annual Energy Review 2004*, U.S. Department of Energy, Energy Information Administration, Office of Energy Markets and End Use, August 2005, http://www.eia.doe.gov/emeu/aer/pdf/aer.pdf (accessed April 5, 2006)

TABLE 9.8

Effective dates of appliance efficiency standards, 1988–2007

Product	1988	1990	1992	1993	1994	1995	2000	2001	2003	2004	2005	2006	2007
Clothes dryers	X				X								
Clothes washers	X				X					X			X
Dishwashers	X				X								
Refrigerators and freezers		X		X				X					
Kitchen ranges and ovens		X											
Room air conditioners		X					X						
Direct heating equipment		X											
Fluorescent lamp ballasts		X									X		
Water heaters		X								X			
Pool heaters		X											
Central air conditioners and heat pumps			X									X	
Furnaces													
Central (>45,000 Btu per hour)			X										
Small (>45,000 Btu per hour)			X										
Mobile home		X											
Boilers			X										
Fluorescent lamps, 8 foot					X								
Fluorescent lamps, 2 and 4 foot (U tube)						X							
Commercial water-cooled air conditioners										X			
Commercial natural gas furnaces										X			
Commercial natural gas water heaters										X			

SOURCE: "Table 2. Effective Dates of Appliance Efficiency Standards, 1988–2007," in *Annual Energy Outlook 2002*, U.S. Department of Energy, Energy Information Administration, Office of Integrated Analysis and Forecasting, December 2001, http://www.eia.doe.gov/oiaf/archive/aeo02/pdf/0383(2002).pdf (accessed May 8, 2006)

In addition to concerns about efficiency, appliance makers—especially those who make refrigerators and air-conditioning systems—are developing alternative cooling techniques to replace chlorofluorocarbons (CFCs), which are ozone-damaging chemicals that can no longer be legally sold in the United States. CFCs were initially substituted with somewhat less dangerous hydrochlorofluorocarbons (HCFCs), but they are now being replaced with hydrofluorocarbons (HFCs), which lack chlorine. In Europe, other substances, such as propane and butane, are being used as refrigerants. Known as "greenfreeze" technology, these materials are rapidly replacing HCFCs.

Lawn and Garden Equipment

In 1994 the Environmental Protection Agency reported that as much as 10% of the nation's air pollution was generated by gasoline-powered lawn and garden equipment, including lawn mowers, chain saws, and golf carts. Carol Browner, who was the agency's administrator at the time, estimated that Americans used eighty-nine million pieces of such equipment, with lawn mowers alone accounting for 5% of the nation's air pollution.

Under Browner, the agency established engine label and warranty requirements, exhaust emissions standards, and test procedures, requiring that engine makers meet the new requirements by 1996. Effective that year, new products were equipped with improved carburetion systems, and additional standards were scheduled for subsequent years. Agency officials reported in 2000 that the new regulations had reduced smog-forming hydrocarbon emissions by 32% and projected that additional standards, to be phased in by 2007, would reduce hydrocarbon emissions an additional 10%.

INTERNATIONAL COMPARISONS OF CONSERVATION EFFORTS

One test of a country's efficiency is the amount of energy it consumes for every dollar of goods and services it produces. According to the Energy Information Administration (*International Energy Annual 2003*, 2005), the United States lags behind some industrialized countries in energy efficiency and conservation efforts, but is also considerably ahead of others. In 2003 the United States consumed 9,521 Btu per dollar (in 2000 U.S. dollars) of gross domestic product compared with 8,269 for France, 7,545 for Germany, and 4,605 for Japan. That same year, however, Canada consumed 17,863 Btu per dollar of gross domestic product; Belgium, 11,465; and Spain, 10,217. If fuel prices continue to increase, countries that are least efficient could face serious economic challenges from those that have embraced efficiency and conservation.

Because the United States is the world's largest producer of greenhouse gases, its emissions per capita are also significantly higher than those in other industrialized countries. For instance, the *International Energy Annual 2003* determined the "carbon intensity" of countries by comparing the metric tons of carbon dioxide they produce per thousand dollars of gross domestic product. The figures for 2003 showed that the carbon intensity of the United States was substantially higher than that of many other industrialized nations. For example, the carbon intensity of the United States was 0.56, while Germany's was 0.45, France's was 0.30, and Japan's was 0.25. Spain's carbon

FIGURE 9.8

Energy use per capita and per dollar of gross domestic product, 1980–2030

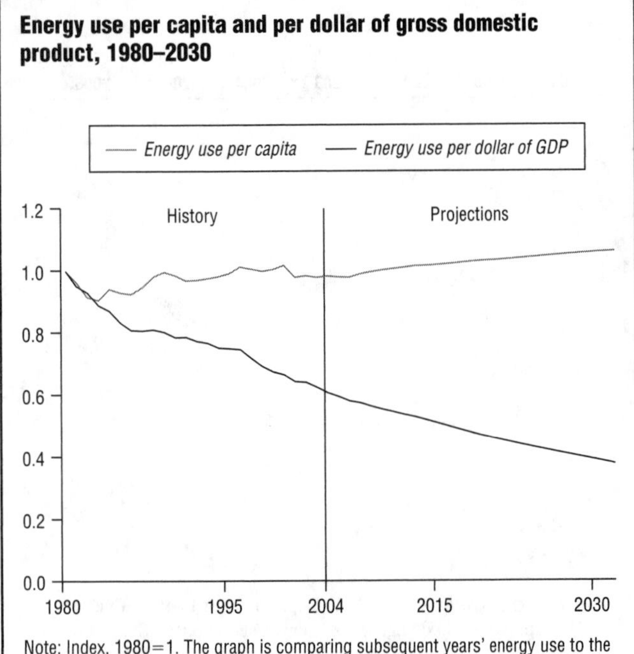

Note: Index, 1980=1. The graph is comparing subsequent years' energy use to the amount of energy that was used in 1980 per person and per dollar of the GDP. That amount of energy use in 1980 is designated as 1.

SOURCE: "Figure 4. Energy Use Per Capita and Per Dollar of Gross Domestic Product, 1980–2030 (Index, 1980=1)," in *Annual Energy Outlook 2006*, U.S. Department of Energy, Energy Information Administration, Office of Integrated Analysis and Forecasting, February 2006, http://www.eia.doe.gov/oiaf/aeo/pdf/0383(2006).pdf (accessed April 5, 2006)

FIGURE 9.9

Past and projected average fuel economy of new light-duty vehicles, 1980–2030

[Miles per gallon]

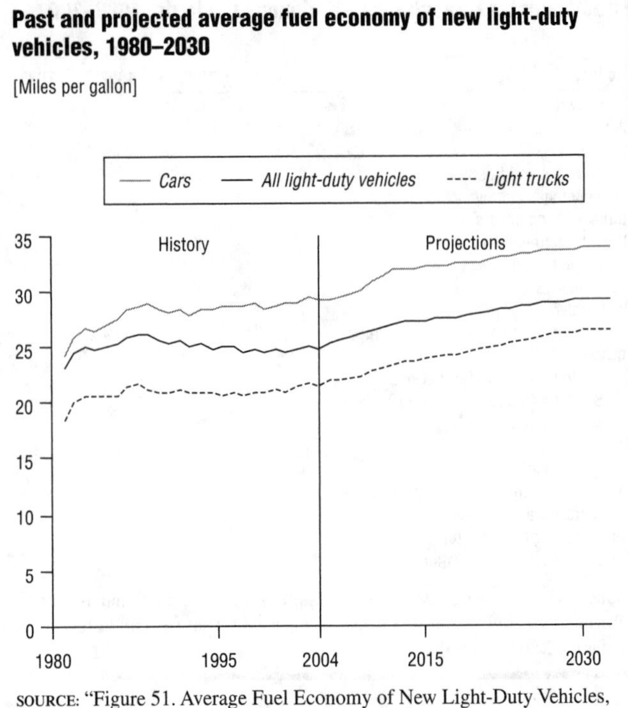

SOURCE: "Figure 51. Average Fuel Economy of New Light-Duty Vehicles, 1980–2030 (Miles per Gallon)," in *Annual Energy Outlook 2006*, U.S. Department of Energy, Energy Information Administration, Office of Integrated Analysis and Forecasting, February 2006, http://www.eia.doe.gov/oiaf/aeo/pdf/0383(2006).pdf (accessed April 5, 2006)

intensity was 0.56, equal to that of the United States, while Canada's was 0.80 and Belgium's was 0.60.

FUTURE TRENDS IN CONSERVATION

According to the *Annual Energy Outlook 2006* (Energy Information Administration, 2006), U.S. total energy consumption is expected to increase at a fairly steady rate from nearly 100 quadrillion Btu in 2004 to about 130 quadrillion Btu in 2030, even with efficiency standards for new equipment taken into consideration. Per capita energy use is expected to increase slightly by 0.3% per year from 2004 through 2030. (See Figure 9.8.)

According to the projections, homes would be larger in 2030, but electricity would be used more efficiently. Higher energy prices would encourage conservation. Although annual personal highway and air travel would increase, efficiency improvements would offset much of that increase. Energy use per dollar of gross domestic product is expected to decrease at an average annual rate of 1.8% from 2004 to 2030 as efficiency gains are made.

The agency also projected that transportation fuel efficiency for light-duty vehicles would improve slowly from 2004 through 2030. The average fuel economy for new light-duty vehicles is expected to rise to 29.2 mpg in 2030. (See Figure 9.9.) It dropped from 1987 through 2004 because the standard for light trucks, which are included in this group, was set at 20.7 mpg, rather than the 27.5 mpg standard for cars. However, the standard for these vehicles (which include sport-utility vehicles, minivans, and pickup trucks) was increased to 21 mpg for model year 2005; to 21.6 mpg for model year 2006; and to 22.2 mpg for model year 2007. Additional increases are proposed for model years 2008 through 2011.

The *Annual Energy Outlook 2004* predicted that the market for alternative-fuel vehicles would grow as a result of federal and state mandates. By 2025, about 3.9 million alternative-fuel vehicles are expected to be sold—about 19% of total light-duty vehicle sales. Alternative fuels could replace about 166,500 barrels of oil per day by 2025, or 2.1% of all light-vehicle consumption.

IMPORTANT NAMES AND ADDRESSES

American Gas Association
400 North Capitol St. NW, Ste. 450
Washington, DC 20001
(202) 824-7000
FAX: (202) 824-7115
URL: http://www.aga.org

American Petroleum Institute
1220 L St. NW
Washington, DC 20005-4070
(202) 682-8000
URL: http://www.api.org

American Wind Energy Association
1101 Fourteenth St. NW, Twelfth Fl.
Washington, DC 20005
(202) 383-2500
FAX: (202) 383-2505
E-mail: windmail@awea.org
URL: http://www.awea.org

Bureau of Land Management
U.S. Department of the Interior
1849 C St. NW
Washington, DC 20240
(202) 452-5125
FAX: (202) 452-5124
E-mail: woinfo@blm.gov
URL: http://www.blm.gov

Council on Environmental Quality
722 Jackson Pl. NW
Washington, DC 20503
(202) 395-5750
FAX: (202) 456-6546
URL: http://www.whitehouse.gov/ceq

Edison Electric Institute
701 Pennsylvania Ave. NW
Washington, DC 20004-2696
(202) 508-5000
FAX: (202) 508-5015
E-mail: electricsolutions@eei.org
URL: http://www.eei.org

Electric Power Research Institute
3420 Hillview Ave.
Palo Alto, CA 94304
1-800-313-3774
URL: http://www.epri.com

Energy Information Administration
U.S. Department of Energy
1000 Independence Ave. SW
Washington, DC 20585
(202) 586-8800
E-mail: infoctr@eia.doe.gov
URL: http://www.eia.doe.gov

Environmental Defense
257 Park Ave. S.
New York, NY 10010
(212) 505-2100
FAX: (212) 505-2375
E-mail: members@environmentaldefense.org
URL: http://www.environmentaldefense.org

Friends of the Earth
1717 Massachusetts Ave. NW, Ste. 600
Washington, DC 20036-2002
1-877-843-8687
FAX: (202) 783-0444
E-mail: foe@foe.org
URL: http://www.foe.org

Greenpeace USA
702 H St. NW
Washington, DC 20001
(202) 462-1177
1-800-326-0959
FAX: (202) 462-4507
E-mail: info@wdc.greenpeace.org
URL: http://www.greenpeaceusa.org

National Mining Association
101 Constitution Ave. NW, Ste. 500 E.
Washington, DC 20001-2133
(202) 463-2600
FAX: (202) 463-2666
E-mail: craulston@nma.org
URL: http://www.nma.org

Natural Gas Supply Association
805 Fifteenth St. NW, Ste. 510
Washington, DC 20005
(202) 326-9300
FAX: (202) 326-9330
URL: http://www.ngsa.org

Natural Resources Defense Council
40 West Twentieth St.
New York, NY 10011
(212) 727-2700
FAX: (212) 727-1773
E-mail: nrdcinfo@nrdc.org
URL: http://www.nrdc.org

Nuclear Energy Institute
1776 I St. NW, Ste. 400
Washington, DC 20006-3708
(202) 739-8000
FAX: (202) 785-4019
E-mail: webmasterp@nei.org
URL: http://www.nei.org

Public Citizen
1600 Twentieth St. NW
Washington, DC 20009
(202) 588-1000
E-mail: member@citizen.org
URL: http://www.citizen.org

Sierra Club
85 Second St., Second Fl.
San Francisco, CA 94105
(415) 977-5500
FAX: (415) 977-5799
E-mail: information@sierraclub.org
URL: http://www.sierraclub.org

Solid Waste Association of North America
PO Box 7219
Silver Spring, MD 20907-7219
1-800-467-9262
FAX: (301) 589-7068
E-mail: info@swana.org
URL: http://www.swana.org

Union of Concerned Scientists
2 Brattle Sq.
Cambridge, MA 02238-9105
(617) 547-5552
FAX: (617) 864-9405
URL: http://www.ucsusa.org

U.S. Department of Energy
1000 Independence Ave. SW
Washington, DC 20585
(202) 586-5000
1-800-342-5363
FAX: (202) 586-4403
E-mail: The.Secretary@hq.doe.gov
URL: http://www.energy.gov

U.S. Environmental Protection Agency
Ariel Rios Bldg.
1200 Pennsylvania Ave. NW
Washington, DC 20460
(202) 272-0167
URL: http://www.epa.gov

U.S. House of Representatives Committee on Resources
1324 Longworth House Office Bldg.
Washington, DC 20515
(202) 225-2761
E-mail: resources.committee@mail.house.gov
URL: http://resourcescommittee.house.gov

U.S. Nuclear Regulatory Commission
Washington, DC 20555
(301) 415-8200
1-800-368-5642
E-mail: opa@nrc.gov
URL: http://www.nrc.gov

U.S. Senate Committee on Energy and Natural Resources
364 Dirksen Senate Office Bldg.
Washington, DC 20510
(202) 224-4971
FAX: (202) 224-6163
URL: http://energy.senate.gov

Waste Isolation Pilot Plant
U.S. Department of Energy
4021 National Parks Hwy.
Carlsbad, NM 88221
1-800-336-WIPP
E-mail: infocntr@wipp.ws
URL: http://www.wipp.energy.gov/

Worldwatch Institute
1776 Massachusetts Ave. NW
Washington, DC 20036-1904
(202) 452-1999
FAX: (202) 296-7365
E-mail: worldwatch@worldwatch.org
URL: http://www.worldwatch.org

RESOURCES

The Energy Information Administration of the U.S. Department of Energy is the major source of energy statistics in the United States. It publishes weekly, monthly, and yearly statistical collections on most types of energy, which are available in libraries and online at http://www.eia.doe.gov. The *Annual Energy Review 2004* (2005) provides a complete statistical overview, while the *Annual Energy Outlook 2006* (2006) projects future developments in the field. The *International Energy Annual 2003* (2005) presents a statistical overview of the world energy situation, while the *International Energy Outlook 2005* (2005) forecasts future industry developments.

The Department of Energy and the Energy Information Administration also provide *Natural Gas Annual 2004* (2005) and *Electric Power Annual 2004* (2005). The *U.S. Crude Oil, Natural Gas, and Natural Gas Liquids Reserves 2004 Annual Report* (2005) discusses reserves of coal, oil, and gas. In addition, the Department of Energy makes available information on the development of alternative vehicles and fuels, renewable energy sources, and electric industry restructuring.

The Energy Policy Act of 2005 (PL 109-58) is available at http://www.fedcenter.gov. The 2006 Advanced Energy Initiative is available at http://www.whitehouse.gov/stateoftheunion/2006/energy/print/index.html.

The U.S. Department of Transportation's Bureau of Transportation Statistics provides transportation information in its *Transportation Statistics Annual Report 2005* (2005).

The U.S. Environmental Protection Agency maintains an Internet page (http://www.epa.gov/radiation/yucca/index.html) for the Yucca Mountain Repository, which was helpful in the preparation of this book. The agency also provides *Light-Duty Automotive Technology and Fuel Economy Trends 1975–2005* (2005) and *Inventory of U.S. Greenhouse Gas Emissions and Sinks: 1990–2004* (2006).

The U.S. Nuclear Regulatory Commission is also an important source of information. It provides the documents *NRC—Regulator of Nuclear Safety* and *Information Digest, 2005–2006 Edition* (2005).

INDEX

Page references in italics refer to photographs. References with the letter t *following them indicate the presence of a table. The letter* f *indicates a figure. If more than one table or figure appears on a particular page, the exact item number for the table or figure being referenced is provided.*

P

R

S

T

U

W

Y